Chinese Handicrafts

Hua Jueming · Li Jinsong · Wang Lianhai
Editors

Chinese Handicrafts

Volume 2
Compiled by Hua Jueming, Li Jinsong, Wang Lianhai,
Guan Xiaowu, Li Hansheng, Luo Xingbo

Editors
Hua Jueming
The Institute for the History of Natural
Sciences, Chinese Academy of Sciences
Beijing, China

Li Jinsong
The Institute for the History of Natural
Sciences, Chinese Academy of Sciences
Beijing, China

Wang Lianhai
Academy of Arts and Design
Tsinghua University
Beijing, China

Translated by
Zhang Weihong
Zhengzhou Business University
Gongyi, China

Jelle Smets
Vancouver, Canada

经典中国国际出版工程
China Classics International

ISBN 978-981-19-5378-1   ISBN 978-981-19-5379-8 (eBook)
https://doi.org/10.1007/978-981-19-5379-8

Jointly published with Elephant Press Co., Ltd
The print edition is not for sale in China (Mainland). Customers from China (Mainland) please order the
print book from: Elephant Press Co., Ltd.
ISBN of the Co-Publisher's edition: 978-7-5347-7329-7

Translation from the Chinese language edition: "Zhongguo Shougong Jiyi" by Hua Jueming et al.,
© Elephant Press Co. Ltd 2014. Published by Elephant Press Co. Ltd. All Rights Reserved.
© Elephant Press Co., Ltd 2022

This work is subject to copyright. All rights are reserved by the Publishers, whether the whole or part of the
material is concerned, specifically the rights of reprinting, reuse of illustrations, recitation, broadcasting,
reproduction on microfilms or in any other physical way, and transmission or information storage and
retrieval, electronic adaptation, computer software, or by similar or dissimilar methodology now known
or hereafter developed.
The use of general descriptive names, registered names, trademarks, service marks, etc. in this publication
does not imply, even in the absence of a specific statement, that such names are exempt from the relevant
protective laws and regulations and therefore free for general use.
The publishers, the authors, and the editors are safe to assume that the advice and information in this book
are believed to be true and accurate at the date of publication. Neither the publishers nor the authors or
the editors give a warranty, expressed or implied, with respect to the material contained herein or for any
errors or omissions that may have been made. The publishers remain neutral with regard to jurisdictional
claims in published maps and institutional affiliations.

This Springer imprint is published by the registered company Springer Nature Singapore Pte Ltd.
The registered company address is: 152 Beach Road, #21-01/04 Gateway East, Singapore 189721,
Singapore

Preface: Re-introducing Handicrafts

China is universally renowned for its traditional handicrafts.

China's traditional handicrafts are characterized by their long history. The fact that it goes back to ancient times, is technologically exquisite, and has rich, distinct features is socially and culturally significant. Its handcrafted cultural relics, that have been either unearthed or handed down, its ancient buildings, and its ancient projects were all created using traditional crafts. To this end alone, we can see how important role handicrafts have played in the national economy and people's livelihoods, as well as in the formation and development of Chinese civilization itself. Like traditional Chinese medicine, they are also a treasure trove of Chinese science and technology and have a far-reaching influence worldwide.

For a long time, due to misconceptions and the influence of the ideology of the "extreme left", there has been a prejudice against handicraft techniques and craftsmanship. Handicraft techniques are often considered outdated and obsolete, as if they should all be replaced by modern technology and only exhibited in museums. They are regarded as dispensable and insignificant skills. Such misconceptions are even popular among all levels of the handicraft industry's administration staff. For a long time, traditional handicrafts have been neglected and their protection, inheritance, and discipline have drawn little attention from the responsible authorities, resulting in many precious techniques becoming endangered or dying out. This worrying situation is intrinsically tied to people's misconceptions about handicrafts.

In light of this inheritance crisis, the lack of further development of handicrafts, as well as the people's misunderstanding of them, it is necessary for us to get rid of the confusion and give a clear revision of their connotations, categories, essential characteristics, value, and prospects in the new historical period of reform and opening up. This way, we can remove obstacles for creating a new era of protection and revitalization of traditional crafts.

Categories of Handicrafts

What are Handicrafts?

In the early days, human beings made objects using their hands (and other limbs) and tools (and simple devices). This kind of labor, aimed at making objects, is called handicrafts or handcrafts.

Handicrafts are both material and spiritual, both technical and artistic, both historical and actual. In the social division of labor and economy, it falls in the category of the handicraft industry. It is also known as craft, handcraft, handwork, manual techniques, and traditional handicraft, all of which are used on different occasions, with slight deviations in meaning.

Handed down since ancient times, handicrafts are humanity's most basic form of labor and lifestyle, and thus will continue to be handed down to future generations. Handicrafts are such a part of humanity that it can be seen as the embodiment of the human essence.

Crafts (traditional handicrafts) is an umbrella term that encompasses 14 different categories, namely, (1) making tools and devices, (2) agricultural and mineral processing, (3) construction, (4) weaving, dyeing, and embroidering, (5) ceramics, (6) metallurgy and metalworking, (7) sculpture, (8) weaving and tying, (9) lacquering, (10) furniture making, (11) making calligrapher's tools (12) printing, (13) carving and painting, and (14) special handicrafts and others.

There are different types within each category, and every type contains various kinds. For example, there are different types of lacquering, such as carved lacquering, inlay lacquering, and bodiless lacquering. Then there are different kinds of inlay lacquering, such as mother-of-pearl inlay and metal inlay. In this book, we will use this three-level classification system to sort all types of traditional Chinese handicrafts.

Some techniques are used to solely make artwork, such as carving and painting. However, some others are used to make tools as well as artwork. For example, there are household ceramics and art ceramics; weaving and dyeing are used to produce not only cloth and brocade but also artwork such as *kesi* silk tapestries and double-sided embroidery.

For a long time, industrial arts have received more attention than handicrafts for making useful items, because they enable foreign exchange through export. In the eyes of some people and even some experts and scholars, traditional crafts are equivalent to arts and crafts (such as paper cuttings and New Year pictures), while other more basic techniques for making tools and devices, oil, salt, sauce, and vinegar, which are more closely related to the national economy and people's livelihoods, are neglected. This is, of course, a great misunderstanding, and in order to better understand, rescue, and preserve traditional handicrafts, it is of the utmost importance to rectify it.

Essential Characteristics of Handicrafts

Practicality

As the saying goes: "Necessity is the mother of invention". All handicrafts are invented to meet people's needs in production and their daily life.

"A handy tool makes a handy man". In order to transform and adapt to nature, human beings have to be able to do all kinds of work. This makes it necessary to create and use a great variety of tools, devices and machines. From the most rudimentary sharp-pointed wooden stick to the *lei* (similar to a spade), *si* (similar to a plow), axe, pickaxe, yardstick, marker, divider, carpenter's square, pestle and mortar, well sweep, water mill, windmill, seismograph, compass, astronomical clock, drawloom, and many more, all of these tools and devices have been widely used in people's daily lives. Due to its fundamental significance to human life, their production techniques have become one of the main categories of traditional handicrafts. For example, the large windmill in Sheyang county, Jiangsu province (as shown in Picture 1) is 8 meters high and has 8 fans, which can rotate from all sides. The fans drive the keel of the waterwheel underneath to pump seawater onto the salt field where the salt is extracted. The windmill was restored by Zhang Baichun, director of the Institute for the History of Natural Sciences, Chinese Academy of Sciences. It is now installed at the gate of the Southern Taiwan University of Science and Technology in Tainan city, Taiwan province. There are two major schools of wind-driven machinery, one

Picture 1 Large windmill, located in Sheyang county, Jiangsu province

in the East and one in the West. The shafts of the large windmills in the Netherlands (West) are horizontal, while those in China (East) are vertical.

Such basic necessities as oil, salt, soy sauce, vinegar, and tea are indispensable in the daily lives of Chinese people. The agricultural and mineral processing techniques like flour-grinding, rice-pounding, oil-extracting, sauce-making, vinegar-brewing, tea-making, salt-extracting, and leather-tanning are closely related to people's livelihoods and are still widely used by Chinese people. Luzhou *Laojiao Baijiu* in Sichuan province is going through aging in storage caves. The more *baijiu* ages, the more fragrant it becomes (as shown in Picture 2). Some people say that brewing techniques should not be categorized as intangible cultural heritage. This is incorrect. Fermentation is applied during the alcohol brewing process, which is one of the earliest forms of bioengineering involving microorganisms. It is of great value and has a place in the history of science and technology as well as the national economy.

If people hope to live and work in peace and contentment, they need to build dwellings and houses. Therefore, geotechnical engineers, masons, carpenters, and bricklayers are necessary. The design and production of various building structures, decorations, urban planning, and garden construction are generally called "construction". Picture 3 shows *tulou*—local pieces of architecture characterized by their strong ethnic and regional features, embodying the culture of the Hakka people in Fujian province.

Clothing is a symbol of civilization. Perhaps because of this, clothing takes precedence over food, housing, and transportation. Cotton, hemp, silk, and wool are woven, printed and dyed, and cut and sewn into clothes, shoes, and hats. China is the birthplace of silk-weaving technology. The Silk Road, named after this highly desired commodity, has played a great role in promoting transportation, trade, and cultural exchanges between China and foreign countries. Picture 4 shows a *da'ao,* a women's coat decorated with brocade from the Qing dynasty. The beautiful patterns and colors are what inspired Bai Juyi (772–846), the famous poet of the Tang dynasty, to write

Picture 2 *Baijiu* for aging

Picture 3 *Tulou*

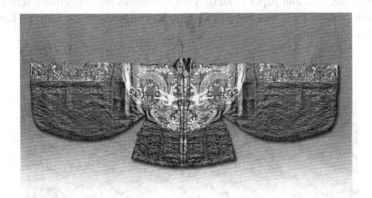

Picture 4 *Da'ao*

in his poem: "The delicate patterns of silk are required by the royal palace… The colors of silk are like those of river water in the south".

Chinese people can't live without pots and jars. The making of porcelain (china) is one of China's great inventions, hence the country's English name. Earthen pots and urns are used for cooking and storing water, while purple clay teapots and porcelain vases are used for drinking tea and arranging flowers, yet having the additional function of being used as an ornament on their own. Colored glazing, glass, and glassware are all categorized as silicate products. Picture 5 shows a woman glazing porcelain in Chenlu town, Shaanxi province.

China had a splendid bronze and iron age and has long been a global frontrunner when it came to the technology of metallurgy and metalworking. Even nowadays, Wang Maizi knives and scissors, Zhang Xiaoquan knives and scissors, gold foil, and iron paintings are still popular among Chinese people. Tibetan, Mongolian, and Husa knives, as well as Bao'an waist knives, are common tools and ornaments among ethnic minorities such as the Tibetan, Mongolian, Hui, Achang, Bao'an, and Salar. Picture 6 shows ironworkers of Jiuxinglu Company in Zhangzi county, Shanxi province, forging gongs with ultra-high-tin bronze (containing 24% tin). The 1-meter-wide gong is the only Chinese musical instrument used in modern symphony orchestras.

Sculpting is one of the oldest artforms. Wood carving, stone carving, jade carving, ivory carving, clay sculpting, and dough modeling are used in the construction and decoration of houses, furniture, bridges, boats, and carriages, but also simply as art, to convey a sense of beauty. Picture 7 is a masterpiece by *Nirenzhang* (Clay Figurine Zhang) of Tianjin city. Two senior farmers are playing chess with pieces of fruit on a self-drawn chessboard in the field.

Grass, hemp, bamboo, rattan, and paper are in themselves quite common materials, but they can be skillfully woven and tied into baskets, mats, tables, chairs, shoes, hats, kites, lanterns, and paper models. Such techniques are collectively referred to as weaving and tying. Picture 8 shows the grand scene of the annual Qinhuai Lantern Market in Nanjin city.

Picture 5 Glazing

Preface: Re-introducing Handicrafts

Picture 6 Forging of a gong

Picture 7 Game of chess

Ten Defects in *Han Feizi* records that Yu the Great, the first king of the Xia dynasty once "painted black on the outside and red on the inside" of sacrificial vessels. Lacquering originated from the need for the protection and decoration of utensils. It later developed into a variety of types such as engraved cinnabar lacquering, engraved cloud lacquering, and ramee-lacquering. Picture 9 is the work "Radiation" by Professor Qiao Shiguang, a master lacquer artist, who stood on a high place and scattered the paint downwards to get the desired visual effect. The materials used are traditional, while the technique is modern.

Picture 8 Qinhuai lantern market

Picture 9 Radiation

Preface: Re-introducing Handicrafts

Picture 10 Alcove bedstead

As early as the Warring States period, there were different patterns of mortise and tenon joint structures, which have become a major feature of making Chinese furniture. The design and making of furniture reached its highest peak in the Ming dynasty, and it is now a museum favorite. In recent years, classical furniture has become more and more popular in China. Picture 10 shows an alcove bedstead commonly used in the past. It contains a large bed with a small cabinet and a wooden urinal beside it, thus forming some private space. Picture 11 is an armchair for women. According to an old folk custom in Zhejiang province, women shouldn't lean against the back of the chair when sitting down. Therefore, the front part of the armrest was purposely missing, so women could only sit on the front half of the chair.

Papermaking and printing are two of China's four great inventions and have played a great role in the inheritance and spread of human civilization. *Xuan* paper, or rice paper, has remained a must-have for Chinese calligraphers and painters, and can in no sense be replaced by regular machine-made paper. Picture 12 shows 18 craftsmen lifting *xuan* paper with a width of one *zhang* and two *chi*, that is, approximately 4 meters. Engraved block printing techniques are still used to print ancient books and Buddhist scriptures. The people in Rui'an city, Zhejiang province, still print their genealogical books with wooden movable types. Picture 13 is the Woodblocks Storage Hall of Dege Parkhang Sutra-Printing House in Garzê Tibetan autonomous

Picture 11 Armchair for women

prefecture, Sichuan province, which is the largest among the three Parkhang Sutra-Printing Houses in the Tibetan region, with 220,000 woodblocks in its possession. Picture 14 shows two Tibetan craftsmen printing Tibetan Buddhist scriptures: one coloring and the other printing.

Paper cuttings, fine-grain paper cuttings, shadow play, New Year Pictures, and inner paintings embody both the richness of humanity and the value of artistry, especially since they all originate from people's living habits and aesthetic needs. Picture 15 shows a *Yangliuqing* New Year Picture, from Tianjin, depicting peace, auspiciousness, and abundance, which are the eternal themes of these types of Pictures.

Some techniques, such as how the Li ethnic group in Hainan province drills wood to make fire, are relics of ancient times and extremely rare (as shown in Picture 16). Drilling wood to make fire is the first significant invention of mankind, which, together with the steam engine, electricity, nuclear power, and computers, is a milestone in human progress. Human beings separated themselves from animals and started walking the path of civilization by learning to make fire and preserve kindling. Nowadays, such an ancient technique is only practiced in areas such as

Picture 12 Lifting *xuan* paper with a width of one *zhang* and two *chi*

Hainan, China, and a few countries and areas in the South Pacific. In Picture 17, a native Hawaiian is demonstrating how to drill wood to make fire at the Polynesian Cultural Center in Honolulu, Hawai'i. Some techniques, such as how the Oroqen ethnic group uses fish skin and birch bark in crafting, are so unique that they cannot be classified into the above categories. Therefore, we describe those and other special techniques in Chap. 14, aiming to include as many remaining handicrafts as possible.

Everything mentioned within the 14 categories of traditional handicrafts listed above originates from common people's needs in production and their daily life. Handicrafts are a way of life. Chinese people love them and cannot do without them. Even today with increasing modernization, various kinds of crafts are still quite popular. In recent years, crafting activities such as pottery and knitting have shown signs of revitalization. This is due to its practicality and its place in the modern lifestyle. Handicrafts embody the essence of human culture and labor. Li Bocong, a famous contemporary Chinese philosopher, said, "I create, therefore I am". By imitating "Je pense, donc je suis", the well-known saying of Descartes, French mathematician and thinker, Li's remarks reveal the significance of handicrafts to human culture and labor. The practicality of handicrafts is its most essential and important feature, as Japanese scholar Sōetsu Yanagi (1889–1961) once said, "Practical crafts are the true crafts".

Picture 13 The woodblocks storage hall of dege parkhang sutra-printing house

Picture 14 Printing

Preface: Re-introducing Handicrafts xvii

Picture 15 *Yangliuqing* New Year picture

Picture 16 Drilling wood to make fire in Baoting county, Hainan province

Picture 17 Demonstration of wood drilling to make fire by an aboriginal Pacific Islander

Rationality

Manual work is one of the main ways for people to understand both the natural and artificial world.

Hegel (1770–1831), a German philosopher, once said that what is rational is actual, and what is actual is rational. What is learnt from handicrafts is concrete and rational knowledge. Rationality, a form of high-level cognition, is the soul not only of human nature and science but also of handicrafts. Whether we are making sharp-pointed wooden sticks or drawlooms, we require at least some rational guidance, like knowledge of materials and mechanics. This kind of rationality may be clear or hazy in a craftsman's mind, but craftsmanship is rational in every sense since it is a purposeful activity. Otherwise, even a single rudimentary sharp-pointed wooden stick couldn't be made, let alone other objects.

Let's take the traditional Chinese lost-wax casting as an example. The composite clay-mold casting technique reached its highest peak and yet also encountered its technological bottleneck during the Warring States period. To meet the demand for surface decoration of bronze wares, lost-wax casting was invented: a brand-new casting technique which could produce permeable decoration in three-dimensional space. In doing so, it made possible such fine bronze creations, known from the Shang and Zhou dynasties, as (1) the Moiré Copper Ban, unearthed in Xichuan county, Henan province, (2) Marquis Yi of Zeng's Bronze *Zun* (wine vessel) and *Pan* (plate), and (3) the Chen Zhang Pot. Although there are no records of the precise production process and there is no way to know the terms and formulas used at that time, there is plenty of physical evidence that such technical concepts as the fusibility and cladding of molds were used, as well as such techniques as plastic wax production and wax molds welding. These concepts and techniques were quite

Picture 18 Gating system of the Moiré Copper Ban unearthed in Xichuan county, Henan province

mature during this era, some of which were, in turn, passed on and developed into lost-wax casting and modern precision casting. Picture 18 shows the gating system of the Moiré Copper Ban unearthed from a tomb of the Chu State in the Spring and Autumn period in Xiasi village, Xichuan county, Henan province. It consists of main runners, sub-runners, and branch runners, whose diameters decrease with every turn, totaling five layers. Decorative coiled chi-dragon patterns are formed at the end (on the outermost layer). This system, though quite complex, conforms to the principles of fluid mechanics and foundry engineering.

All forms of handicrafts are based on people's rational cognition of the natural and artificial world and must conform to the objective laws of physics, chemistry, biology, and other scientific disciplines, either by understanding them but not knowing why or understanding them and knowing why. No matter which of the two groups people belong to, rational cognition is essential. That is the rationality of handicrafts.

Aesthetics

Manual work is a combination of wisdom and strength. Even a bone spindle from ancient times (as shown in Picture 19) shows its beauty in its simplicity as well as its harmony between shape and function.

However, not all handicrafts are beautiful. Every period in history inevitably produces some shoddy or overcomplicated handiwork, but these do not represent the majority of all craftsmanship.

In our daily life, the handicraft products we see or use are either handy or pleasing to the eyes, but they are all beautiful. Who is willing to acquire them if they are ugly or inconvenient to use?

Aesthetics is one of the essential characteristics of manual work. As Marina Tsvetaeva (1892–1941), a Russian poet, once said, "Venus was hand made". Great artists use their hands to imbue their creations with great spirit. *"The Thinker"* (as shown in Picture 20) by Auguste Rodin, a French sculptor, is a typical example of this.

Picture 19 Bone spindle

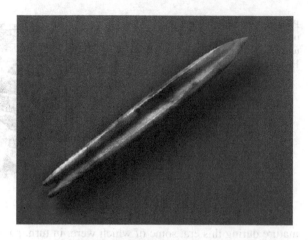

Picture 20 *"The Thinker"* by Auguste Rodin

Humanity

Just like home-brewed rice wine can invoke homesickness, so can a hand-knitted sweater still envelop us with the warmth of our loving mother after many years.

Cloth shoes with multi-layered soles may look a bit clumsy, but they are comfortable. Though in order to get these comfortable shoes, they need to be meticulously made by hand, stitch by stitch, layer by layer.

Preface: Re-introducing Handicrafts

The cloth tigers in Picture 21 were made by an elderly woman in Shaanxi province. They are vivid and full of life. She said she made them by imitating the appearance and facial expressions of her grandson. Surprisingly, the cloth tigers actually resemble the boy.

An old Tibetan man was carefully engraving decorative patterns on a pottery jar. He said that he wanted to make this pottery jar out of his desire to wish his family peace and good fortune ("tashi delek" in Tibetan). He didn't plan to sell it. He did

Picture 21 Cotton tiger

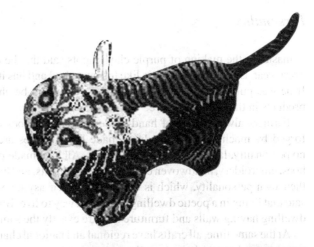

Picture 22 A dressed-up little girl of the Miao ethnic group

not care about how much time he had already spent on the pottery jar, nor how much time he would still spend on it.

Handicrafts embody humanity.

Picture 22 shows a dressed-up little girl of the Miao ethnic group. The Miao people are especially fond of silver. Brides will wear complete sets of silver accessories, such as silver headwear, earrings, neck ornaments, and chest ornaments.

Personality

A master in the making of purple clay teapots said that he only made a few teapots every year. Every one of them is like his own child and has its own distinct character. If he was put in front of a pile of teapots, he would be able to recognize his own products in the blink of an eye.

Farmers always say that hand-made sickles and hoes are different from those forged by machines. The products made by machines are almost the same, with no personality, lacking character and spirit. All handmade products, such as sickles, hoes, embroidery, handwoven cloth, silver ornaments, and hand-made furniture, have their own personality, which is conveyed in their asymmetry and irregularity. It is said that living in a poetic dwelling is the best way to live. But can you imagine such a dwelling having walls and furniture that are exactly the same as any other dwelling?

At the same time, all crafts have regional and national characteristics, which reflect their personality and their roots. *Xuan* paper, or rice paper, is made of blue sandalwood bark, long-stalk straw, and carambola vine juice produced in Jingxian county, Anhui province. All the other kinds of paper made in other areas are not permitted to be called *xuan* paper. The Naxi people in Yunnan province make *dongba* paper with a fixed mold, which is used to print *jiama* (paper charms) and record *dongba* scriptures. Tibetan paper, made of chamaejasme, is mothproof and used to print Tibetan Buddhist scriptures. These kinds of paper are so closely related to the traditional culture, living customs, religious beliefs, and ethnic pride that they cannot, in any way, be replaced by machine-made paper. The essence of their handicrafts is deeply rooted in each culture's personality and location. Shown in Picture 23 are window paper cuttings from Yuxian county, Hebei province, with a style typical of northern China, while shown in Picture 24 is a Dai-style paper cutting, and in Picture 25 a fine-grained one from Yueqing city, Zhejiang province with a style typical of southern China.

Preface: Re-introducing Handicrafts   xxiii

Picture 23 Paper cuttings from Yuxian county, Hebei province

Picture 24 Dai-style paper cutting

Picture 25 Fine-grained paper cutting from Yueqing city, Zhejiang province

Combining Variance and Invariance

Handicrafts are dynamic and always changing with time.

There are no handicrafts in the world that do not undergo change and simply remain the same. Porcelain made in the Ming and Qing dynasties are different from those of the Song and Yuan dynasties, which, in turn, are different from those of the Han and Tang dynasties; likewise, celadon made in the Han dynasty is different from the proto-porcelains of the Western Zhou and Shang dynasties. However, all of them are considered porcelain, which exemplifies the fact that variance and invariance are closely related. No one would fault the porcelain of the Song and Yuan dynasties for their differences from those of the Han and Tang dynasties, or those of the Ming and Qing dynasties for their differences from those of the Song and Yuan dynasties. Then who can guarantee that the handicrafts we know today will be the same in 100 years, or even force those handicrafts of today to remain static and unchanged?

The technique of drilling wood to make fire has not changed much since ancient times, although it takes different forms in different parts of the world. If a technique fails to keep pace with the times due to its innate limitations, it can only decline, die out, and be replaced by other more advanced techniques. Isn't it quite natural and reasonable to replace wood with flint, flint with matches, and matches with a lighter to make fire?

It is often said that those traditional handicrafts which have been recognized as intangible cultural heritage should be preserved in their original state. Of course, this kind of advice is positive to some extent, since they should be preserved conditionally and selectively. To meet the needs of mass production and market competition, the Anhui Jingxian Xuan Paper Making Mill has used machinery, instead of manual labor, to pulp paper and steam, instead of fire, to dry paper. No one has raised any objections to this change, as it is, in fact, impossible to pass on the tradition and keep it intact. However, to preserve the original techniques—techniques used about one century ago instead of those in the Ming dynasty—the mill has built a paper-making production line which uses only traditional techniques. The superior *xuan* paper that is made in this traditional way is then sold at a higher price. A wise decision, indeed. Some handicraft techniques are highly complex and cannot be understood in a short time. For example, the brewing of distilled spirits is a kind of biological engineering. Scientific analysis shows that some kinds of *baijiu* contain more than 400 types of beneficial microorganisms. The effect of these microorganisms on the quality of spirits and the way they act and react with the raw materials, yeast, water, and production procedures have not yet been closely examined. In this case, most of the traditional production procedures have been preserved, such as the spreading, airing, and loading of fermented materials, assessing the quality of *baijiu*, and blending, because those techniques heavily depend on the artisans' senses and judgement. Only procedures like the handling of raw material have been replaced by machinery. Also a wise decision.

It is widely recognized that Japan has done the best in preserving their intangible cultural heritage. Moreover, Japan is quite entrepreneurial when it comes to the inheritance and development of traditional handicrafts. As early as the 1980s, or even earlier, Japanese traditional metal crafts and lacquering adapted to the needs of modern life and the changes in people's aesthetic preferences. Since then, there have been many innovative changes when it came to shape, color, and patterns, whereas they maintained the natural materials and traditional techniques used in production. No one can deny that these are still authentic traditional craft products, and no one can accuse such practices of parting with tradition, because keeping pace with the times is in fact the tradition of handicrafts. China lags behind Japan both in theory and practice in this respect, so we still have a long way to go. Nevertheless, it is comforting to see that experts and scholars in some Chinese colleges and institutions are working towards this very goal. For example, outstanding progress has been made in ceramics and lacquer painting. If we use the correct concepts and take appropriate measures, then traditional handicraft techniques will surely become better and more effective in combining both variance and invariance.

Eternality

Handicrafts go hand in hand with human beings, who have changed the world and themselves using their own two hands. Long-term labor makes manual labor an instinctive need in people's lives. There are even famous people who liked handicrafts. For example, among those who loved woodworking, there was the Chinese

Emperor Zhezong of Song (1077–1100), the great Russian writer Lev Tolstoy (1828–1910), and former U.S. President Jimmy Carter. In India, Mahatma Gandhi (1869–1948) (as shown in Picture 26) was a big fan of the spinning wheel. There are also many common people who are fond of handicrafts and love them purely as hobbies.

Handicrafts that aim to create things that are useful are also eternal. Many people say that the pre-modern era was the era of manual labor, the industrial era was the era of machines, and the post-modern era of automation, information, or digitalization. It is not very accurate, but also not very inaccurate. Like Professor He Xuntian, a contemporary Chinese musician, once said, "It is an era where manual and non-manual labor coexist",

Nearly all original work requires a manual touch, especially in manufacturing. At present, there is a shortage of 100,000 senior technicians in China, which is related to our common misunderstanding of what manual labor entails. People who have a bit of common sense when it comes to manufacturing all understand the extreme importance of mold making for machine manufacturing, and mold making is inseparable from manual labor. Similarly, the modification of all cars is based on car models, which are made by hand. The best clothes, hats, leather shoes, and jewelry around the world are all handmade. The handmade leather shoes of the French Massaro family cost $3,000 per pair. How can machine-made ones even begin to compare? Academician Wu Mengchao (1922–2021), a famous Chinese surgeon who won the Highest Science and Technology Award in 2005 (as shown in Picture 27), said that all his operations could be made public. You might know in theory how he did the operation, but you would not be able to replicate it in practice, because you do not possess the same skills. That is why capable surgeons are called "Golden Fingers" and why they can all proudly call themselves craftsmen or artisans.

Picture 26 Mahatma Gandhi is spinning

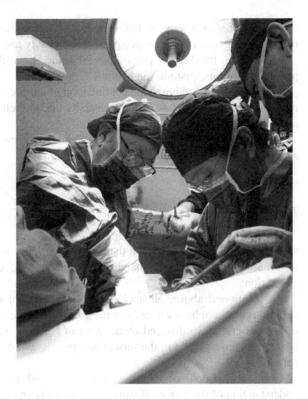

Picture 27 Dr. Wu Mengchao during an operation

Therefore, it is definitely a great misunderstanding to call the current era "the era of machines or automation".

Where there are humans, there are handicrafts, there is a fondness of participating in craft activities, and there is a love for handmade products. Handicrafts are eternal.

The Value of Handicrafts

Inherent Value

The value of handicrafts is determined by their core characteristics.

Their practicality determines that they are of great value both in people's daily lives (practical value) and on the market (exchange value), both on a personal and national level.

Their rational foundation determines their academic value. There are scientific principles behind many kinds of traditional handicrafts and they deserve to be preserved, just like wildlife.

Their aesthetics determine their general and educational artistic value. Handmade products can give us different perspectives on beauty and mold our temperament.

Meanwhile, handicrafts possess social value, since they are closely linked to people's emotions, customs, and beliefs.

They are of great value to the continuation of culture and the preservation of our national identity as well as cultural diversity due to their intrinsic regional, ethnic, and local characteristics.

Because handicrafts are eternal and dynamic, they are a testimony to humanity's shared universal values and historical continuity.

Historical Value

Chinese people are famous for their ingenuity. This is the natural result of the influence and cultivation of handicrafts being passed down from generation to generation.

As mentioned above, all ancient buildings, ancient projects, and handcrafted cultural relics, that have either been unearthed or handed down, were created using traditional crafts. To this end alone, we can see how important a role traditional handicrafts have played in the formation, growth, and development of the Chinese nation.

Even in the field of historiography, it is still necessary to gain a deeper understanding of the historical value of handicrafts and appraise them properly.

Modern Value

This is a topic that has hardly been talked about or simply avoided, be it intentionally or unintentionally. After understanding more about the essential characteristics and value of handicrafts in this chapter, I believe you, my dear readers, will be able to draw your own conclusions. The main thing that we, as the authors, would like to convey is the following:

Traditional crafts are one of China's treasures as well as a treasure trove of Chinese science and technology comparable to traditional Chinese medicine.

Much of what we eat with, wear, use and play with every day are still handmade products, such as axe, chisel, hoe, pickaxe, black tea, green tea, yellow rice wine, *baijiu*, woven bamboo ware, rattan plaited articles, wood carvings, jade carvings, clay sculpting, dough modeling, tie-dyed cloth, batik cloth, brocade, embroidery, gold leaf ornaments, silver ornaments, celadon ware, purple clay ware, paper cuttings, New Year Pictures, Tongrentang traditional Chinese medicine, Quanjude roast duck, Wang Mazi knives and scissors, Zhang Xiaoquan knives and scissors, cloisonné enamelware, Chinese knots, oil, salt, soy sauce, vinegar, clothes, shoes, hats, fireworks, firecrackers, brush, ink, paper, inkstone, and the list goes on and on. With the improvement of people's living standards, their yearning for returning to nature and their roots, and the new perspective in regards to their aesthetic preferences, there will be more people asking for handmade products.

Today, in China, we can see that traditional handicrafts are still of great value in people's daily lives and of great value to the economy, academic research, artistic appreciation, and culture. Handicrafts will play a vital role in the further development of the central and western regions in China as well as the agricultural, rural areas. Handicrafts will prove their value, not only by improving people's living standards but also by improving the national economy. However, we still need to delve deeper into their academic value. It needs further study and exploration, which will hopefully result in great achievements that will attract worldwide attention. At the same time, their artistic value will steadily garner more and more recognition and interest. While facing the impact of globalization and the dominance of Western culture, the cultural value of Chinese traditional handicrafts is especially important and shall draw ever more attention.

Protecting Handicrafts: Global Consensus

It has taken nearly half a century for the world to reach a consensus on the protection of handicrafts.

Japan formulated and promulgated the *Law on the Protection of Cultural Property* in 1950, which was the first step in protecting intangible cultural heritage in the world. Intangible cultural heritage means non-physical cultural heritage. Traditional handicrafts, such as inlaying gold and silver, Japanese shippo, and the making of Japanese blades, swords, and magic mirrors, account for a considerable proportion of Japan's intangible cultural heritage.

In 1982, UNESCO established the Section for the Non-Physical Heritage, and in 1989, the 25th General Conference adopted the *Recommendation on the Safeguarding of Traditional Culture and Folklore*.

In 1997, UNESCO launched the program of *Proclamation of Masterpieces of the Oral and Intangible Heritage of Humanity*, in which handicrafts are listed as one of the five forms of intangible cultural heritage. This program was officially announced and launched the following year.

In 2001, the *Universal Declaration on Cultural Diversity* was adopted by the 31st General Conference, which emphasized the significance of intangible cultural heritage in maintaining human cultural diversity. In the same year, the first 19 *Masterpieces of the Oral and Intangible Heritage of Humanity* were recorded and Chinese *Kunqu* Opera was one of them.

In 2002, the *Istanbul Declaration* was adopted by the *Third Roundtable of Ministers of Culture of the United Nations*, which emphasized the importance of intangible cultural heritage.

In 2003, UNESCO launched the *Convention for the Safeguarding of the Intangible Cultural Heritage*. China joined this convention the following year. At present, the convention has entered into force, after being ratified by nearly 30 countries.

In 2004 and 2006, UNESCO added more elements to the *Representative List of the Oral and Intangible Heritage of Humanity*. Until then there had been a total of 90 elements listed, among which were the *guqin* (the Chinese zither), the Xinjiang Uyghur *Muqam* (folk melodies), and Mongolian *Urtiin Duu* (long folk songs).

So far, the importance of intangible cultural heritage, including handicrafts, and the necessity and urgency of its preservation has become an international consensus. Especially in developing countries where intangible cultural heritage is still well preserved and relatively complete, it usually receives financial support from both the government and NGOs.

The Protection and Revitalization of Traditional Handicrafts are Part of Chinese Modernization

Chinese handicrafts have long been neglected and improperly protected. The reasons for this are closely related to profound social and ideological misconceptions. For a long time, under the influence of the ideology of the "extreme left", cutting off history, ignoring tradition, and even acting against tradition have become a new kind of "tradition". At the same time, modernization has caused many people to become prejudiced against traditional handicrafts due to a decline in knowledge and understanding, especially in the humanities.

However, since China joined the UNESCO *Convention for the Safeguarding of the Intangible Cultural Heritage*, the preservation and inheritance of intangible cultural heritage have been officially put on the Chinese government's agenda. In 2006 and 2008, the State Council approved and published *List of the First and Second Batch of National Intangible Cultural Heritage* respectively. Among the more than 1,200 pieces of national intangible cultural heritage, accounting for more than a quarter, there are more than 300 traditional handicrafts. The importance of handicrafts and the necessity and urgency of their preservation and inheritance have become the consensus of all sectors of society. With the joint efforts of the government, communities, enterprises, artisans, and experts, a number of precious handicrafts are expected to be protected, inherited, developed, and revitalized. When we have a better and more objective understanding of traditional handicrafts and recognize their value and importance in modern society, we should naturally think of their protection and revitalization as part of China's modernization, and by no means as dispensable or a burden. The protection, revitalization, and development of traditional handicrafts should be put forth as a national decree signed by the executive branch of the central government, in which their short-, medium-, and long-term goals are formulated, worked out, and put into practice. All in all, realizing that goal will require all of our joint efforts (Picture 28).

Picture 28 *Painting of Gusu, the Flourishing City*

How to Regard and Live with Handicrafts?

Handicrafts are a Treasure

Please regard them with respect and develop alongside them.

There is a Chinese TV series by CCTV called *Liuzhu Shouyi*, which means *Retaining Handicrafts*. It is rich in content, practical, and simply made really well. In the series, they interview Japanese scholar Shiono Yonematsu. In fact, the name of the program *Retaining Handicrafts* is exactly the same as Shiono Yonematsu's book. Just like the title of the book and TV series, I agree that we should "retain handicrafts". However, I also think that "developing alongside handicrafts" sounds more positive and reflects reality more accurately, because we do not only retain handicrafts but also develop and revitalize them. Since people will not and cannot live without handicrafts, they are eternal and inseparable from human life. This is why we grow with them, why we develop alongside them, as one.

Beijing, China

Hua Jueming
Li Jinsong
Wang Lianhai

Translators' Preface to "Chinese Handicrafts"

As one of the four ancient civilizations, China has a long history of material and spiritual culture. Its traditional handicrafts have played a crucial role in the formation and development of its civilization, and have also made outstanding contributions to the development of the rest of the world, continuously promoting the exchange and mutual appreciation of Chinese and foreign civilizations. For example, in terms of science and technology, the *Four Great Inventions of China*, namely, paper, the printing press, the compass, and gunpowder, have been an immense contribution to global development. Furthermore, in terms of trade, Chinese silk, tea, and porcelain have been popular and exported in large quantities to Central Asia and Europe for a long time.

Chinese traditional handicraft techniques are the crystallization of thousands of years of hard work and wisdom of folk artisans and represent the cultural heritage and lineage of the Chinese nation. These techniques mainly include (1) making tools and devices, (2) agricultural and mineral processing, (3) construction, (4) weaving, dyeing, and embroidering, (5) ceramics, (6) metallurgy and metalworking, (7) sculpture, (8) weaving and tying, (9) lacquering, (10) furniture making, (11) making calligrapher's tools, (12) printing, (13) carving and painting, and (14) special handicrafts. A detailed introduction to all of the above is provided in this book. The editors are professional scholars in various industry fields, with a solid grasp of the content of each sector, supplemented by credible field research and empirical studies, making this book a professional and readable monograph on traditional Chinese handicraft techniques with rich content, detailed data, and illustrations.

The English translation and publication of Chinese Handicrafts in other countries will be of great value to the world in understanding the history and contemporary protection of Chinese traditional handicrafts, its contribution to Chinese civilization and world civilization, and its contemporary significance.

The two translators of this book are a native Chinese and an English speaker. Zhang Weihong is a senior scholarly translator with a solid background in traditional Chinese culture, as well as profound basic Chinese and English language skills and over twenty years of teaching and practical experience in translation. Jelle Smets has a strong interest in traditional Chinese culture, has lived and worked in China for many years, and is specialized in Chinese-English language and cultural translation. Our collaboration ensured an accurate understanding of the language and issues of the original text, as well as the accuracy and fluency of the translation.

When we saw the nearly 400,000-word Chinese Handicrafts for the first time, we were very proud to be working on its translation and doing our part for the exchange of Chinese and foreign cultures, but at the same time we were under great pressure. In addition to the time constraint and the heavy workload, the most troublesome problem was that there were many technical terms related to the materials, production, and usage of traditional Chinese handicraft techniques in the book, many of which did not have ready English equivalents and needed to be translated by the translator according to the actual situation. Therefore, after reading through the entire book before actually translating it, we developed our own translation principles and strategies for translating terminology.

Our main purpose of translating this book is twofold: first, to preserve the expressions and connotations of traditional Chinese culture as much as possible in its original form; second, to focus on the fluency and acceptability of the translated text for readers. Therefore, when translating the terminology, we have extensively consulted a large number of materials and decided that if there is a corresponding English expression, we will adopt it directly; if not, we will adopt the method of transliteration plus free translation, which means adding the corresponding explanatory translation besides the Chinese *pinyin* to facilitate readers' understanding, and occasionally we will directly adopt the free translation method according to the specific situation. In terms of annotation, in order not to add extra burden to the readers and not to disturb their reading fluency, we have incorporated most of the explanatory text directly into the translation, and only added a table of Chinese dynasties and corresponding Western chronology in the appendix, as well as the English-Chinese comparison of Chinese classical works, so that those readers who are interested can cross-reference these themselves. In addition, all the photos in the original text have been retained in the translation.

We hope that our translation efforts will be well received by our readers. We also sincerely hope that our readers will enjoy reading this book and that it will enhance their understanding of traditional Chinese technology, especially handicrafts, and ultimately promote mutual understanding and communication among world civilizations.

Gongyi, China	Zhang Weihong
Vancouver, Canada	Jelle Smets

Contents

1. **Making Tools and Devices** 1
   Guan Xiaowu and Feng Lisheng

2. **Agricultural and Mineral Processing** 115
   Li Jinsong

3. **Construction** 227
   Luo Xingbo and An Peijun

4. **Spinning, Dyeing, and Embroidering** 297
   Wang Lianhai and Zhao Hansheng

5. **Ceramics** 425
   Hua Jueming and Qiu Gengyu

6. **Metallurgy and Metalworking** 521
   Hua Jueming

7. **Sculpture** 637
   Luo Xingbo

8. **Weaving and Tying** 693
   Wang Lianhai

9. **Lacquering** 757
   Zhou Jianshi and Hua Jueming

10. **Furniture Making** 809
    Hua Jueming

11. **Making Calligrapher's Tools** 875
    Hua Jueming and Guan Xiaowu

12. **Printing** 935
    Guan Xiaowu and Fang Xiaoyan

13	**Carving and Painting** 1007	
	Wang Lianhai	
14	**Special Handicrafts and Others** 1053	
	Yang Yuan and Li Jinsong	
15	**Protection, Inheritance, and Revitalization of Traditional Crafts** 1119	
	Hua Jueming	

Conclusion: Destiny of Traditional Crafts in Contemporary Times 1135

Postscript 1141

Appendix I 1143

Appendix II 1145

Chapter 6
Metallurgy and Metalworking

Hua Jueming

Metal materials have such a range of properties that they have played an extremely important role throughout history. The point at which a civilization is actually considered "civilized" is the invention of metallurgy.

There are many metals found in nature, such as gold, copper, silver, and meteorite. People always use readily available materials to create tools and devices first. Therefore, the use of natural metals predates that of artificially smelted metals, in other words, the technology of metalworking predates that of metal smelting.

There are eight kinds of metals that were smelted and used in ancient China: bronze, gold, lead, tin, iron, silver, mercury, and zinc. Antimony may have been used in the Ming dynasty, but this has not been verified yet. Xu Shen's *Discussing Writing and Explaining Characters* writes that "every type of metal can be considered *jin* (金)". This is the reason why the Chinese character *jin* (金), which originally only meant "gold", also became the umbrella term for all types of metal.

Of the eight metals mentioned above, copper and iron are the most important. Archaeology categorizes civilizations according to the materials they used. In China, green copper (bronze) and black iron each led the way for 2,000 years, creating a splendid bronze age during the Shang and Western Zhou dynasties and an iron age from the ancient times to the recent past (as shown in Pictures 6.1 and 6.2).

In a certain sense, we can say that China is a civilization built on bronze and iron, and many precious traditional metal crafts are the inheritance and extension of these beginnings.

H. Jueming (✉)
The Institute for the History of Natural Sciences, Chinese Academy of Sciences, Beijing, China
e-mail: huajueming@163.com

© Elephant Press Co., Ltd 2022
H. Jueming et al. (eds.), *Chinese Handicrafts*,
https://doi.org/10.1007/978-981-19-5379-8_6

Picture 6.1 Simuwu *Ding* of the Shang dynasty, unearthed in Yin Ruins in Anyang city

Picture 6.2 Iron Lion of Cangzhou in Cangzhou, Hebei province, dating back to the Five Dynasties period

6.1 The Bronze Age

6.1.1 The Bronze Smelting and Casting Industry in the Xia, Shang, and Zhou Dynasties

The Xia, Shang, and Zhou dynasties are called "the great three dynasties", which encompass "Bronze Age of China".

Bronze smelting and casting require thorough technical know-how and skill, as well as a lot of time to produce. To be able to support the creation of a smelting and casting industry, a society needs to be sufficiently advanced in agriculture; animal husbandry; transportation; and the production of stone wares, jade wares, and pottery. The relationship between bronze casting and pottery is particularly close. The materials for preparing pottery molds and furnaces, the methods and tools for making molds and cores, and the high-temperature technology needed for bronze smelting and casting are all inherited from and developed by the pottery industry. *Duke Wen of Teng I* in *Meng Zi* states: "Using millet to exchange kitchenware is not harmful for pottery smelting industry". In ancient times, the Chinese ancestors used to smelt and cast pottery. It can be concluded that the bronze smelting and casting industry came into being from the pottery-making industry.

The oldest metal relics discovered in China up to now are the brass sheet and tube unearthed at the Jiangzhai site in Lintong, Shaanxi province, which can be traced back to 5,000 years ago (as shown in Picture 6.3). During this period, bronze smelting appeared sporadically, seemingly disappearing and then popping up again, which indicates it being only in its embryonic state.

From the late Longshan culture to the early Xia dynasty, bronze smelting and casting technology was still in its infancy. *Gengzhu* in *Mo Zi* states: "The emperor of Xia once ordered Feilian to mine gold on mountains an in rivers, and Feilian smelted bronze wares on Kunwu Mountain". Once the Xia dynasty entered the Bronze Age, metal handicrafts became a stand-alone discipline. The first bronze alloys were smelted by using symbiotic ores or mixing different types of ore. Occasionally, they would use natural copper. The shape of bronze wares was simple (as shown in Picture 6.4). The casting process mainly consisted of single-sided and double-sided pottery molds. Stone molds were also widely used. Technically, they were still experimenting with molds and the best versions hadn't been determined yet. However, the early techniques of metal processing, which were mainly based on

Picture 6.3 The Yangshao culture's brass sheet and tubes, unearthed in Lintong, Shaanxi province

Picture 6.4 The Qijia culture's early copper wares, unearthed in Yongjing, Gansu province

casting, were perfected during that time. The appearance of gold, silver, and lead utensils indicates that they discovered that lead could be smelted and natural gold and silver could be cast as well.

In the late Xia dynasty, composite pottery was used to cast utensils such as the *jue* (wine vessel) (as shown in Picture 6.5), *jia* (wine cup), *ding* (cooking vessel), and *zhi* (drinking vessel), which were used in sacrificial ceremonies and rituals. Weapons and tools such as the *ge* (dagger-axe), *qi* (battle-axe), adze, and axe were large in size and solid for use. The bronze plate, shown in Picture 6.6, with turquoise inlay in the shape of a *taotie* (Chinese mythological creature), showcases the artistry and skill of that time. Based on findings at the ruins of Shang City in Zhengzhou, Henan Province, and Panlong city, Huangpi, Hubei province, the people of the early Shang dynasty were able to cast large-scale devices. The ruins show that there was a clear division of labor between the workshops. The bronze alloy used was smelted from copper, tin, and lead in one workshop, and the bronze items made from that alloy were cast in another. During the Shang and Zhou dynasties, composite pottery molds advanced at such a rate that it caused the formation and development of bronze smelting and casting. This was referred to as "the formation period".

After Pan Geng, a king of the late Shang dynasty, moved his capital to Yin, the state was relatively stable and the economy prospered. With Yin (nowadays Anyang) as the center, the bronze smelting and casting industry made great progress and its technology reached maturity. Its influence extended to Liaoning in the north, Huguang in the south, Haibin in the east, and Shaanxi and Gansu in the west. The skillful application of the separate casting method made it possible to constantly cast a large number of elegant and complicated bronze wares. The casting process became standardized, which made it possible to smelt different forms of bronze alloys and use them to cast pieces that required those different compositions. Small objects made of gold and objects covered with gold leaves became more and more popular, as well as funerary objects cast from lead. The early Western Zhou dynasty

Picture 6.5 *Jue* (wine vessel) of the Xia dynasty, unearthed in Yanshi, Henan province

Picture 6.6 Turquoise inlaid bronze plate in the shape of a *taotie* of the Xia dynasty

inherited the bronze casting methods of the Shang dynasty. Halfway through the Zhou dynasty, the bronze smelting workshops belonging to the royal court and its vassals created their own styles, with new kinds of vessels emerging such as the *fu* (vessel for boiled grain), *xu* (food vessel), *pan* (tray), and *yi* (gourd-shaped vessel). During this pinnacle of bronze smelting and casting, ritual vessels, mainly composed of *ding* and *gui*, completed the perfect set of vessels used for sacrificial ceremonies.

However, due to the decline of the ritual tradition from the late Western Zhou Dynasty to the middle of the Spring and Autumn period, some bronze wares were not further refined and remained coarse in shape. However, the bronze smelting and casting industry still developed on a larger scale and in a wider range. The wide application of long inscriptions, mold reproduction, and core bracing showed that expanding smelting and casting technology as deeply and broadly as possible was the main characteristic of this era. Because of this, it was called "the extension period".

The great change took place from the middle and late Spring and Autumn period to the beginning of the Warring States period. Great social changes brought new impetus to bronze smelting and casting. The casting of bronze wares was changed from the previous, relatively simple, single clay-mold casting technique to a series of new metalworking techniques such as one time casting, split casting, lost-wax casting, pewter welding, rivet weld casting, inlay, gold plating, engraving and hollowing, and forging; smelting techniques, however, remained the same. During this transition period, representative works, such as the Moiré Copper Ban, unearthed in Xichuan county, Henan province, Bianzhong (a set of bronze bells), Zun and Pan (set of wine vessels), the Square Desk with Patterns of Dragons and Phoenixes (as shown in Picture 6.7), and the Chen Zhang Pot, surpassed the creations from the previous period with their colorful and novel shapes, and gorgeous and fine decorative patterns, marking a technological breakthrough. The ancient mining and metallurgy sites at Tonglushan in Daye, Hubei province and in southern Anhui province showed the production scale and technical level of copper mining and smelting during this period. During the Warring States period, the number of descriptions for mining, smelting, casting, alloy preparation, and device manufacturing in books, such as *Book of Diverse Crafts* and *Classic of Mountains and Seas,* increased significantly, reflecting the deepening understanding of metallurgy and hinting at the prosperity the development of metal handicrafts brought (as shown in Picture 6.8). During this period, the smelting and metalworking techniques for lead, tin, gold, silver, mercury, and other non-ferrous metals developed simultaneously (as shown in Picture 6.9). Silver smelting adopted the ash-blowing method, also known as the lead-silver separation method, mercury refining and amalgam preparation reached a higher level, and the production of bronze currency, copper mirrors, hooks, and other devices for daily use increased exponentially. The momentum of technological progress was strengthened by the commercialization of bronze wares and the growth of private workshops. At this point in time, China entered the Iron Age. Iron and steel tools were already in use for making bronze wares at the beginning of the Warring States period, but their use was mainly restricted to agriculture and the handicraft industry. The Bronze Age passed its pinnacle and ended, to be replaced by the age of iron and steel.

It was possible that China had had a division of labor or expertise in the handicraft industry among tribes since the late Neolithic Age. *Commentary of Zuo* and *Selections from Lu's Commentaries of History* record that Xi Zhong was the Minister of Carts in the Xia dynasty, and the Kunwu people, who were good at pottery and smelting, were named Xia Bo (the leader of all the vessels of Xia), which could be used as evidence. After being defeated, some conquered or imprisoned tribes became workers and

6 Metallurgy and Metalworking

Picture 6.7 Inlaying gold and silver square desk with patterns of dragons and phoenixes, unearthed in Pingshan, Hebei province

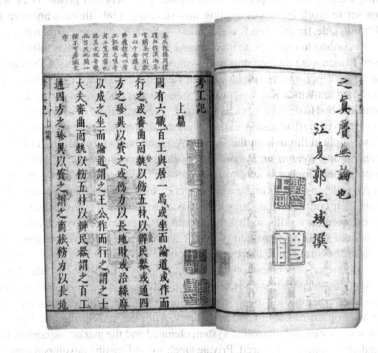

Picture 6.8 Photocopy of *Book of Diverse Crafts*

Picture 6.9 Golden wine cup, unearthed in Suizhou, Hubei province

slaves, for example, when the Zhou dynasty destroyed the Shang dynasty in Lu, the conquered Yin people were divided into seven ethnic groups and set to work, among which the Tao were set to make pottery and the Suo to make ropes. Archaeological excavations showed that copper casting workshops serving the royal family were permanently located near the palace. Both regular prisoners and prisoners of war were often set to work in copper mining and smelting. Only those who were in charge of "gold, jade, tin and stone" were called "human" (*Office of Earth* in *Rites of the Zhou*). According to oracle bone inscriptions from the Shang dynasty, officials in charge of craftsmen were divided into "left workers" and "right workers". Craftsmen were called "workers", and those below them were slaves of workers and smelters. Although skilled craftsmen had the status of free citizens, they still belonged to the government, and had to do hard labor as required. They only lived in order to do their jobs. This system of craftsmanship, called "industry and commerce only to serve the officials", was upheld throughout the Bronze Age and had a far-reaching influence in later generations. *Record on the Subject of Education* in *Classic of Rites* says: "The son of a good smelter is sure to learn how to make a *qiu* (bellow); that of a good bowyer to learn how to make a *ji* (quiver)". Learning art from an early age and working their craft for a long time enabled craftsmen who inherited their trades to achieve extraordinary skills so that they could perform almost instinctively. This was the benefit of skills being passed on within a family and one of the reasons why bronze wares in the Shang and Zhou dynasties could reach a very high level of quality. However, the personal attachment of craftsmen to the royal family and nobles led to their humble status and poor life, and those with low status sometimes became victims of blood sacrifice when casting tools, which seriously restricted the development and technological progress of mining and metallurgy. After the mid-Spring and Autumn period, when the social and economic system changed and the market expanded, the private handicraft industry appeared. Private smelting and casting developed rapidly during the Warring States period. For example, as mentioned in the literature, "casting passengers" were traveling artisans who walked among the people and were hired by workshops.

What is worth noting of this period was the rise and formation of the Five Elements Theory. Ancient civilizations such as Mesopotamia, Egypt, and Greece, all took natural objects, such as water, air, earth, and fire, as the elements of the formation of all things in the universe. Only the Chinese Five Elements Theory regarded the "metal" obtained from artificial smelting as a basic material that was part of the formation of the world. It states that the five elements of metal, wood, water, fire, and earth are interrelated and endless. The formation of this ideology was partly a reflection of the historical fact that bronze production tools and other devices had been widely used by ancestors and had played an important role in the development of society and life, all possible under the impetus of advanced bronze smelting and casting techniques. It was no accident that the Five Elements originated in the Shang and Zhou dynasties and became a theory used to explain the origin of the universe and the operation of all things during the Warring States period. This showed that the highly developed bronze smelting and casting techniques of the Shang and Zhou Dynasties had a profound impact on later societies' daily life, economy, and culture, including people's thoughts and cognition.

6.1.2 Copper Mining and Smelting Technology

Like other ancient civilizations in the world, China's early copper mining started with open mining and then pit mining. When mining, vertical shafts, inclined shafts, and flat roadways were often used together (as shown in Picture 6.10).

Common mining tools were a stone hammer, a copper axe, and an iron punch. The big copper axe produced at the Tonglushan Mine in the Spring and Autumn period weighs more than 16 kg (as shown in Picture 6.11). Miners would swing them overhead and use the inertia to hit the rock surface. Some ancient mine walls show traces of smoke, which are the remains of mining under high temperatures.

Picture 6.10 The Tonglushan mining site in Daye city, Hubei province

Picture 6.11 Big copper axe of the Spring and Autumn period, unearthed from Tongxianshan, Daye city, Hubei province

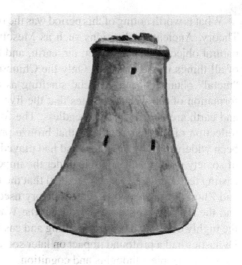

The copper ore belt in the middle and lower reaches of the Yangtze River belonged to the ancient Jingzhou, with soft surrounding rocks and highly developed wooden structure supports in the shafts and roadways. The supporting structure consisted of a bamboo and wood frame with back plates, which were mostly fabricated on the ground beforehand and put in place while digging (as shown in Picture 6.12). The roadway structure of the Shang dynasty only allowed people to crawl through, while the roadway in the Western Zhou dynasty and the Spring and Autumn period allowed people to squat and bend, and miners in the Warring States period were able to work upright.

Underground lighting was often achieved by lighting bamboo strips or oil in bamboo buckets. Ventilation depended on the natural airflow, sometimes by connecting multiple tunnels or repaired wind barriers to control the direction of the airflow and its volume. Groundwater was collected at the water warehouse through drains and drainage channels, and lifted out of the well. Ore was packed in bamboo baskets and dragged back by hand. The exit of the shaft was equipped with pedals or wooden ladders to facilitate going up and down. Well sweeps, well windlasses, and pulley components were found in the Tongling, Tonglushan, and Nanling mining sites in Ruichang, Jiangxi province, which were used to lift and transport materials. There were two methods of ore processing: hand selection and elutriation. Elutriation method was commonly used, that was, using different specific gravity of minerals and rocks in water medium to separate useful minerals. Ruichang's Tongling site has a wooden chute, tail sand pool, and water filtration platform. The ore materials in the chute were washed with water. Mud and rock particles were washed into the tail sand pool, while coarse ore particles were deposited at the bottom of the tank and finer ore particles moved forward along the bottom of the tank. This way ore was recovered and distributed by levels of size, and the waste was discharged through the tail sand pool.

6 Metallurgy and Metalworking

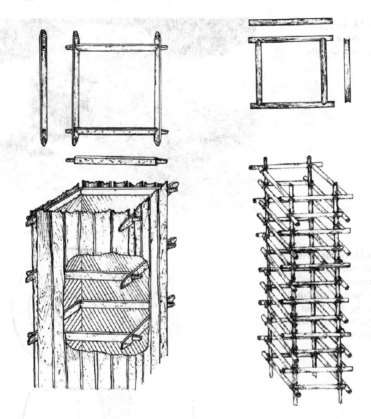

Picture 6.12 Frame structure of a shaft and roadway, dating back to the Western Zhou dynasty and the Warring States period

Smelting operations were mostly active within the same mining area. A copper smelting furnace consisted of a furnace base, a hearth, and a shaft with a wind ditch to prevent the bottom from freezing (as shown in Pictures 6.13 and 6.14). The furnace wall was equipped with a "metal gate" for discharging slag and copper.

Copper oxide ore (as shown in Pictures 6.15 and 6.16) was mostly used in copper smelting. Charcoal was used as a reducing agent and as fuel, and iron minerals were used as flux. A shaft furnace produced about 200 kg of crude copper per day. Smelting slag was mostly black and flake-shaped, with a smooth and corrugated upper surface, which indicated that the smelting furnace ran normally and the fluidity of slag was good. The obtained crude copper contained between 93 and 94 percent copper.

Copper deposits found in nature were mostly sulfide deposits, and the upper parts became oxidized due to being exposed to the elements. With the passage of time, after the oxide ore was exhausted, it was necessary to smelt copper with sulfide. The analysis and detection of rhombic copper ingots and smelting slag unearthed in Guichi, Tongling, and Nanling in Anhui Province showed that the "sulfide ore-copper sulfide-copper" method of smelting, desulfurization, and repeated refining

Picture 6.13 Copper smelting furnace at the Tonglushan mining site

Picture 6.14 Section view of a copper smelting shaft furnace (restoration view).
1. furnace base 2. wind ditch 3. golden gate 4. exhaust hole 5. tuyere 6. inside wall 7. workbench 8. furnace wall 9. floor

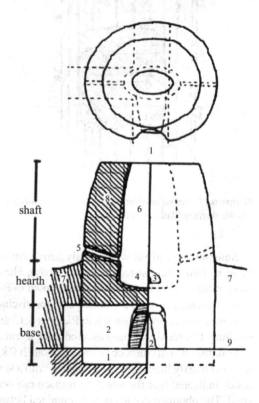

had been adopted since the late Western Zhou dynasty (as shown in Picture 6.17). This is a great event in the history of mining and metallurgy, and explains why the bronze smelting and casting industry in the Shang and Zhou dynasties had such a continuous and abundant supply of copper.

Picture 6.15 Native copper from Tonglushan, Daye city

Picture 6.16 Malachite from Tonglushan, Daye city

6.1.3 Preparation of Bronze Alloys

Book of Diverse Crafts lists the copper-tin ratio of alloys used in six kinds of utensils: cauldrons, axes, spears, swords, arrowheads, and bronze mirrors. This shows that there was a good understanding of changing alloy properties during the Warring States period. After physical inspection, it can be seen that the alloy ratio of several bronze ritual vessels, musical instruments, tools, swords, and cutting tools are similar to that of the above-mentioned six utensils. For example, the tin content of Marquis Yi of Zeng's Bianzhong (a set of bronze bells) is about 14%, which is similar to the balance of 60% copper and 10% tin. The tin content of many pieces of copper unearthed in Guangdong and Henan provinces was as high as 28% to 30%, which

Picture 6.17 Diamond copper ingot and its metallographic structure (lower part)

was similar to the composition of the above-mentioned arrowheads with 50% copper and 20% tin. It can be seen that the mix of 60% copper was indeed the culmination of craftsmen's long-term practical experience with smelting and casting and a major achievement in alloy research. However, in practice, the alloy composition of many bronze wares did not consistently contain 60% copper, which showcases the difficulty and reality of trying to synchronize bronze smelting and casting techniques in the Warring States period. *Book of Diverse Crafts* had been compiled by later generations, and the existing version could have been altered. For example, it is doubtful that it includes two devices with very different performance requirements, namely, cutting and killing vectors, in the same mix.

One of the technical characteristics of creating bronze alloy in the Shang and Zhou dynasties was the serialization of tin bronze, which was unique among the various ancient civilizations. Another was the wide use of copper-tin–lead ternary alloy, which was also rare in the ancient world. The ancient names for metals are common in bronze inscriptions. With the formation of standard alloy preparation, the inscriptions became more stylized and were added to the cast items. For example, the typical format could be written as "*Xuan Liu Chi*", which means that the item was made of black tin and red copper. Most of the bronze wares in the Shang and Zhou dynasties tested, especially the royal treasures and pewter bells, were pure in composition and had few impurities, which showed that the alloy preparation of bronze wares in this period was indeed quite advanced.

6.1.4 Clay-Mold Casting

The foundation of the highly developed composite clay-mold casting techniques of the Shang and Zhou dynasties was laid between the late Xia dynasty and the middle Shang dynasty. The Rectangle *Ding*, an ancient cooking vessel unearthed in Zhengzhou, was made by multiple casting. This technique matured in the Shang dynasty, as can be seen from *ding*, the ancient cooking vessels found at the Yin Ruins, such as *Simuwu Ding*, Oxen *Ding*, and Deer *Ding*. The molds used for casting were made of clay, fine sand, and soil. Once made, the mold was turned over, dried in the shade, and baked in the kiln. The movable furnace consisted of a furnace seat and a furnace ring. The diameter of the fixed furnace ranged between 0.6 and 0.7 m. Charcoal was used as fuel and animal skins for bellows. After the metal had melted in the furnace, it was injected into the ladle for pouring, and large-scale devices were injected into the mold troughs in multiple furnaces. After the metal was stripped, burrs and flash were removed, and then a stone was used to sharpen and polish the metal again.

During the Shang dynasty, the key to obtaining highly complex shapes was the use of separate additional casting procedures, which included the post-casting procedure and the pre-casting procedure (as shown in Pictures 6.18 and 6.19). The multi-step casting technique, however, was more complicated (as shown in Picture 6.20).

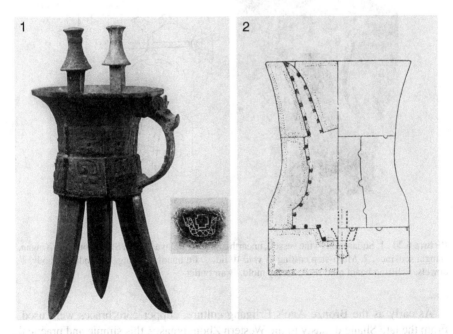

Picture 6.18 1. *Yaqi Jia* (wine cup) of the Shang dynasty, unearthed at the Yin Ruins, Anyang city, Henan province; 2. *Jia* casting mold

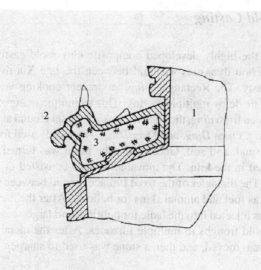

Picture 6.19 Additional casting of a square *lei* (wine jar) 1. body 2. attachments 3. soil core

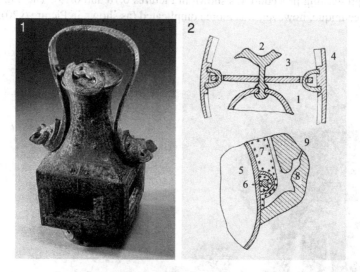

Picture 6.20 1. Square *you* (wine vessel), unearthed from the Dayangzhou Shang tombs in Xingan, Jiangxi province; 2. Multi-step casting of *you:* ① lid, ② lid handle, ③ ring, ④ handle, ⑤ body, ⑥ dowels, ⑦ lifting beam mud core, ⑧ clay mold, ⑨ air outlet

As early as the Bronze Age's Erligang culture, copper core braces were used. From the late Shang dynasty to the Western Zhou dynasty, this simple and practical technology gradually spread, which played an important role in improving the yield of castings.

6 Metallurgy and Metalworking

Picture 6.21 1. Great Round *Ding* of the Western Zhou dynasty, unearthed in Chunhua, Shaanxi province; 2. Rubbings of the inscriptions on the *Hu Ding*, which writes that five slaves are worth as much as a horse and ten hanks of silk (according to Guo Moruo's textual research)

The ritual vessels from the Western Zhou dynasty are characterized by their long inscriptions, which concerned many aspects, such as enfeoffment, giving life, sacrifice, conquering, meritorious services, banquets, marriages, and civil affairs (as shown in Picture 6.21). In the Shang dynasty, characters were engraved in intaglio in the mold, then turned over so they function as characters in relief, to finally be cast as characters in intaglio. Long inscriptions in the Western Zhou dynasty were either attached with clay pieces or turned over with core boxes. During the Spring and Autumn period, the technique of printing words by using impressions appeared.

The movable block molds and movable block patterns were already used in the early Western Zhou dynasty. Complex shapes could be created this way, which would have been difficult to achieve using conventional modeling practices (as shown in Picture 6.22). The progress of casting techniques mainly manifested itself in the way molds were remade and reused. In order to meet the needs of mass production, the Houma copper casting site in Shanxi province used the first casting mold again as a second-generation mold and then as a second-generation pattern, forming a complicated and highly mature technological system (as shown in Picture 6.23).

During the Warring States period, the outstanding achievements in terms of clay molds are mostly exemplified in the casting of the large bells in Marquis Yi of Zeng's *Bianzhong*, which is an ancient musical instrument consisting of bronze bells (as shown in Picture 6.24). This set of bells consist of eight groups, totaling 64 bells. The inscriptions on the inlaid gold of the bells contain more than 2,800 words, marking the notes and temperament. Each bell can produce two notes, giving the whole set a range as high as five octaves, which can move in the 12-pitch scale. All the bells were made using composite pottery molds. The *yong* part of the bell was cast together with the body, while the ornamentation of each part of the body was made

Picture 6.22 Part of the *Bo Gui* (food container), unearthed in Baoji, Shaanxi province

Picture 6.23 Box with patterns of animals, unearthed at the Houma copper casting site in Shanxi province

with separate-mold casting (as shown in Picture 6.25). The sounding mechanism of the two tones of the *Bianzhong* could be revealed by using laser holography (as shown in Pictures 6.26 and 6.27). This large-scale Bianzhong combines the achievements of the pre-Qin bronze smelting and casting techniques with the achievements made in acoustics and melody. In the past 2,500 years, a wonder like this has not often been witnessed.

6 Metallurgy and Metalworking

Picture 6.24 Marquis Yi of Zeng's *Bianzhong* (set of bronze bells)

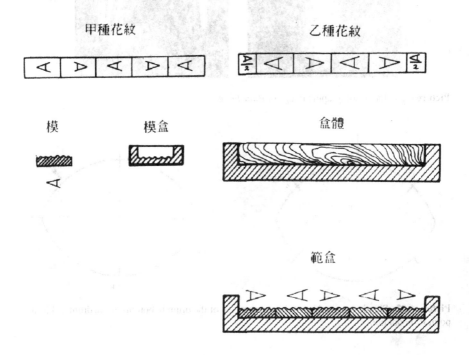

Picture 6.25 Making and molding of strips of characters as ornaments

The bronze sword of the Wuyue kingdom, also known as the Sword of Goujian (as shown in Picture 6.28), is another piece that exemplifies Shang and Zhou dynasty bronze wares. Its forward is restrained, curved twice, and there are blood grooves in two clusters. The sword is scientifically shaped. The body of the sword is decorated with diamond-shaped dark patterns, and there is a bird seal inscription near the grid: "the self-made sword of Goujian, king of Yue"; the sword grid is inlaid with blue glazing and turquoise on both sides. The end of the sword's pommel is decorated with narrow and deep concentric circles. The central ridge of the sword was cast first and

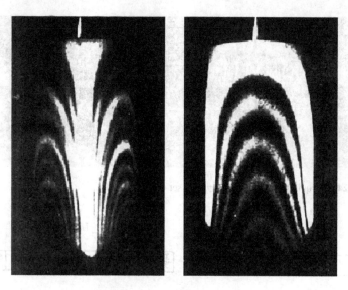

Picture 6.26 Laser holographic image of *Bianzhong*

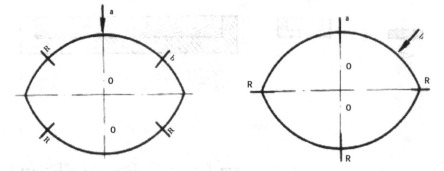

Picture 6.27 Phonic mechanism of the drum: a. top of the drum b. bottom of the drum ↓. hitting point R. nodal line

then the blade. The former contains a lower tin content, while the latter has a higher tin content. The composite sword thus formed is rigid on the outside and soft on the inside, which makes it harder to break (as shown in Picture 6.29). The concentric circles on the pommel were made by scraping them into the mold. The dark diamond pattern was treated with a paste with a high tin content, which formed the pattern over a long period of time. *On Boardland* in *Xun Zi* writes, "With a standard mold, high quality gold and tin, the correct use of fire, and a skilled blacksmith, it is easy to make a sword". Wuyue had a technological tradition of making high-quality bronze swords. The Sword of Goujian represents the highest level of bronze sword casting.

6 Metallurgy and Metalworking

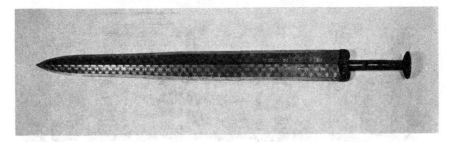

Picture 6.28 The Sword of Goujian, unearthed in Jiangling, Hubei province

Picture 6.29 Section of the compound sword

6.1.5 New Metalworking Techniques

The casting of bronze wares changed from the previous, relatively simple clay-mold casting technique of the Spring and Autumn period to a series of new metalworking techniques in the Warring States period, such as one time casting, split casting, lost-wax casting, pewter welding, rivet weld casting, inlay, gold plating, engraving and following, and forging. Below we will briefly introduce each technique.

6.1.5.1 Lost-Wax Casting

The earliest known product made by lost-wax casting is King of Chu's *Yu* (water vessel), which is now in the Metropolitan Museum of New York. The button on the lid is only 2 cm. It is composed of three copper rings and several copper stalks,

Picture 6.30 The Moiré Copper Ban, unearthed in Xichuan county, Henan province

which shows the characteristics of wax strip kneading. This broad-mouthed vessel was owned by Xiong Shen, the king of Chu, and was cast in 560 BC or even earlier.

One of the most famous metal artifacts is the Moiré Copper Ban unearthed from the tomb of Zigeng, the Minister of Chu state, in Xichuan, Henan province, which is 131 cm long and weighs 94 kg (as shown in Picture 6.30). Its entire surface is comprised of openwork patterns. Its gating system consists of four layers. It was cast in 552 BC or later.

Another artifact that was cast around the same time as the Moiré Copper Ban is one with openwork coiled chi-dragon patterns unearthed from Xu Gongning's tomb in Yexian county, Henan province. After inspecting it with an industrial CT scan, no traces of welding nor multi-step casting were found, which confirmed that it was made by lost-wax casting (as shown in Picture 6.31).

Marquis Yi of Zeng's Bronze *Zun* (wine vessel) and *Pan* (plate) unearthed in Suizhou, Hubei Province is one of the Shang and Zhou dynasties' bronze treasures (as shown in Picture 6.32). Both the front edge and the openwork accessories around the *Zun* were cast using the lost-wax method. There are 19 kinds of pattern units on the edge of the *Zun*, which form a carefully arranged, gorgeous, and elaborate wreath (as shown in Picture 6.33). Up to now, more than ten items made by lost-wax casting have been discovered dating back to the pre-Qin period, which proves that this technique had already reached a high level since the middle of Spring and Autumn period at the latest, and its origin dates back to an even earlier time. All these show that the lost-wax casting method with Chinese characteristics is a brilliant creation and great technical achievement of pre-Qin casters, and the judgement such as "lost-wax casting is not the choice of Chinese Bronze Age" or "there is no example of lost-wax casting (in pre-Qin period)" is groundless.

6.1.5.2 Rivet Weld Casting

The most common form of pewter welding in the Spring and Autumn period and the Warring States period was to pre-cast tenons on the wall of the vessel, inject molten solder into the accessory's cavity, and then connect them with the vessel's body after

Picture 6.31 Artifact with coiled *chi*-dragon patterns dating back to the middle-to-late Spring and Autumn period, unearthed in Yexian county, Henan province. The bottom image is the industrial CT scan of the artifact

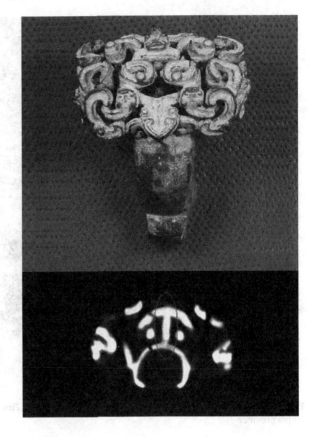

it has solidified. Later devices, such as *ding* and *jian* (mirror) were often pre-cast with mushroom-shaped tenons to make the connection stronger, such as the Wang Ziwu *Ding* found in the Chu tomb in Xichuan, Henan province. During the Warring States period, there were more forms of rivet weld casting, as can be seen on the bronze mirror of Marquis Yi of Zeng's tomb, which had holes in the handles where solder was used to connect them. For the Marquis Yi of Zeng's Bronze *Zun* and *Pan*, the copper sleeve of the bell beam was filled with lead–tin alloy, and the bronze ring foot accessories were welded by rivet weld casting (as shown in Picture 6.34).

Copper brazing was also in use in the late Spring and Autumn period. For example, the handles of the *Wang Ziwu Ding* were inserted into the cavity and then brazed with liquid copper. The base of Marquis Yi of Zeng's drum and the dragons on it were cast and welded with tin and bronze. All these are different from modern welding techniques. They are unique techniques created under the technical conditions of the times.

Picture 6.32 Marquis Yi of Zeng's Bronze *Zun* (wine vessel) and *Pan* (plate), unearthed in Suizhou, Hubei province

Picture 6.33 Structure of the openwork patterns on the *Zun*

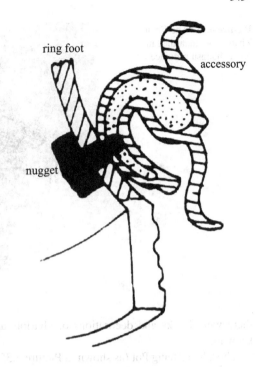

Picture 6.34 Rivet weld casting of ring foot accessories Marquis Yi of Zeng's bronze *Zun*

6.1.5.3 Inlay

During the Warring States period, staggered inlay ornaments depicting realistic fishing and hunting, feasting, warfare, and fighting prevailed. In addition to pressing red copper sheets and strips into grooves and adding staggered inlays, some sheet of red copper ornamentation is pre-cast and then put into the mold to be cast into the wall of the vessel, such as the *guanfou* (water vessel) unearthed in Marquis Yi of Zeng's tombs.

The *Luanshufou* (water or wine vessel) was the first to make staggered inlays in gold, other famous similar artifacts were The Golden Tally to Monarch Qi of the Er found in Shouxian county, Anhui Province, and the Square Desk with Patterns of Dragons and Phoenixes found in the tomb of the Prince of Zhongshan.

6.1.5.4 Gold Plating

The amalgam used for gold plating is made of gold foil and mercury. The making process is like as follows. The clean surface of artifacts is first coated with gold amalgam. Then it is heated, making the mercury evaporate and the gold cling to the surface. To make the gold layer denser, this process is often repeated three to six times. The earliest known gold-plated artifacts were found at the turn of the Spring and Autumn period and the Warring States period. During the Warring States Period,

Picture 6.35 The Chen Zhang Pot, unearthed in Xuyi, Jiangsu province

there were hooks and decorations of chariots and horses made in Changsha and Luoyang.

The Chen Zhang Pot (as shown in Picture 6.35) is the most representative artifact to showcase the application of various techniques, thus fully proving the outstanding achievements and level of metallurgy in this period.

6.1.6 Influence of Bronze Smelting and Casting on Ancient Technical Concepts and Social Cognition

In most pre-Qin ancient books, smelting and casting meant the same, and sometimes casting was used to refer to smelting, from which we can see that the meaning of "casting" was broader than that understood in modern times. The fact reflects the unique technical concept of ancient Chinese ancestors. Casting was considered the "big smelting" and forging the "small smelting". *Great Master* in *Zhuang Zi* writes: "Take heaven and earth as the big furnace and nature as the big smelting". The influence of this idea derived from the casting-based technological tradition was so long that there were still some words like "casting knives" and "casting swords" in literature long after knives and swords were forged instead of being cast.

The inscription on the Qiang Basin, a bronze ritual vessel of the Western Zhou Dynasty, contains such a sentence "井帅宇诲" (meaning "implementing uniform laws for people under heaven to follow"). "井" (meaning literally "well" in Chinese) is the same as "型", meaning "teaching by serving as a role model". The everyday Chinese words like "mofan" (meaning "model"), "fanwei" (meaning "scope"), "fanchou" (meaning "category"), "guimio (meaning "scale"), and "guifan" (meaning "standard") were originally terms of smelting and casting industry, which had gained universal meanings for a long time and changed into words such as "shifan" (meaning "teacher training"), "mohu" (meaning "vague"), "xingxiao" (meaning "marketing"), and "xiaoshou" (meaning "sell").

6.2 Metal Civilization in the Iron Age

Iron was first used in the Shang Dynasty. The evidence of coahuilite being cast with bronze indicated that iron-making techniques were produced in the bronze smelting and casting (as shown in Picture 6.36). Iron-making techniques appeared in the late Western Zhou dynasty (as shown in Picture 6.37), and the iron age began at the Spring and Autumn period and the Warring States period. Over 2,400 years, mining and metal techniques evolved on the basis of iron-making techniques and the metallurgical industry.

6.2.1 Metal Handicraft Industry in the Iron Age

The time between Warring States period and the mid-Western Han dynasty was not a complete Iron Age. *King of Sea* in *Guan Zi* says: "A woman must have a needle and a knife" and "a tiller must have a *lei* (similar to a spade), a rake and a cymbal", which was the portrayal of the popularization and application of iron devices. *Biography of Merchants* in *Records of the Historian* records that Guo Zong in Handan, Zhuo's family in Zhao state, Kong's family in Wei state, and Cao Li's family in Lu state

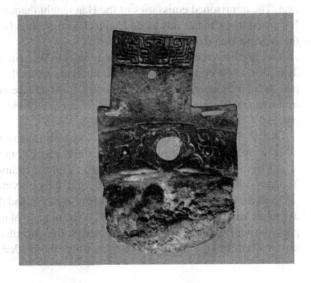

Picture 6.36 Battle-axe with iron blade and a bronze handle, unearthed in Gaocheng city, Hebei province

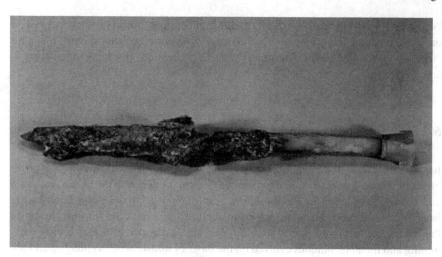

Picture 6.37 Iron sword with a jade handle

were all famous for metallurgy (iron smelting). During the Warring States period, the flourishing economy and the contention of a hundred schools of thought were based on the metallurgical production.

Metallurgy industry in the Han dynasty really had a national scale, and 49 iron officials in the Western Han dynasty were set in Shandong, Henan, Jiangsu, Hebei, Shaanxi, Shanxi, Sichuan, Gansu, Liao, Anhui, Hunan, and other provinces. Remote and ethnic minority areas such as Qiuci, Nuoqiang in Western Regions, and Dianchi Lake in Yunnan also mastered iron smelting technology (as shown in Picture 6.38). *Biography of Western Regions* in *History of Han* says: "From the west of Yuan (an ancient country in Central Asia) to the Parthia...people there didn't know how to cast iron. The imprisoned emissaries of the Han taught them to cast iron into weapons". It showed that China's advanced iron technology already spread to Central Asia in the Western Han dynasty.

The time between late Western Han dynasty to Wei, Jin, the Northern and Southern dynasties was a complete iron age. Iron frying technology came into being in the mid-Western Han dynasty at the latest. With the increase of wrought iron devices, all iron weapons replaced bronze weapons, marking the basic completion of ironization process.

During this period, bronze smelting and casting currency and mirrors were the most prosperous. Ethnic minorities in Southwest China were able to cast bronze drums from the Spring and Autumn period, and the technology gradually spread to Yunnan, Guizhou, Guangxi, Guangdong, and other provinces after the Han dynasty.

From the late Western Han dynasty to the Wei, Jin, and the Northern and Southern dynasties, steel making developed greatly. The method of making steel by combining pig iron with wrought iron was invented at the end of the Eastern Han dynasty, and evolved into the method of *sutie* casting (the predecessor of pouring steel) in

Picture 6.38 Ironwares in the Warring States period and the Han dynasty: 1. hexagonal hoe in the Warring States period. 2. iron pickaxe in the Western Han dynasty. 3. iron fork in the Eastern Han dynasty. 4. iron fork in the Eastern Han dynasty

the Northern and Southern dynasties, which could obtain high-quality medium and high-carbon steel.

Iron-making technology had made great progress by that time, and advanced water-driven bellows had been used. *Commentary on the Water Classic* quotes *History of Xicheng of Shi's Clan* as saying that "there is a mountain 200 miles north of Quci, and there are fires at night, but smoke in day on the mountain. People take carboniferous rocks from the mountain and smelt them into iron, which are enough for the supply of 36 countries". The passage indicates that coal was used for iron making in the Jin dynasty at the latest.

In the Han dynasty, prisoners and soldiers worked in mining and metallurgy, which continued in Wei, Jin, and the Northern and Southern dynasties. However, craftsmen were restricted and hur as usual. It is recorded in the imperial edict of in the fifth year of the Taiping Zhenjun period (444) in *Biography of Shizu* in *Book of Wei*: "The servants and descendants of hundreds of workers must learn the craft from their masters and forefathers instead of receiving ordinary learning; otherwise, offenders will be punished and put to death". These outdated conventions severely oppressed the craftsmen and seriously restricted the progress of metal technology.

The time between the Sui and Tang dynasties to the Ming and Qing dynasties belonged to the late iron age. At the beginning of the Tang dynasty, there were 168 iron pits, which increased to 271 in the mid-Tang. According to *Treaties of Foods and Commodities* in *New Book of Tang*, the annual output of iron in the Yuanhe period (806–820) during the reign of Xianzong reached more than 5,000 tons. After the mid-Tang, wars began to occur frequently, and the mining and metallurgy industry in the Yangtze River Basin began to surpass that in the northern region.

After a long period of integration and selection, the great inventions of iron smelting in the Warring States period and the Western Han dynasty gradually tended to be finalized in Tang and Song dynasties, forming an iron and steel process system with Chinese characteristics, which was the most significant technological achievement in this period (as shown in Picture 6.39).

Rich and exquisite gold and silver wares are important achievements of metal technology in the Tang dynasty. Their shapes and ornamentation were obviously influenced by Central Asia and West Asia, showing the characteristics of opening to the outside world in the Tang dynasty.

The scale of official mining and metallurgy in the Northern Song dynasty was far larger than that in previous dynasties, and more than half of them were in Jiangnan area. There was a great growth in private mining and metallurgy industry. For example, according to volume II of *Dongpo's Memorial*, there were "36 smelters, which were large-scale, and each one had more than 100 workers" in Liguo Iron Mine in Xuzhou.

Bile copper method was an outstanding technological invention in the Song dynasty, which reached 15% to 20% of the total copper output in the Northern Song dynasty, and was the earliest scale production of hydrometallurgy in the world. There were 19 workshops in the Imperial Household Department, such as copper, iron, tin, grinding, dragging, and gold plating, which showed the fine division of labor and the specialization of skills.

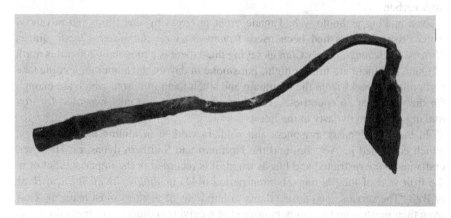

Picture 6.39 Iron hoe of the Northern Song dynasty, 63.7 cm long, marking the transformation of farm tools from casting to forging

6 Metallurgy and Metalworking

In the Yuan dynasty, the General Office of Mining and Smelting was set up to govern the whole country's pit smelting. The Wax Production Bureau, Drum Casting Bureau, and Wrought Iron Bureau were in charge of the production of court utensils and weapons; Smelting Office, Silver Office, and Gold Panning Office were set up to levy taxes from craftsmen. This structural retrogression in economy and the prohibition of holding iron blades among the people seriously hindered the development of metal handicraft industry.

At the beginning of the Ming dynasty, the annual income of iron was as high as more than 9,000 tons. Therefore, official metallurgy was closed, and private enterprises were allowed. The tax rate was reduced to 1 out of 15, and the mining and metallurgy industry grew greatly. Hunan and Hubei Provinces received more than one-third of the national iron. Foshan, Guangdong province had become an iron smelting center by the middle of the Ming dynasty. It is recorded in *New Writings of Guangdong* by Qu Dajun of the Qing dynasty that there were nearly 1,000 craftsmen in a smelting workshop with a daily output of more than 20 tons of iron. The main producing areas of copper were Sichuan, Yunnan, and Guizhou provinces. Tin was produced in Hexian and Nandan, Guangxi province; Hengyang, Hunan province; and Chuxiong, Yunnan province. Mercury mostly came from Guizhou province. Zinc reached mass production in the middle of the Ming dynasty, and brass was widely used for casting currency and various utensils (as shown in Picture 6.40). With the expansion of market economy and the formation of civil society, folk workshops flourished and guild organizations were established. In the tenth year of the Chongzhen period (1637), *Heavenly Creations* written by Song Yingxing of the Ming dynasty was published, in which *Smelting and Casting*, *Forging* and *Hardware* accounted for about a quarter.

In the Ming dynasty, the autocratic rule was strengthened, and the government even set up a smelting tax-levying department to collect the tax, which caused strong resistance from the people. The mining and metallurgy industry that once flourished in Yunnan Province and other places soon tended to decline. It was quite insightful to describe the disadvantages of mining administration in *Treaties of Foods and Commodities* in *History of Ming* as "those wise men know that this is sign of the fall of the Ming dynasty".

In the Qing dynasty, mining and metallurgy were mainly private, which were divided into the balance of 20% and 80% for the government and the owners. In the second year of the Shunzhi period (1645), the craftsman's membership was abolished, which lifted the personal attachment and labor of craftsmen to the government for a long time, and promoted the mining and metallurgy industry to exceed the level of the middle Ming dynasty. During Kangxi period (1662–1722), copper mines in Dongchuan and other places in Yunnan province were developed on a large scale, with an annual output of more than 7,000 tons, which was called Dian Copper (Yunnan Copper). Large copper factories hired thousands of craftsmen to mine ore, and the profit was divided equally among the government, miners, and craftsmen. Some craftsmen were paid monthly and free to leave or stay. This mode of operation had the nature of wage labor. The Internal Affairs Office, which served the imperial family, had copper workshops, iron workshops, knife workshops, and enamel workshops.

Picture 6.40 Acupuncture bronze man of the Ming dynasty, imitating that of the Song dynasty

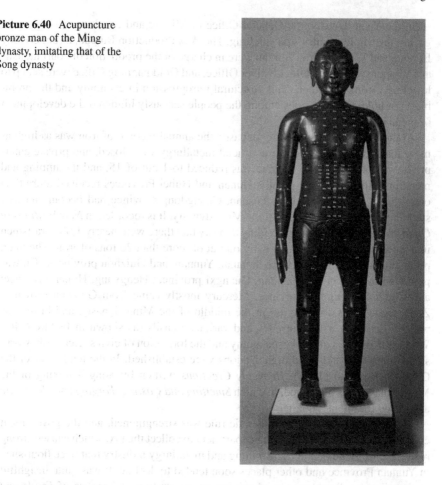

The materials, techniques, and prices of materials had been standardized. During the Qianlong period (1736–1795), Benxi in north China became the center of coal mining and iron making. There were many shaft furnaces and thousands of craftsmen in Tieluchuan, Hanzhong, Shaanxi Province. Folk workshops were more developed than before. In the late Qing dynasty, there were nearly 100 workshops for copper casting beating copper inside and outside Beijing. Advanced western metal technology was introduced to China after the Opium Wars at the end of the nineteenth century. The drastic changes in society made it impossible for traditional metal technology to change to modern metal technology.

6.2.2 Pig Iron Smelting and Casting

Iron smelting in China originated from bronze smelting and casting. The early metal process tradition, which was dominated by casting, made casters instinctively seek ways to obtain liquid iron and cast it into shape. Block smelting furnaces had been used in Europe from ancient times to the Middle Ages, producing spongy wrought iron containing slag, and pig iron only appeared in the fourteenth century. In early China, the shaft and furnace temperature of iron-making shaft furnaces were quite high, which enabled the reduced iron to melt into more carbon, thus lowering the melting point and allowing the iron discharged as liquid (as shown in Pictures 6.41 and 6.42). This is the biggest feature and advantage of traditional Chinese iron-making techniques, from which a series of great inventions and creations were derived, such as softening of cast iron, iron-mold casting, and iron frying, which were 1,600 to 2,000 years earlier than those in Europe.

6.2.2.1 Pig Iron Smelting

The iron-making shaft furnace in the Warring States period was about 1.5 m high and had a furnace capacity of about 0.4 cubic meters. In the Han dynasty, the furnace shaft was increased to 2–3 m, and the ore contained 50% iron. Charcoal was used as fuel and reducing agent, and limestone was used as flux.

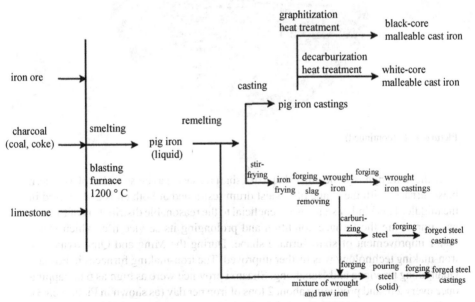

The iron and steel technology route in ancient China

Picture 6.41 Metallographic structure of pig iron

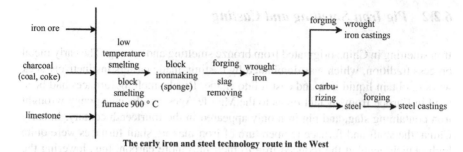

The early iron and steel technology route in the West

Picture 6.41 (continued)

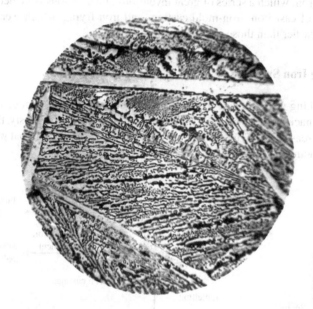

Picture 6.41 (continued)

In the Song dynasty, the inner shape of a shaft furnace was close to that of a modern blast furnace, with the shape of a waist drum restrained at both ends and relaxed in the middle. This kind of shape was beneficial to the reasonable distribution of furnace gas, improving the furnace condition and prolonging its service life, which was a major improvement of shaft furnace shape. During the Ming and Qing dynasties, iron-making technology was further improved. The iron-making furnaces in Foshan, Guangdong province and Hanzhong, Shaanxi province were as high as 6 m, tapping once every 2 h, and producing about 2 tons of iron per day (as shown in Picture 6.43).

Crucible iron making was an original creation in China and prevailed in Shanxi, Henan, Shandong, and Liaoning provinces. The furnace was square, 1-m-high and

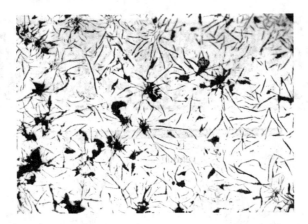

Picture 6.42 Metallographic structure of gray cast iron

Picture 6.43 Traditional iron-making shaft furnace in Yunnan province

large enough to hold 100 to 300 crucibles, which were filled with ore, anthracite and "black soil" (inferior pulverized coal, used as flux). Iron could be produced during 24 h after ignition. According to the statistics in 1899, Jincheng, Yangcheng, and Taiyuan in Shanxi province had an annual output of 50,000 tons of crucible iron, accounting for about a quarter of the national iron production.

As the saying goes: "Where there is air, there is iron" and "good air makes good iron", Blast technology was of great importance to iron smelting. The ox hide bellows in the Han dynasty were simple in structure and reliable in operation, and could be used for large smelting furnaces (as shown in Picture 6.44). It was a great innovation for blast technology to develop from manpower drive to animal power and hydraulic drive. According to *Biography of Du Shi* in *History of Later Han*, when Du Shi was

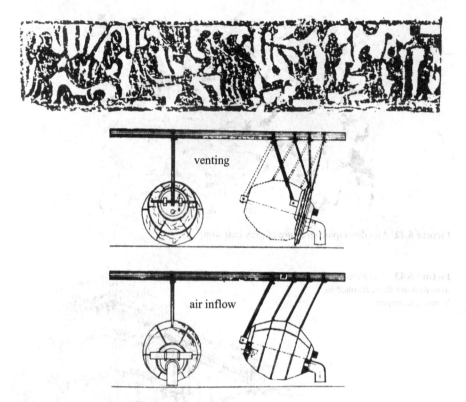

Picture 6.44 Rubbing of the stone relief of iron smelting and casting in the Eastern Han dynasty in Hongdao Academy in Tengzhou, Shandong province and restoration of the ox hide bellow

the prefect of Nanyang, he was "invented the water-driven bellows to cast agricultural tools. This type of bellows was labor-saving and efficient, which was quite convenient". The blast device invented by Du is still widely used in many provinces in modern China.

6.2.2.2 Softening Technique of Cast Iron

Pig iron was brittle and not resistant to bumping. Pig iron could be refined in Europe during Roman times, but it was regarded as useless because it was broken when forged. Chinese casters adhered to the craft tradition handed down from generation to generation, and creatively adopted the method of softening treatment at high temperature to solve this problem.

As early as the Warring States period, decarburization heat treatment of cast iron was carried out in oxidizing atmosphere, and white-core ductile cast iron and decarburized steel of cast iron were obtained. The latter could be forged into various

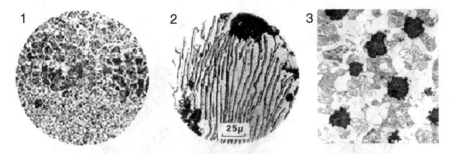

Picture 6.45 Metallographic structure of ductile cast iron. 1. white-core malleable iron 2. black-core malleable iron 3. spherical graphite malleable iron

devices because of more complete decarburization. The other way was the graphitization heat treatment in neutral atmosphere, which could produce black-core ductile cast iron and ductile cast iron with spheroidal graphite (as shown in Picture 6.45).

High-temperature annealing of ductile cast iron was carried out in batches in a kiln, and the annealing time was as long as 4 or 5 days and nights. The iron thus treated had higher strength, better plasticity, and impact toughness. Most of the iron tools in the Warring States period and Western Han dynasty were softened. For a long time, it had been recognized that ductile cast iron was invented by Leomuel of France in 1722, and black-heart ductile cast iron was invented by Sith.

Bowden of America in 1831. No one could have thought that ductile cast iron originated from China more than 2,000 years ago. This great technological achievement, which was full of originality and has been used for thousands of years in China, has been lost in oblivion for 2,000 years. Now, when we look back at history of this period, we should give high praise to the outstanding creation of ancient Chinese craftsmen.

6.2.2.3 Iron-Mold Casting

Iron-mold casting was another great creation of iron smelting in early China. It was only in the fifteenth century that pig iron shells with simple shape were cast with iron mold in Europe.

In 1953, 86 iron molds of the Warring States period were unearthed in Xinglong, Hebei province, including hoes, sickles, axes, and chariots. Their common characteristics were that the profile of the molds was consistent with the shape of the casting, the wall thickness was uniform, which was beneficial to heat dissipation and could prolong the service life; some iron molds were equipped with iron cores, demonstrating a high technological level. The material of iron molds was hypereutectic cast iron, with carbon content of about 4.45%. From the Warring States period to the Northern and Southern dynasties, the use of iron molds gradually expanded, and its shapes included shares, scuffle hoes, hammers, boards, plows, arrowheads, and

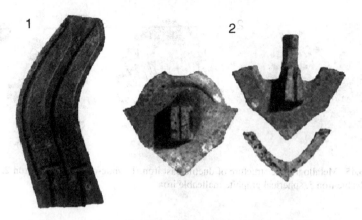

Picture 6.46 Iron molds: 1. mold of double sickles of the Warring States period, unearthed in Xinglong, Hebei province 2. mold of the share of the Northern Wei dynasty, unearthed in Mianchi, Henan province

sickles (as shown in Picture 6.46). Most of the materials were gray cast iron with good thermal stability.

Iron-mold casting was not clearly recorded in ancient books. We think that *Biography of Dong Zhongshu* in *History of Han* records metal molds (including iron molds) as "…like the metal in the mold to be treated by the smelter".

6.2.3 Steel Making

The properties of steel depend on its carbon content. The carbon content of wrought iron is less than 0.1%, that of low-carbon steel is 0.1–0.25%, that of medium carbon steel is 0.25–0.6%, that of high-carbon steel is higher than 0.6%, and that of pig iron is between 2.0 and 4.5%.

Wrought iron is soft, while pig iron is brittle and hard. Steel has both high strength and good toughness, and its microstructure and properties can be adjusted in a wide range by heat treatment (as shown in Picture 6.47), which made it the best material in ancient times.

6.2.3.1 Steel Making with Carburizing

There were many inclusions in the high-carbon layer of the steel sword (M44: 100) made in the Warring States period in Yanxiadu, Yixian county, Hebei province, which was formed by carburizing. From the Warring States period to the early Western Han dynasty, steel weapons and tools were mostly made of wrought iron by carburizing and repeated folding and forging. This method of making steel was widely used.

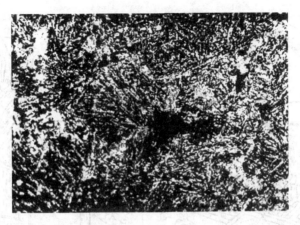

Picture 6.47 Metallographic structure of the iron sword (martensite formed after quenching) of the Western Han dynasty, unearthed from Liaoyang, Liaoning province

The needle-making method recorded in *Heavenly Creations—Forging* was to forge wrought iron into thin strips, draw them into filaments with a mold, and process them into needles. The original description is "...put the needles in a cauldron and roast them with slow fire. After this, cover them with earth, mixed with pine wood charcoal powder, and bean jam and heated from beneath. Set two or three needles outside to test the proper time and temperature. When the needles outside can be twiddled into powder by using fingers, it is time to uncover the needles underneath and to quench them in water". This was a comprehensive method using cold forging, wire-drawing, annealing, shearing, filing, drilling, carburizing, and quenching. Dayang Town, Jincheng City, Shanxi Province was the center of needle-making industry in the Ming and Qing dynasties, known as "City of Needles", and gradually declined until a large number of foreign needles were dumped in the late Qing dynasty and the early Republic of China. The Hezhe people still use this method to make fishhooks in modern times, which can catch big fish weighing a thousand *jin*.

6.2.3.2 Iron Frying and Making Multiple-refined Steel

Iron frying technique was invented in China as early as the beginning of the common era. There was the earliest iron frying furnace at Tieshenggou Iron Smelting Site in Gongyi city, Henan province. It took pig iron pieces and blocks as raw materials, heated them to semi-molten state, and made carbon, silicon, and other elements oxidize violently after stir-frying, so that the temperature rose and the carbon content decreased, thus obtaining solid wrought iron. As iron frying is concerned, it was easy to get raw materials and operate, with high productivity, which could provide a large number of high-quality and cheap steel-making raw materials for the society. This was of key significance for the transition from the early Iron Age to the complete Iron Age. Picture 6.48 shows the furnace for smelting pig iron and wrought iron in

Picture 6.48 Furnace for smelting pig iron and wrought iron

Heavenly Creations. It saved labor and materials by connecting the two processes, and was advanced and novel in technical thought.

Multiple-refined steel was an outstanding creation of blacksmiths in ancient China. Since Liu Kun wrote the popular verses "I intended to be as brave as multiple-refined steel, only to find myself in a helpless situation" in the Western Jin dynasty, more words related with steel have become daily expressions, such as "severe training and hammering" and "making steel after hard work", both of which mean "achieving sth. through hard work". The 30-*jian* (meaning that the knife was folded and hammered 30 times when being made) knife (as shown in Picture 6.49) unearthed in Cangshan, Shandong province was made in the sixth year of the Yongchu period (112) in the Eastern Han dynasty. The blade was composed of pearlite and ferrite, with a carbon content of about 0.5% to 0.7%. There was a small amount of martensite in the blade, which had been quenched. The 50-*jian* steel sword unearthed in Xuzhou, Jiangsu province was forged by the Xigong Office of Shu prefecture in the second year of the Jianchu period (77) in the Eastern Han dynasty, and was made of medium- and high-carbon steel. *Swords Inscription* in *Classical Treaties* by Cao Pi, the first emperor of Wei kingdom in the Three Kingdoms period, writes: "Choose fine metal, and order the civil craftsmen to make steel knives by multiple folding and hammering". The swords and knives thus made show beautiful texture "as colorful as ribbons" and "as shiny as the rising sun". According to *Records of Swords and Knives* by Tao Hongjing (456–536), Liu Bei, the Emperor of Han Kingdom in the Three Kingdoms period, ordered Pu Yuan to make 5,000 knives, all of which were 72-*jian*. This technique

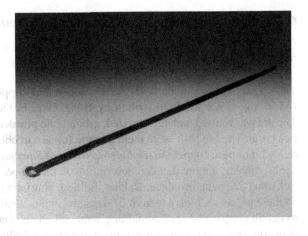

Picture 6.49 30-*jian* Knife made in the sixth year of Yongchu (112) in the Eastern Han dynasty

was quite mature in the Three Kingdoms period, and has been used in successive dynasties since then, becoming a widely accepted and excellent type of steel.

6.2.3.3 Steel Pouring

At the end of the Eastern Han dynasty, Wang Can describes in *Knife Inscription* that the knife is "folded and poured many times", which is the earliest recording of steel-pouring. Zhang Xie of the Western Jin dynasty also says in his article *Qi Ming* that "refine the steel by hammering and pouring a thousand times". *History of the Northern Dynasties* records that Qiwu Huaiwen made *sutie* iron knives by pouring molten pig iron onto wrought iron, thus producing steel. With this steel as the blade and wrought iron as the ridge, 30 layers of an armor can be cut off after urine quenching and oil quenching.

Steel pouring became a conventional process during the Song dynasty. According to *Dream Pool Essays* by Shen Kuo of the Yuan dynasty, "The way to make steel is to curl up wrought iron with pig iron in the middle, cover them with mud, and refine them in the furnace. This method is called 'draw steel' (*tuangang*) or 'pouring steel'". The steel thus made is called "plastering steel" (*mogang*) and "Jiangsu steel", which still prevail in Jiangsu, Anhui, Hubei, Hunan, Sichuan, and Fujian provinces in modern times. The carbon content of pouring steel was mostly between 0.6% and 1%, which belonged to high-quality medium and high-carbon steel. As an original steel-making technology in China, it occupies an important historical position in the world metallurgical history.

6.2.4 Smelting and Alloy Preparation of Non-ferrous Metals

6.2.4.1 Smelting of Copper by Fire and Water

The copper-making method of "sulfide ore–matte–copper" was first described in *Ode to the Great Smelting* written by Hong Zikui in the Southern Song dynasty. The ore was roasted and desulfurized, ground into powder, and made into balls with rice paste. The ore balls were melted in the furnace to obtain matte, which was then refined into pure copper. *Miscellaneous Notes of Shuyuan* by Lu Rong of the Ming dynasty makes a more detailed description on the copper smelting from sulfide ore in Lishui, Zhejiang province. Baking, milling, elutriating, briqueting, and refining, all these processes took a total of 27 days and nights. Lead was added to raw copper to obtain copper and lead containing gold and silver. The former was refined into pure copper, while the latter was refined into gold and silver by ash blowing method.

Hydrometallurgical copper processing was invented in China, which was commonly called "*dantong* method". The whole process was to place iron in chalcanthite (copper sulfate) solution, so that copper was replaced by iron to generate copper powder, and then refined to obtain pure copper. In the Northern Song dynasty, there were three major hydrometallurgical copper processing sites: (1) Qianshan county, Xinzhou city, Jiangxi province; (2) Dexing county, Raozhou city, Jiangxi province; (3) Censhui, Shaozhou city (nowadays Shaoguan), Guangdong province. During the Daguan period (1107–1110), 6.6 million *jin* (3.3 kg) of copper was made, including more than one million *jin* of *dantong*.

6.2.4.2 Mining and Smelting of Gold and Silver

Liu Yuxi, a famous poet in the Tang dynasty, describes gold panning in his poem, "Mist opens as the red sun shines over the shoal, and gold panning women are crowded along the river bend; Jewelry of the beauty and seals of the noble, all are made of gold from the river sand". Wooden plates, chutes, and gold panning beds were used to form gold panning. Gold particles were obtained by baking mercury and gold chips in a crucible, similar to amalgamation in modern times (as shown in Picture 6.50).

The detailed records of silver mining and refining can be found in Zhao Yanwei's *Casual Notes from the Cloudy Foothill* in the Song dynasty: "The method silver mining and refining are as follows. Where there is black lines on the stone wall, there is silver ore. Chisel holes along the lines... The silver ore mined is gravel, which will be mashed with a pestle, ground, refined with silk, and then scoured in water. Discard the yellow stones and keep the black silver, which will be battered. Put lead in it and calcine silver particles into large pieces by fire" (as shown in Picture 6.51).

Picture 6.50 Horseshoe gold in the Han dynasty

Picture 6.51 "*Chengan Baohuo*" silver ingot

6.2.4.3 Smelting of Lead, Tin, and Mercury

Lead has a low melting point (327 °C) and is easy to smelt. It is called "the ancestor of metals" by Li Shizhen, a famous medical scientist of the Ming dynasty, because it must be added during tin smelting and silver smelting by ash blowing method. There were three methods of lead smelting: (1) remove silver and then sink it to the bottom of the furnace; (2) remove lead before putting copper when smelting in the furnace; (3) elutriate, crush, and refine galena.

Most of tin exists in nature in the form of an oxide (cassiterite). Its famous producing areas were Guiyang in Hunan province; De'an and Dayu in Jiangxi province; Hezhou and Hechi in Guangxi province; Chuxiong, Baoshan, and Gejiu in Yunnan province.

China abounds in antimony deposits, and Xikuangshan (Tin Ore Mountain, thus named because antimony there was mistaken as tin) site in Xinhua, Hunan province is the most famous. It was once mined in the Ming dynasty and mined on a large scale in modern times.

As early as the Spring and Autumn period, elemental mercury could be extracted from cinnabar. Qing, a widow in Ba area (nowadays Sichuan province) became rich by smelting mercury. After the Han and Jin dynasties, mercury was made by being airtight and pyrolysis. There were mainly three ways, namely, (1) lower fire and upper condensation, (2) upper fire and lower condensation, and (3) distillation. Similar mercury smelting methods are still used in Guizhou and other places in modern times, which are called *"Tianguo Dizao"*.

6.2.4.4 Smelting of Zinc

Zinc was called Japanese lead in ancient China, and its smelting process can be found in *Heavenly Creations—Metallurgy*: "It is extracted from calamine. The Taihangshan Mountains in Shanxi province is the top producer of zinc, followed by Jingzhou and Hengzhou. To produce zinc, put ten *jin* (5 kg) of calamine in an earthen jar. Seal it tightly with mud and smooth the exterior. The jar is then allowed to dry. Caution must be taken to keep it away from fire lest the heat cracks the mud exterior. When the jars are ready, coal cakes together with firewood are piled up underneath them to make a fire. The molten material inside the jars is allowed to cool for the zinc to take shape. The jars are then broken in order to get the metal out. Every ten *jin* of calamine produces eight *jin* of zinc, i.e. Japanese lead". The boiling point of zinc is 907 °C, which is very close to the reduction temperature (904 °C). Elementary zinc can be obtained only when it is recovered by condensing device, which is the difficulty of zinc smelting. The invention of zinc smelting in India predated that in China. The lower coagulation method was adopted in India, while the upper coagulation method was used in China. It is most likely that the two methods were invented and developed independently. In 1745, a Chinese freighter sank in Port of Gothenburg, Sweden, and the zinc ingots loaded on it were salvaged in 1842, with a purity of 98.99%. The export of zinc ingots from China undoubtedly promoted the rise of zinc smelting industry in Europe.

6.2.4.5 Preparation of Brass and Cupronickel

Brass was an alloy of copper and zinc, which was called "stone" in ancient China. Refining brass from calamine was quite common in the Song dynasty. *Compendium of Materia Medica* quotes *Waidan Materia Medica* by Cui Fang of the Song dynasty as saying: "With a kilo of copper and a kilo of calamine, they can be refined into a kilo and a half of stone". *True Purpose of Sanyuan Dadan Secret Garden* written during the Jiajing period (1522–1566) in the Ming dynasty records that "Japanese lead with white gas" was produced in Fujian, "Japanese lead with green gas" was produced in Henan, and "Japanese lead with yellow gas" was produced in Shanxi, all of which could be used to make brass. Brass coins were first cast during the Tianqi period in the Ming dynasty (1621–1627), which showed that zinc was mass produced.

Cupronickel is an alloy of copper and nickel or copper and arsenic. Chang Qu of the Eastern Jin dynasty records in *Annals of Huayang State* that Tanglang county (nowadays Huize and Qiaojia counties in Yunnan province) produced cupronickel. Crude cupronickel was obtained by melting "green ore" (nickel ore) and "yellow ore" (copper ore) in Huili, Sichuan Province, and then copper and zinc were added to prepare cupronickel. Because of its white color and difficulty in rusting, the devices such as ink cartridges and hookah bags were very popular among the people. Cupronickel was introduced to Europe in the eighteenth century and was called "Chinese silver". In 1776, Swedish scientist Engistrom analyzed Yunnan cupronickel ware and confirmed that it was copper-nickel alloy. In 1823, Thompson, an Englishman, and Hanninger brothers in Germany, copied cupronickel, which was widely used in industry afterwards.

The preparation of arsenic cupronickel was first seen in *The Master Who Embraces Simplicity* written by Ge Hong in the Western Jin dynasty. He Yuan of the Northern Song dynasty writes about the preparation of cupronickel from arsenic powder by Xue Tuo, a copper-making expert, in *Chunzhu Documentary*. Xue's method was to make pills with arsenic powder and jujube meat, throw them into the copper melting furnace to make the arsenic sulfide (As2S2) reduced to arsenic, which would melt with copper. Then mirabilite (Na_2SO_4) was added to make slag, which was arsenic cupronickel.

6.2.5 Metal Processing Technology

6.2.5.1 Casting

Cheng Kong, a monk in the Sui dynasty, devoted his whole life to casting the Jinyang Iron Buddha Figure with a height of 70 *chi* (23 m). *Collection of Wonders* by Xue Yongruo of the Tang dynasty records that the big Buddha figure was cast several times but failed every time, and it was made only when the monk threw himself into the furnace. There were folk smelters in China, who regarded Golden Flower Empress as the furnace goddess. According to legend, her father was a smelter, who had to pay taxes on time, but the furnace couldn't produce iron. In the end, his daughter jumped into the furnace and the iron was eventually produced. There were similar legends in ancient Europe.

Picture 6.52 Iron man and iron ox at Pujin ferry, cast in the Tang dynasty

During Wu Zetian's reign in the Tang dynasty, Nepalese craftsman Mao Boro presided over the casting of *Tianshu*, a column commemorating Wu's greatness, which is 105 *chi* (35 m) high and made of two million *jin* of copper and iron.

The famous Iron Ox in Pujin, unearthed in 1989, was cast in the 12th year of the Kaiyuan period (724) in the Tang dynasty to maintain the floating bridge at Pujin ferry in Yongji, Shanxi province. It was 3.3 m long and 2.5 m high, with an iron base under its abdomen and a 3-m-long iron column (as shown in Picture 6.52). This giant ox manifested the highly developed iron smelting industry and its technical level in the Tang dynasty, which could be called a world miracle.

Iron Lion of Cangzhou was cast in the third year of the Guangshun period (953) in the Later Zhou dynasty during the Five Dynasties period, with a length of 5.3 m, a height of 5.4 m, and a weight of about 40 tons. There was the inscription "Lion King" under the lion's head and neck, and the inscription "Made by Li Yun, Shandong" on its left rib. The body was formed by combining and casting more than 360 pieces of mud molds. Dangyang Iron Tower was located outside the mountain gate of Yuquan Temple (as shown in Picture 6.53), which was built in the sixth year of the Jiayou period (1061) in the Northern Song dynasty. It was about 18 m high and was made of nearly 40,000 kg of iron. It was made of 44 whole cast pieces. *Compendium of Materia Medica* by Li Shizhen quotes *Treasure Store Treatise* as saying: "There are five kinds of iron, with Jing iron produced in Dangyang city the best, which is purple and strong, followed by Shangrao iron". The iron tower built in Dangyang, Hubei province was a symbol of the developed metallurgical industry in South China during the Tang and Song dynasties.

Zhengding Bronze Buddha is located in Dabei Pavilion of Longxing Temple, Zhengding, Hebei province, with a height of 22 m and a weight of about 36 tons (as shown in Picture 6.54). The temple was built in the sixth year of Sui Dynasty's Kaihuang period (586) and rebuilt in the fourth year of Song dynasty's Kaibao period

Picture 6.53 Dangyang Iron Tower

(971). There were 3,000 craftsmen involved in this project, and the base was cast with pig iron, which was 2 m deep. The part from the lotus pedestal to the Buddha's head was cast seven times upwards.

Yongle Bell was cast during the Yongle period (about 1418–1422) in the Ming dynasty, with a height of 6.75 m, a diameter of 3.3 m, and a weight of 46.5 tons (as shown in Picture 6.55). There are 227,000 words of Buddhist scriptures cast inside and outside the bell body. The mud molds of the body consists of seven sections. The alloy composition of the bell is 80.54% copper, 16.40% tin, and 1.12% lead.

Copper and iron buildings have been built since the Tang dynasty. In the second year of the Dali period (767) in the Tang dynasty, Jinge Temple was built on Wutai Mountain, and its roof was covered with gold-plated copper tiles. Similar golden tops include Jokhang Temple in Lhasa, Tibet, Ta'er Lamasery in Xining, Qinghai province, and Sumifushou Temple in Chengde, Hebei province. The ancillary hall of Bixia Temple on Mount Tai and Guangxiang Temple on Mount Emei are covered with iron tiles. The golden hall on Wudang Mountain was built in the 14th year of the Yongle period (1416) in the Ming dynasty, with a height of 5.5 m and a width of 5.8 m. The golden hall on Mingfeng Mountain in Kunming, Yunan province was built in during the Wanli period (1573–1620) in the Ming dynasty, with a height of 6.7 m, a width and a depth of 6.2 m.

In the pre-Qin period, cloth coins, knife coins, and circular coins were mostly cast in mud molds, while ant-nosed coins in the Chu State were cast in copper molds (as shown in Picture 6.56). In the Western Han dynasty, the above two kinds of molds were used together to cast coins. From the Three Kingdoms period to the

Picture 6.54 Zhengding Giant Buddha, made in the Northern Song dynasty

Picture 6.55 Yongle Bell, made in the early Ming dynasty

Picture 6.56 Copper molds of ant-nosed coins, found in Fanchang, Anhui province

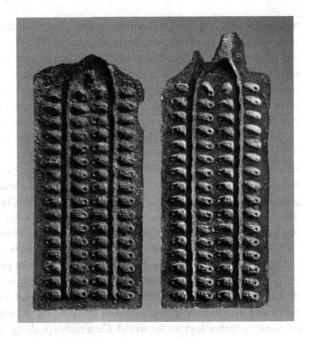

Southern and Northern dynasties, the stack casting had always been an important coin-casting method, and some copper-mold boxes have been handed down from ancient times, such as "Daquan Dangqian" (coins made in Wu Kingdom during the Three Kingdoms period) and "Changping Wuzhu" (coins made in Northern Qi) (as shown in Picture 6.57). The scale of coin-casting in the Tang and Song dynasties was far larger than that in previous dynasties, which promoted great technological changes. In the Song dynasty, it became a conventional method to make coins by casting the mother-coin molds in sand. The whole process included mold making, coin grinding and arranging (as shown in Picture 6.58).

The method of lost-wax casting gradually gained popularity since the Han dynasty, and its technological process was first recorded in *Record of the Pure Registers of the Cavern Heaven* written by Zhao Xihu in the Southern Song dynasty. In the Yuan dynasty, the Wax Bureau was set up to take charge of lost-wax casting. Lady Jiang was one of the well-known artisans. In the Ming dynasty, Xuande Incense Burner was cast in *dingyi* (a kind of ritual vessel) wax-mold made of yellow wax. In the Qing dynasty, the Copper Workshop set up in the Office of Internal Affairs also used the lost-wax casting as the main means to make artistic pieces (as shown in Picture 6.59).

6.2.5.2 Forging, Brazing, Cold Working, and Decoration Techniques

In the Tang and Song dynasties, wrought iron farm tools gradually replaced cast iron farm tools. This milestone event not only contributed to the establishment of China's traditional iron and steel technology system, but also changed its overall composition and landscape of the metallurgical industry. The blacksmith group rose rapidly, and almost every village had a blacksmith workshop. *Heavenly Creations* quotes the proverb that "all tools take pliers as their ancestors", which was a very accurate statement. Chinese traditional forging technology had one defect in its process composition in that it paid more attention to free forging than die forging, and there was only one example of large item forging, i.e. the anchor weighing 1,000 *jin*.

Ancient brazing was divided into tin brazing, copper brazing, and silver brazing. The commonly used fluxes were rosin, borax, halogen sand, halide salt, and diversifolious poplar resin. Diversifolious poplar resin, first recorded in *Biography of Western Regions* in *History of Han*, was a high-temperature flux and could be used for brazing gold, silver, and copper. The craft is still used by Uygur silversmiths in modern times. The native place of poon was the Euphrates River Basin, and diversifolious poplar resin might have been introduced from West Asia as a brazing flux.

The earliest known example of metal lathing was the bronze boat (water container) of the Han Dynasty unearthed in Nanyang city, Henan province. The silver box of the Tang dynasty unearthed in Hejia village, Xi'an city, had close circle lines formed by cutting both inside and outside the box bottom, and the circle lines on the outer surface of the silver bowl were formed by spinning, which was the earliest example of spinning technology in the world. Generally speaking, metal cutting was not very developed in ancient China, and its application scope was narrow.

Surface decoration skills such as gold coating, gilding, inlaying copper, inlaying gold and silver, tin plating, gold plating, and engraving and hollowing reached a high level in the Bronze Age, and the technology has become more refined since the Iron Age. According to *Six Codes of the Tang Dynasty*, the gold and silver workshop in

Picture 6.57 Copper molds of "Daquan Wushi" and the cast coins of the Han dynasty, now collected in the National Museum of China

Picture 6.58 Coin casting by mother-coin mold method

Picture 6.59 Bronze pavilion in the Summer Palace

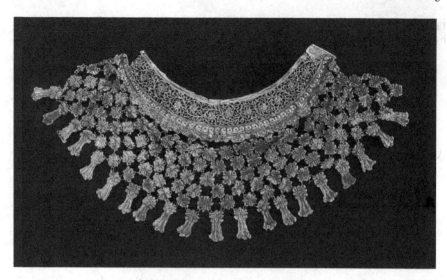

Picture 6.60 Flower-patterned gold headdress, handed down from ancient times

Minor Treasury applied many techniques, such as gold melting, gold pasting, gold plating, gold weaving, gold grinding, gold coating, gold painting, gold hollowing, gold twisting, gold engraving, gold couching, gold laying, gold embedding, and gold wrapping.

Gold ornaments in the Song dynasty, Xuande Incense Burner, cloisonné enamel, and filigree inlay in the Ming and Qing dynasties were popular for a while (as shown in Picture 6.60). The shape, scale, and pattern of some extremely exquisite works were created by craftsmen's understanding and skills trained over the years.

6.3 Traditional Precious Metal Technology

6.3.1 Zhaoyuan Metallurgy

Heavenly Creations records that the price of superior gold was 16,000 times that of black iron. Since ancient times, about 100,000 tons of gold have been mined worldwide. In 1848 and 1851, gold mines were discovered in San Francisco, USA, and Melbourne, Australia, which triggered the gold rush. The Chinese names for San Francisco (*Jiujinshan*, meaning "Old Golden Mountain"), and for Melbourne (*Xinjinshan*, meaning "New Golden Mountain") came from this (as shown in Picture 6.61).

Zhaoyuan in Shandong province is a famous gold producing area in China, especially Fushan and Linglong, which are known as "the crown of gold mines in Asia". Judging from the burn marks on ancient potholes and relics such as iron chisels and

6 Metallurgy and Metalworking

Picture 6.61 Gold panning in Honduras

hammers, Zhaoyuan used pyrotechnic methods to mine gold as early as the Warring States period. In the fourth year of the Jingde period (1007) in the Song dynasty, Minister Pan Mei supervised the gold mine in Zhaoyuan and used gunpowder for mining. During the Jiajing period (1522–1566) in the Ming dynasty, gold mines in Zhaoyuan flourished even more.

Zhaoyuan's traditional gold mining and metallurgical techniques are still well preserved in Jiangjia village, Jiuqu, Fushan town. The main processes will be described below.

6.3.1.1 Mining and Crushing

After the ore is mined, it is crushed with a sledgehammer into smaller pieces with a diameter of about 1 cm.

6.3.1.2 Grinding

The ore particles are mixed with water and ground into ore powder (as shown in Picture 6.62). The beneficiation rate depends on the purity of the ore. Manual grinding is very hard, and people often hum some tunes or tell stories to relieve or distract them from their fatigue. The lower limbs of some women who have been grinding ore for many years actually deform because of the labor.

Picture 6.62 Grinding

6.3.1.3 *Laliu*

Laliu (pulling water) is the common name for mineral processing. The slide board used for this was 1-m wide and 2.5 m long. Willow board with rough and hairy surface that easily catches gold grains is usually used at an inclination of 15–18°.

First, workers spread the ore on the upper end of the slide board with a shovel and press it to form a shallow ditch with a rake. When the water flows down the ditch, the ore will loosen and flow down with the water. The concentrate will accumulate downstream, at the upper end of the plate (as shown in Picture 6.63), and the gangue (the unwanted materials), being lighter, would sink to the bottom of the water. Workers will sweep the gangue into the sand pool with a broom. Concentrate, also known as gold mud, is packed in a mud bowl to dry. Once dry, the gold mud is brushed onto paper with rabbit legs (because gold particles don't stick on them) and stored for refining.

6.3.1.4 Smelting

It is commonly referred to as "*lahuo*" (pulling fire). After the stone stove is ignited, the paper bag containing the gold mud is put into the crucible. After melting, the impurities gradually evaporate, and mirabilite and borax are added to make slag. Then it is purified and poured into a mold to form ingots (as shown in Picture 6.64). All these processes are controlled by experienced goldsmiths and are a typical example of intangible cultural heritage.

Picture 6.63 *Laliu*

Picture 6.64 Ingot casting

Traditional gold mining and smelting techniques have been passed on within families. There are two seventh-generation descendants among the goldsmiths in Jiangjia village of Jiuqu, namely, Wang Jinyong and Chi Mengwen, whose inherited techniques can be traced back all the way to the 1870s and 80s.

Wang Jinyong, aged 47, studied the craft with his father at the age of 18, graduated from university and became a senior engineer. He is currently the Director of the Production Technology Research Institute of Gold Mining Company, and belongs to a new generation of highly educated inheritors of traditional handicrafts.

Zhaoyuan is still the largest gold base in China, with an annual output of one-seventh of the whole country. It was named "China Gold Capital" in 2002.

Since the 1960s and 1970s, traditional gold mining and smelting techniques have been gradually replaced by machinery. Most goldsmiths holding on to traditional goldsmithing practices are nearly 60 years old now. It is of the utmost importance to protect and inherit this precious cultural heritage under the leadership of the government and through the joint efforts of artisans, enterprises, communities, and experts.

6.3.2 Zinc Smelting in Hezhang

Traditional zinc smelting is concentrated in Guizhou, Yunnan, and Hunan provinces, and mostly in Hezhang county, Guizhou. According to Xu Li's research on Hezhang's Magu area in the 1980s, the raw materials used in zinc smelting came from local zinc-lead ore deposits. These had a mineral composition that was mainly smithsonite (calamine), but also heteropolar ore (H_2ZnSiO_5) and zinc sulfate ($ZnSO_4$), with a zinc content of 16% to 20%. The distillation pot was made of refractory mud and clinker, with a height of 80 cm. The upper part was made of a bucket shell for recovering zinc vapor (as shown in Picture 6.65). Zinc smelting furnaces were rectangular, 10 m long and 1.5 m wide, commonly known as manger furnaces, which could hold 120 distillation pots at a time. The ore and coal were crushed, mixed, and put into a tank.

Slag and coal were put in the furnace first, then a distillation pot would be placed on top of them, honeycomb briquettes around it, slag on top of it, and then mud was used to cover it. After igniting the furnace and heating it up, zinc ore would be reduced at a high temperature and waste gas would escape from the exhaust hole of the pot cover and ignite. This would cause zinc vapor to condense at the bottom of the bucket shell, which was taken out after the furnace was put out. The purity of zinc obtained this way was 97% to 98.7%, containing only a small amount of lead and iron. The residue could also be packed into a crucible to recover zinc. About 700 kg of ore was loaded in each furnace. The metal recovery rate was 85% to 90%. The production took about one day and one night, requiring 16 craftsmen per furnace (as shown in Picture 6.66).

Both *Annals of Taiding Prefecture during the Daoguang Period* and *Annals of Weining County*, compiled in the Qing dynasty, state that Tianqiao Yinchanggou (now Shashi Township, Magu District) produces zinc, and its mining era began in the Tianfu period of the Five Dynasties period (936–947). This provides an important clue for determining the invention time of zinc smelting technology in China, which can be further proved by *Treasure Theory* (written by Xuanyuan Shu in the second

Picture 6.65 Distillation pots

year of the Ganheng period, that is, 918) quoted in *Compendium of Materia Medica*. The book says that "gold can be smelted from Japanese lead (zinc)".

From the perspective of technical history, the existing traditional zinc smelting processes in Hezhang and Huize do not meet the modern safety standards and regulations, since they are extremely polluted and thus not suitable for continuation. However, they are still of great value to the study of the origin, spread, and technological history of zinc smelting. Therefore, it is still necessary to protect it, but more as data.

6.3.3 The Stone-Mold Casting Plowshare in Qujing

Stone molds were replaced by pottery molds as early as the Bronze Age. However, Yunnan and Sichuan provinces still retain this ancient skill and use it to cast iron plowshares. These can be called living fossils of ancient techniques and miracles of technological history. Wang Dadao and Li Xiaocen researched the stone-mold

Picture 6.66 Operating a furnace

casting in Dongjia village, Zhujie town, Qujing city, Yunnan province, which had attracted great attention in academic circles. Their research is as follows.

In Dongjia village, the Yuan family, whose ancestors were from Henan and were banished to Yunnan, where their hometown was Zhaizikou township, Chengguan town, Fuyuan county, moved to the present site 100 years ago. Their technique has always been passed down from father to son, not daughter. In 1992, the inheritor Yuan Decheng died, and his brother-in-law Xu Zhongheng succeeded him. He studied the art from Yuan's grandfather in 1949. In the 1980s, plowshares were cast in Yingshang, Dahe, Huangnihe, Yuwang, and Laochang in Fuyuan county. Farmers paid 100 yuan a month to the craftsmen each and provide them with free food and shelter. They were able to cast 500 plowshares a month, which were still in short supply. Each plowshare could be sold for 1.5 yuan. Four people could make 40 pieces a day, earning 20 yuan per person per day. At present, there are still 10 stone molds in his family, but new plowshares have not been made since 1996.

How are these stone-mold casting plowshares made? We will take a quick look.

1. Making and revising molds

Stone molds require fire resistance; otherwise, they might crack or explode. The white sandstone and red sandstone used come from Shenjia village at the junction of Zhanyi and Xuanwei. Those with laminated striation cannot be used.

When making a mold, one flat side is cut first, a rectangular concave hole chiseled on each side for holding and lifting. On the other side, a plow shape is chiseled. The mold weighs about 40 kg and should be stored in a dry place.

The core skeleton is made of iron sheets and the lower part of the head is coated with refractory mud mixed with 80% coke powder and 20% white cement in multiple layers. Drying needs to be done after every coating, and this process is repeated many times before the mold is trimmed and formed.

After casting, the coating must be repaired with white cement. After a long period of casting, large defects in the mold surface will require more thorough repairs. When repairing the mold, one needs to make up the raw materials first, dry it and flatten it, and then brush it on with the ash obtained by burning dragon claw (a kind of plant). Stone molds can be used for 50 years if they are well maintained, but only half a year if they are not well maintained.

2. Mold baking and mixing

When baking the mold, it is put on the iron frame and wood burned below. It is baked for 2–3 h and the moisture is removed before casting, otherwise the stone mold will crack. The core must be repeatedly coated with black mud while baking.

Before casting iron plowshares, the molds must be jointed. First, put the mold flat on the ground, with the casting face facing up, and put the core into the groove of the lower stone mold. The placement of the core is related to the thickness and quality of the castings, and the operator must be experienced. After placing the mud core brace, the upper mold is joined with the lower mold. The core brace is flattened and bonded with the core and the upper mold. The gap between the core and the mold does not only show the thickness of the plow, but it is also used as the filling gate. After that, the iron sheet is hooped at the waist of the mold, a crowbar is inserted into the gap between the stone mold and the iron hoop, and then they are fastened with a wedge.

Before casting, the mold must be tilted at an angle of about 80 degrees with the filling gate facing up. If the stone mold is straightened too early, its own gravity will displace the core brace, resulting in a casting failure.

3. Casting

The retort furnace used for stone-mold casting plows has its own characteristics. It is also known as the *Bagua* furnace. The crucible is lined up with the bottom of the furnace. The joint is coated with yellow mud to make it seamless and dried with a small fire. After preheating for about 20 min, workers put the broken iron into the layer of carbon and blast it. Pig iron melts and flows into the crucible. An iron rod with a tenon is inserted into the square handle of the crucible to lift it and strip the slag, which is then ready to be cast. Casting should first be fast and then slow. In order to avoid the buoyancy of molten iron to lift the core, a wooden stick is used to hold the core in place. About one minute after casting, workers insert a wooden stick into the core mortise to pull it out. They then take out the mold in two or three minutes, clip out the casting, knock off the burrs, and get the finished product.

The whole casting process should be done by at least three to four people. The iron used for a plowshare costs about 3 yuan per kilogram. When opening the furnace, you should choose even-numbered days and auspicious days. First, you should pay your respects to *Taishang Laojun* (Supreme Venerable Lord) and ask the ancestor to bless you. Then you offer chicken, wine, and dishes; burn yellow paper; and light three incense sticks.

6.3.4 The Pig Iron Smelting and Casting Techniques in Yangcheng

Yangcheng in Shanxi province has always been an important iron industry town, with their own complete sets of technologies such as the square furnace (as shown in Picture 6.67), plow furnace iron making, iron frying, iron casting, iron molds, and plowshare casting. Among them, plow furnace iron making, iron molds, and plowshare casting techniques were used until the 1990s. They could be called the representative works and living pieces of pig iron smelting and casting history in China (as shown in Picture 6.68).

Picture 6.67 Smelting with a square furnace

Picture 6.68 Night view of a plowshare workshop

6.3.4.1 Plow Furnace Iron Making

The plow furnace consists of upper, middle, and lower sections, with a height of about 3 m and a curved inner wall (as shown in Picture 6.69). It is made of the "gem" processed from quartz sandstone and uses charcoal as fuel. Charcoal firing requires a balance of 30% wood and 70% charcoal, which has high strength and is not easy to crush. Iron ore is usually hematite that is rich in iron. This must be crushed and roasted at a high temperature to remove sulfides before entering the furnace.

Workers add iron ore and charcoal to the plow furnace after making a fire, resulting in iron after 4 h. It takes 6–7 h for the furnace to be in a normal condition, and every half hour an amount of 10–15 kg of molten iron can be obtained. The control of molten iron composition in the furnace is commonly known as "watching the color of fire" and "watching the color of molten iron". Looking at the fire color means judging whether the furnace condition and molten iron composition are qualified by looking at the flame ejected from the taphole and the condition of molten iron in the furnace. If the flame is bright and white, the furnace condition is normal. The color of the molten iron can be checked by scooping it out with a small iron spoon and gently blowing on its surface. If the surface of molten iron is red and gray, it indicates that graphite floats and the carbon content is on the high side, which is called *"rangshui"*. The worker who is responsible for watching the color of fire and the color of the molten iron is called the "fire master". This traditional furnace checking technique is purely based on experience, but is simple, accurate, fast, and efficient, and has its own clever and unique features.

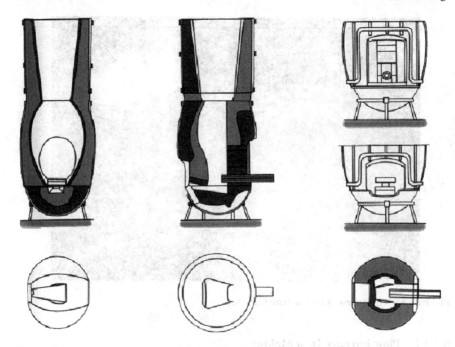

Picture 6.69 Cross-sectional view of a plow furnace

6.3.4.2 The Iron-Mold Casting Plowshare

The iron mold used for casting plowshare is commonly known as "plow box" or "*hezi*", which was previously exclusively supplied by Li family in Shangqin village in Yangcheng. It uses real objects as molds and obtains sand molds by duplicating many times. After casting, it can obtain a pair of iron mold. The technique is simple and easy, and it is quite ingenious. A good plow box can be used for more than 10 years and be cast over 20,000 times. The iron used is mostly gray cast iron, because of its good thermal stability, strength, and toughness, which gives it a long service life.

Before casting, the iron mold must be preheated and coated with Guanjing charcoal. A craftsman is required to skillfully put his foot on the iron mold to prevent fire from leaking (as shown in Picture 6.70). Craftsmen open the mold immediately after casting is done, take out the plowshare (as shown in Pictures 6.71 and 6.72), and knock out the flash and burr to get the finished product. The plowshare is made of white cast iron with a high carbon and low silicon content. This makes it brittle and hard, but also non-sticky and cheap, making it popular with farmers (as shown in Picture 6.73). There are more than 100 kinds of plowshares that are frequently produced. This is because the soil quality, crops, and farming techniques vary from place to place, and therefore various specifications must be prepared to meet these different needs.

Picture 6.70 Plowshare casting

Plowshares process the recognized acceptance criteria of ten non-acceptance, namely, black tendons, pockmarked surfaces, clear noses, hot frying, round edges, uneven mouths, crisp sound, insufficient pouring, holes, and so on. Because both producers and sellers stick to these quality standards, Yangcheng plowshares have become a famous brand known for their high quality and low price.

With the social transformation and the popularization of agricultural machinery, the demand for plowshares sharply dropped, resulting in a sharp contraction of the industry. In the late 1980s, a small number of plow furnaces were still in use in Henghe, Sanglin, and Majia villiges in Shanxi, all of which have since closed down. In 2006, this skill was added to *List of the First Batch of National Intangible Cultural Heritage* (No. 385). Under the leadership of the Yangcheng County Government, Institute for the History of Natural Sciences of Chinese Academy of Sciences cooperated with Tsinghua University Academy of Arts and Design to establish a Yangcheng Pig Iron Smelting and Casting Techniques Exhibition. It did not only keep the cultural memory alive, but it was also well received by the masses. The replacement of skills is a historical necessity; however, the scientific and technological building blocks contained in these replaced skills, the professionalism and the great contributions made by craftsmen are indelible and should not be forgotten.

Picture 6.71 Red plowshare fresh out of the mold

6.3.5 Wrought Iron

Song Yingxing's *Heavenly Creations* contains a folk proverb: "Ten thousand utensils are made in the crucible". Throughout the Tang and Song dynasties, blacksmith shops and mobile blacksmiths spread all over the country. Besides forging various farm tools and hand tools, they also made daily iron utensils such as kitchen knives, scissors, and razors. They played an extremely important role in the national economy and people's livelihoods. In recent years, due to industrialization, iron is mostly produced by machines, causing the wrought iron industry to shrink. However, in rural and remote areas, wrought iron craftsmen are still alive and kicking. Some of them are so closely entwined with local people's livelihoods and customs that they aren't in danger of suddenly becoming obsolete.

According to Zheng Tao's research in Jinan's rural areas, the brothers Ma Chengjun and Ma Xiangzheng have made a living by forging iron since they were in their teens. For more than 40 years, they have travelled through villages and lanes in

Picture 6.72 Iron mold and plowshare

Picture 6.73 Yangcheng Tieshan plow with pimpled plowshare

Changping, Beijing, to make farm tools for farmers, and are still doing so. They don't have their own shops, so they carry dozens of pounds of iron-making tools on their backs. Because they have worked around here for so long, they know the villagers and have even lost their native accent. They only return to their hometown for a short period of time every 6 months. Nowadays, farm tools are produced in batches by factories and are spread all over the rural areas. The Ma brothers, who have worked hard most of their lives, have to carry heavy outfits to the remote, high-altitude areas where there is still a demand for mobile blacksmiths.

Countless iron materials have passed under the hammers of the Ma brothers in nearly half a century. They are really good at what they do. Besides common ironware such as kitchen knives, sickles, axes, and harpoons, they also make iron clips and furnace bars for villagers. Some former colleagues have gotten good jobs in the city once they grew up. They come back from Haidian, Yanqing, and other places every now and then to meet up with the two brothers. "The farm tools made by hand are better than those produced in the factory, and they are random in size and convenient", which is the villagers' evaluation of Ma brothers' products.

Life is hard. Iron materials bought from waste recycling stations cost 4.8 yuan per kilogram and coal costs 0.45 yuan per kilogram, while one sickle only gets them 2 yuan for half a day's work. Life is not as good as before. Their children are unwilling to inherit their father's business and even disdain it. But the old brothers say that they will continue to struggle as long as they can. If not, it would be such a loss not to do this craft.

For a long time, the lives of ordinary Chinese people have been hard. Broken pots and pans need to be repaired by mobile craftsmen, who are more commonly known as tinkers. It is recorded in Lu You's *Laoxue'an Notes*, "In the market, those who tinker copper and iron utensils are called *gulu*". Zhang Bangmu's *Casual Records of Mo Zi and Zhuang Zi* says: "Tinkering the leakage (锢漏) is commonly known as *gulu* (骨路)". *Discussing Writing and Explaining Characters* interprets "锢" as casting plugs to tinker the leakage, with "金" as its radical, and pronounced the same as "固". From the above literature, we can see that the origin of the term dates very far back. According to Hao Erxu's verification, there are more than ten records of "*gulu* (古路)" and "*gu* (钴)" in the sutras found in caves for preserving Buddhist sutras Dunhuang. And they also record that those engaged in this line were *xiejiang* (casting craftsmen), and *xie* means casting. There are many descriptions of "*gulu ju* (casting to mend)" in the notebook novels of Song and Yuan Dynasties, which showed that they were widely distributed everywhere. Until 1970s and 1980s, tinkers still existed in many places, but now they only exist in remote villages and mountainous areas (as shown in Picture 6.74).

6.3.5.1 Rubbing

According to *Forging* in *Heavenly Creations* by Song Yingxing in the Ming dynasty, "Hoes for plowing fields and planting and hoes with broad mouths are forged from wrought iron. Then sprinkle melted pig iron onto the hoe mouths so that the hoes will become hard and tough after being quenched in water. The best way is that every one *jin* (0.5 kg) of shovel or hoe should be sprinkled with three *qian* (15 g) of pig iron, less than which the product will not be hard enough and more than which the tools will be so hard that they will be easily broken". Earlier than Song, *Compilation of Martial Arts* written by Tang Shunzhi during Ming's Jiajing period also says: "Cast

Picture 6.74 *Casting to Mend*, copper sculpture in Changsha, Hunan province

with pig iron and wrought iron. When the wrought iron is extremely well done and pig iron is about to flow, workers use pig iron on wrought iron and rub it in". Both the recordings refer to the same process.

Ling Yeqin did research in the rural areas of northern China and eastern China, and heard many old farmers speak highly of tools they used, such as hoes, lutetium, and picks. They would rather spend more money on rubbed brand-name goods than steel ones. Some old farmers have used their rubbed tools for 20 or 30 years and cherish them very much. The reason is that these kinds of farm tools have sharp edges and are so handy to use that they will leave no mud unturned, no matter if it is dry land or in paddy field. These tools have durable resistance, self-sharpening, and high farming efficiency.

Rubbing is developed on the basis of the steel-pouring techniques of raw and wrought iron, which is used for the surface treatment of wrought iron billets. This skill has spread almost all over the country in various ways. For example, in the Yanbei and Yixian areas of Shanxi, Beijing and Shandong, high-carbon white cast iron sheets and blocks, such as plowshare iron and pot iron, are "paved" (*pusheng*); in Yangquan and Pingding in Shanxi, Luquan, Yixian, and Jiaohe in Hebei, high-carbon gray cast iron are rubbed; hoes and shovels produced in Wenzhou and Haining, Zhejiang province are also drenched with pig iron; the situation in Northeast China is slightly the same as that in North China. Yangquan, Shanxi province, is famous for producing iron, with a special factory with an annual output of more than 300,000 rubbed shovels and hoes, which are sold to Henan, Hebei, Shandong, and Inner

Mongolia. This technique also spread to neighboring countries, such as Japan. By comparison, it is considered to be superior to the techniques of clamping steel, sticking steel, stamping, and carburizing.

In addition to the above-mentioned plowshare iron and pot iron, these rubbed materials were also widely used in ancient bells, ancient iron columns, iron tiles, and iron weights (all made of white iron). In Yangquan, Shanxi province, gray cast iron strips with a high carbon content were obtained using a crucible furnace. There were many different names for such technique in different places, such as infiltration, *guangtie*, grits, and ice iron. This was a secret technique in the old days, when it was passed on from generation to generation but did not leave the home. Sampling analysis shows that all these raw materials have a high carbon content and low silicon and sulfur components, which results in a binary ferroalloy with high carbon and low silicon. It is characterized by: a low melting point (about 1,130 °C), good fluidity, strong carburization, and difficult graphitization, so that the rubbing process is easily achieved and the product quality is guaranteed.

Taking an iron hoe as an example, the production process of rubbing farm tools can be divided into cogging, upper nose (installing mouth), rubbing, flattening, cold hammering, and quenching. The thickness of the rubbing layer is key. If it is too thick, the carburization effect will be too strong, so that the body metal from wrought iron or low-carbon steel into high-carbon steel will become hard and brittle, and will crack during forging and use. If it is too thin, the carburization effect will be too weak, there will be no denaturation or insufficient denaturation of the surface layer, the hoe board will have poor strength, and it will easily curl and bend. This results in the so-called "less means weak, and more means brittle and cracking" by Song Yingxing. According to Ling Yeqin's research in Xiaoguan Iron Industry Community, Beijing, it is known that the amount of raw materials used for hoe rubbing is actually more than the amount stated in *Heavenly Creations*. The reason may be that only the blade of the old farm tools was rubbed, while the current hoe rubbing cover the entire board surface, which increases the amount of materials needed.

The temperature and time of rubbing and quenching have great influence on product quality. Field measurements show that heating low-carbon steel plates to 1,200 °C is more suitable. The time metal can be rubbed is about 4–5 min, in which the active rubbing time is only 20–30 s, which must be done agilely and accurately. Uneven rubbing or an uneven surface will affect the quality of the hoe board. Before quenching, the workpiece must be cooled to 750–800 °C, the color a cherry red. It is quenched in water for about 5 s, then trimmed and the edge cut to get the finished product.

From the metallographic structure, the outermost layer of the rubbed *Shuangzaohua* (a famous brand) hoe board is the cladding layer of white cast pig iron, the matrix structure is cementite and pearlite, and the inner layer is a hypereutectoid layer, which is a high-carbon steel material. The innermost layer is called the eutectoid layer and the transition to the bulk metal is called the hypoeutectoid layer (as shown in Picture 6.75). It can be seen that the high-carbon rubbing material acts on the wrought iron body with low-carbon concentration at high temperature, which makes the carbon diffuse rapidly, forming a carburized layer as well as the topmost

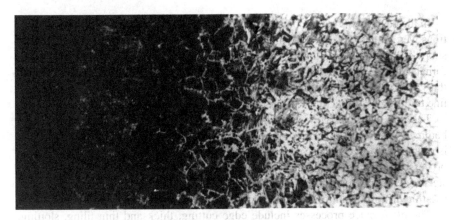

Picture 6.75 Metallographic structure of rubbed workpiece

pig iron cladding layer. These layers on the flexible body form a rigid and flexible steel composite material, which is beneficial to farming, does not break easily, and can sharpen itself, thus becoming a farmer's preferred tool. With the popularization of mass-produced agricultural machinery, rubbed-iron techniques are now rare. It is said that there are still artisans engaged in this industry in rural areas of Shandong and Shaanxi provinces. As a precious craft contained in *Heavenly Creations* and still existing in the world, it should be listed in the list of traditional handicrafts for preservation.

The following introduces two wrought iron skills that have been listed in *List of National Intangible Cultural Heritage*.

6.3.5.2 Wang Mazi and Zhang Xiaoquan Scissors

In China, there is a folk proverb that says the north has Wang Mazi and the south has Zhang Xiaoquan as household names, well known by women and children alike. Wang Mazi scissors came from Shanxi in the early Qing dynasty, named after the shopkeeper with the surname of Wang, who had a pockmarked face. *Complement of Dumen Bamboo Branch Poems* published in the 24th year of the Jiaqing period (1819) reads: "Shops of different owners are all named after Mazi", indicating that the shops for knives and scissors at that time had already been named after Wang Mazi to attract customers.

In the 21st year of Jiaqing (1816), two businessmen, Wang and Ji from Taiyuan, Shanxi Province opened Jutai Scissors Shop in Xuanwumenwai Street, and hung up the sign that reads: "three generations of Wang Mazi", "the real deal and no bargaining". They insisted on quality standards when purchasing finished products from workshops, and all products were engraved with the mark of "Wang Mazi", which had a lifelong guarantee and free replacements. These practices made the quality of Wang Mazi scissors stable for a long time and made them famous brand products.

In the past time, the ones passing on Wang Mazi's techniques to the next generations were honest and rigorous in their teaching. During the period of manual workshops, an apprentice started his career and asked the master to preside over the first furnace; otherwise, he would not be recognized by his peers, and no one would buy the scissors he made. This is the practice of every family in this industry, and it is a fine tradition of respecting teachers and respecting morality.

The tools used to forge Wang Mazi scissors include forging furnace, bellows, anvil, hammer, iron tongs, pliers, steel files, and thick and thin grindstones. Production is divided into two major sequences: on-furnace and off-furnace. The on-furnace processes include material selection, iron flattening, steel cutting, initial forging, steel sticking, cogging, firing, forging and compounding, forming, grooving, strands twisting, trimming and shaping and flattening (as shown in Picture 6.76).

The off-furnace processes include edge cutting, thick and thin filing, slotting, riveting, rough grinding of the edge, drilling, smearing, quenching, fine grinding of the edge, strands circling, marking, and revitalizing (as shown in Picture 6.77). Usually, a forge is equipped with two craftsmen, who are responsible for on-furnace and off-furnace work, respectively.

Wang Mazi scissors stick steel on wrought iron bodies, which makes the configuration of steel and iron optimal, the cost low and the grinding easy. They forge it to make it compact, quench it to make it rigid, then sharpen the blade and make it so that it is prevented from rolling or collapsing. All of the above is something ordinary scissors cannot match.

Traditional forging techniques run the whole on-furnace process. The temperature must be perfectly mastered when heating and forging. As the saying goes, "When the steel sweats, the forging is just right". When quenching, animal hoof powder and

Picture 6.76 Forging

Picture 6.77 Revitalizing

salt are applied to the scissors for carburization, quenching, and tempering, which is called "wiping medicine" by craftsmen. After quenching, the scissors have an optimal hardness and shine. All these unique skills are accumulated by craftsmen's long-term working experience, which can only be mastered by in-person teaching and practice.

Zhang Xiaoquan scissors in Hangzhou, which enjoys the same fame as Wang Mazi, was founded in 1663 by Zhang Sijia, a blacksmith from Yi county, Anhui province. Zhang Xiaoquan, the descendant of Zhang Sijia, moved to the foot of Wushan Mountain in Hangzhou to set up a shop and a stove. He chose high-quality steel from Longquan and Yunhe in Zhejiang province to make scissors, and through the meticulous work his business boomed.

For more than 300 years, the descendants of Zhang Xiaoquan have adhered to the ancestral precept of "good steel and fine work". The technological process claims to have 72 working procedures, and the products include three series: civil scissors, clothing scissors, and industrial-and-agricultural garden scissors.

Over the years, Zhang Xiaoquan scissors have won numerous awards at home and abroad, including the Silver Award at the Nanyang Industrial Exposition in 1910 and the Silver Award at Panama Pacific International Exposition in 1915. Famous inheritors include Xu Weiyi, Xie Rushan, Qi Yongxing, and Ding Chenghong. Among them, Qi Yongxing has a college education and Ding Chenghong has a postgraduate degree, which marks the new inheritance pattern of traditional skills in modern enterprises and is a gratifying phenomenon under the new historical conditions.

6.3.5.3 Bao'an Waist Knives

The main settlements of the Bao'an people are Ganhetan, Meipo, and Dadun villages in Jishishan Bao'an, Dongxiang, and Salar autonomous county in Gansu province, commonly known as the "three villages of the Bao'an". These areas are abundant in copper deposits. Waist knife forging mostly takes place in Ganhetan village. There are more than 620 Bao'an knife makers in total, with an annual output of more than 400,000 waist knives, which occupies an important position in the economic life of this ethnic minority.

The Bao'an waist knife is exquisite in craftsmanship, sharp in blade, and durable. It is a living utensil and ornament used by the Bao'an, Dongxiang, Salar, Tibetan, and Tu people. During the Cultural Revolution in 1966–1976, waist knife production was regarded as a sign of capitalism, so forging furnaces were destroyed and tools were confiscated. However, many artisans were unwilling to lose their ancestral skills and still produced knives in secret. Since the reform and opening up, the Bao'an waist knife has evolved into more than 10 varieties, among which the most famous one is the *Boriji* knife, and the most beautiful one is the *Shiyangjin* knife.

The waist knife is carefully made of high-quality steel, and the charcoal is fired with black thorn tree and birch. Taking *Shiyangjin* knife as an example, its manufacturing process is divided into three parts: knife workblank, knife handle, and knife scabbard. Knife workblank making includes steel cutting, heating and forging, slotting and clamping, forming, bore engraving, primary grinding, edge engraving, character engraving, grain engraving, fine grinding, drilling, quenching, precision forging, and polishing (as shown in Pictures 6.78 and 6.79).

The waist knife is named depending on the different combinations of knife handles. Whether the knife handle is beautiful and neat is an important symbol to judge the craftsman's level. The main processes are cutting horns, fixing, making hand guards, snail handles, customizing copper covers, sharpening knife handles, customizing plum patterns, and fine grinding. The scabbard is made of and coiled with copper sheets, finely ground on the scabbard surface. An iron bar is heated and then inserted into the scabbard to make it become golden yellow.

There are some taboos in the production of waist knives. For example, it takes 2 to 3 years to learn techniques from the masters. In the first year, only chores are done, and then skills are taught and wages are paid. If you start your own stove, you must get the master's permission, otherwise the master has the right to destroy your stove. In the old days, it was forbidden for women to be present, especially those who had just given birth or had their period, and it was forbidden for women to mount and cross tools and equipment. At present, in addition to the Bao'an people, there are also Han commoners who practice the craft. For example, the brothers Liu Wenzhong and Liu Wenji in Ganhetan Village have been making horn knives for more than 10 years.

"The steel knife with ten brocade handles, the sheath under the silver bag, and the tweezers laid by bronze are very attractive to wear". This ballad aptly shows the customs and aesthetics of the Bao'an people. For this reason, artisans carefully make knives and handles to be solemn and beautiful, integrating practicality and aesthetics. They have become symbols and markers to arouse people's feeling and sense of identity.

Picture 6.78 Clamping steel

Picture 6.79 Forging the waist knife

The forging skill of Bao'an waist knives was added to *List of the First Batch of National Intangible Cultural Heritage* (No. 392) in June, 2006. With the joint efforts of the government, community, artisans, and experts, this precious skill will continue to develop and be carried forward, as Ma Xuewu, a Bao'an poet, writes: "Ah, Bao'an waist knives. You exist, and therefore, there exists a nation. Bao'an waist knives, you are our symbol and our soul".

6.3.6 Lost-Wax Casting

The Chinese traditional smelting and casting industry is famous for lost-wax casting in Beijing, Wutai in Shanxi Province, Duolun in Inner Mongolia, Weifang in Shandong province, Suzhou in Jiangsu province, Chengdu in Sichuan province, Baoshan in Yunnan province, Lhasa in Tibet, and Foshan in Guangdong province. In the winter of 1965, Hua Jueming and Wang Ancai made a round trip to study the traditional lost-wax casting method, and found Buddha statue casting master Men Dianpu, who had changed his profession many years ago, in Beijing Micromotor factory. Under his guidance, they copied a Guanyin in the style of Northern Wei dynasty with lost-wax casting.

The process of this technique is as follows.

6.3.6.1 Making the Core

The mud core is made of sculpted pulp clay and the internal iron wire is used as the core bone (as shown in Picture 6.80).

6.3.6.2 Making the Wax Mold

The wax is made of paraffin, rosin, and soybean oil. After melting, it is slightly cooled. After repeated pulling, it becomes a wax with excellent plasticity. After rolled into a wax sheet, it is attached to the core, and then molded with a plectrum (as shown in Picture 6.81). The copper brace is inserted in the appropriate position.

6.3.6.3 Preparing the Mold

Coat the surface layer with refractory mud, and then coat it after drying. Pulp clay is used as fabric, and *madao* mud is used as the back material. After drying in the shade, it is heated and dewaxed, and then roasted in a kiln.

6 Metallurgy and Metalworking

Picture 6.80 Mud core

Picture 6.81 Wax mold being molded by Men Dianpu, a Buddha statue casting master

6.3.6.4 Melt Casting

The yellow miscellaneous copper is melted in a crucible, the mold is heated to about 300 °C in a kiln, and then taken out and buried in a bunker for hot casting.

6.3.6.5 Post-casting and Coloring

After the mold is cooled, remove the mold, core, flash and burr, and finely process and polish the face, hair lines and streamers with a chisel. The coloring method is often called *qianghuang*, which is the ancestral handicraft of Zheng Guanghe, a master who lived in Gongzi Hutong, Beijing, the operation of which is secret. It is said that the copper pieces are boiled in the liquid after decocted with three traditional Chinese medicines. After coloring, the copper Guanyin statue showed a beautiful golden yellow color, which was better in color and texture than electroplating. The surface layer seemed to be attached with colloid, which could be preserved for a long time (as shown in Picture 6.82). This technique is now lost to oblivion.

Men Dianpu died of illness in 1973, and now there are no descendants in Beijing. According to recent visits, only Foshan, Guangdong province and Deqin, Yunnan province still have some artisans who can perform this craft, but most of them have not practiced it for a long time. Since 2007, the Chinese Traditional Handicraft Research Association and Hangzhou Junhua Sculpture Company have cooperated successfully to copy the lost-wax castings of the pre-Qin period by lost-wax casting technique. This precious skill can be inherited and carried forward under the new historical conditions by imitating and creating techniques such as reproducing lost-wax.

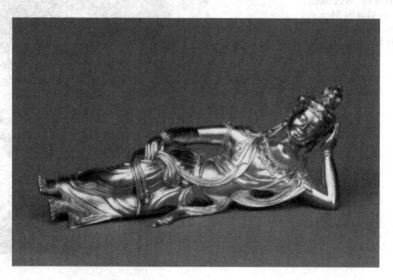

Picture 6.82 Colored copper Buddha

6 Metallurgy and Metalworking

Picture 6.83 Casting copper workblank

6.3.7 Gongs Forging in Changzi

Changzi county, Shanxi province, is the earliest place to forge musical instruments, more specifically metal bells, in China. According to records, in the first year of Tang dynasty's Zhenguan period (627), the copper gongs made in Chengcun village, in the southwest of the county, already gained a good reputation. Its main processes are as follows.

6.3.7.1 Making the Blank with Chemical Materials (as Shown in Picture 6.83)

The alloy ratio of the copper gong is: red copper 77% and pure tin 23%, which are cast into billet by iron mold. To melt the material, use pit stoves with two crucibles placed on in each one.

6.3.7.2 Hot Forging (as Shown in Picture 6.84)

The copper billet is heated to 600 °C, taken it out, hot forge it to make it thinner, and stretch it, which needs to be repeated many times. Hot forging should not be done

Picture 6.84 Demonstration of the operation of forging gongs by a senior craftsman

at a temperature above 600 °C, the quenching temperature cannot exceed 450 °C, otherwise the copper billet will become brittle and can no longer be forged.

6.3.7.3 Forging into a Rough Shape

Take the copper gong as an example, steps include starting with forging the gong's edge, annealing, hammering, and cutting the edge for many times to make the contour regular.

6.3.7.4 Quenching

Heat the rough shape of the musical instrument to 450 °C and quench it with water.

6.3.7.5 Cold Forging and Sound Correction

Musical instruments of different shapes and sizes have their own fixed scales, which must be hammered and calibrated constantly to determine the tone and sound quality.

6.3.7.6 Polishing

Use a scraper to flatten the front and back sides of the musical instrument, commonly known as "scraping it smooth". In this process, the tone is fixed for the first time, beating each part with a hammer, adjusting the thickness of the material, and then forging it with a hammer to produce a loud timbre.

6.3.7.7 Punching

The tether makes it easy to carry yet does not produce an echo.

6.3.7.8 Second Sone Setting

This process is commonly known as "setting the tone with a hammer". After polishing, continue to adjust the timbre, sound quality and tone, so as to achieve the same sound effect of similar musical instruments. Setting the tone is the most critical and technical process in the production of copper musical instruments. This is referred to as "beat and shape the gongs a thousand times with the hammer, but set the tone with one beat". Craftsmen who can set the tone are referred to as "good-handle master" and "full-handle master".

There are dozens of copper musical instruments such as flat tune gongs, high-pitch gongs, tiger-sound gongs, clearing-the-way (*kaidao*) gongs, big cymbals, and small cymbals (as shown in Picture 6.85). The inheritance history of this skill can be traced back to 1858 through musical ancestry records. At present, there are only three folk craftsmen who are proficient in a full set of this handicraft in Chengcun Village, one of whom is 81 years old and the other two are over 70 years old. The inheritance of skills is an urgent problem that needs to be solved.

6.3.8 Gold Foil Making

The traditional gold foil-making technique is called "thin gold", and the relevant records can be found in ancient books such as *Heavenly Creations* and *Trivial Talking about Painting*. The technique of making thin gold is still found in Beijing, Nanjing, as well as Zhejiang, Fujian and Guangdong provinces in modern times. Nanjing is currently the main production area.

Longtan, Qixia district, Nanjing, is the birthplace of gold foil making. According to folklore, this technique was invented by Ge Hong, an alchemist, who competed with Lv Dongbin, one of the Eight Immortals in Chinese legend, and tried to hammer gold as foil in order to put gold on Buddha statues in Longtan. His technique was better than that of Lv, so, later, the technique was handed down.

Multiple processes are necessary to produce gold foil:

Picture 6.85 A large clearing-the-way gong with a diameter of one meter

6.3.8.1 Melting and Casting

The fineness of gold foil is 98 gold, 88 gold, 77 gold, and 74 gold. Taking 98 gold as an example, 49 gold (with a gold purity of 99.99%) is used to start, mixed with 2% silver and copper, heated and melted in a crucible. Then borax is added to make slag, and then gold bars are cast in an iron mold.

6.3.8.2 Forging Workblank and Blade

The gold bars are forged into blanks about 8 *si* (0.08 cm) thick, and then forged and cut into 16.cm gold leaves. Heating is needed during forging to eliminate metalwork hardening. 120 of the obtained gold leaves are called one "*zuo*".

6.3.8.3 Making a Ribbon

Cut the gold leaves into ribbons with a width of 1 cm with a bamboo knife, with 2,048 ribbons for each "*zuo*".

6.3.8.4 Gold Extension

The black gold paper with a square of 10 cm is heated, so that it can be extended quickly after the ribbon is put into it.

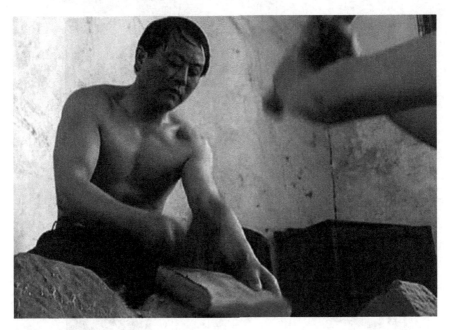

Picture 6.86 Beating the "*kaizi*"

6.3.8.5 Putting the Ribbon

Put the ribbon into the interlayer of black gold paper with fingertips or tweezers, and seal it with paper.

6.3.8.6 Beating the "*Kaizi*"

Beat the "*kaizi*", i.e. the ribbon wrapped in black gold paper, thinner (as shown in Picture 6.86).

6.3.8.7 Putting the "*Kaizi*"

The "*kaizi*" is very thin and fragile, so it is necessary to pick up the ribbon with goose feather by blowing and put it into black gold paper about four times as large as the "*kaizi*", commonly known as the "*jiasheng*".

6.3.8.8 Heating and Temperature Control

Put the "*jiasheng*" into the furnace to control the temperature for about half an hour, so as to prevent the metal from hardening and avoid the influence of external temperature.

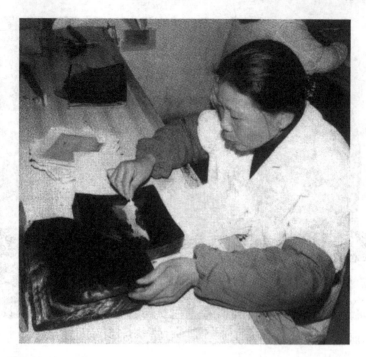

Picture 6.87 Moving gold foil

6.3.8.9 Thinning

Continue to hammer the "*jiasheng*", to change the thickness of the layer and avoid direct contact with the gold foil in the middle. At this time, the gold foil has been extended to a 0.12-micron-thick piece with a width of 10 cm.

6.3.8.10 Moving Gold Foil

Use a goose feather to pick up the gold foil by blowing and put it on Maotai paper (as shown in Picture 6.87).

6.3.8.11 Cutting Gold Foil

Cut the foil into squares with a width of about 3.3 cm with a bamboo knife.

6.3.8.12 Packaging

10,000 pieces of gold foil weighs a total of only 25 g.

Black gold paper is the key material for forging gold foil, and its specifications are extremely strict. It is derived from a bamboo leaf in the mountainous area of Zhejiang province, which must be buried underground for several years. After being taken out, it is smashed with a stone hammer to make paper. Tiles are smoked and baked under soybean oil lamps. After becoming carbon black, the carbon is scraped off, mixed with plant gum, used to coat the paper, dried, and rolled flat. The paper thus made has a very smooth and shiny surface, and gold sheets can be well extended when sandwiched between it. At present, the areas where black gold paper is produced is still limited to Nanjing and Shaoxing, Zhejiang Province.

The hammer art of gold foil is exquisite. To start the process, two craftsmen must sit opposite each other, with a small hammer in their left hand and a sledgehammer in their right. The technical standard of making the "*kaizi*" is to beat it evenly until it takes the shape of a crab shell.

The "*jiasheng*" is beaten on a stone pier, whose surface is angular. During the process of beating, unwrap the "*jiasheng*" every 45 min, cool for about 20 min, switch the upper part and the lower part, rewrap and tighten it, and repeat the process four times. Then the beating and unwrapping process should be done another three times, but the "*jiasheng*" should be unwrapped every half hour to cool down since it has become thinner. In this way, a pack of "*jiasheng*" will be beat 30,000 times in total, which shows the hard work of craftsmen.

According to the survey, the famous foil artisans in the early Republic of China were Han Xinggui, Yin Fucheng and Yin Fujia; Liu Xingguo, Guo Yifa and Guo Yishun in the 1920s and 1930s, and Mei Zhenghua, Wu Tingkui, and Gu Guangfu in the 1950s.

The production of gold foil has many forging processes and strict technical requirements, with low wages and heavy physical work. At present, the industry lacks young artisans. Mechanized gold foil making has a higher productivity and lower price, which has greatly impacted the traditional skills. This skill was added to *List of the First Batch of National Intangible Cultural Heritage* in June 2006. With the support of the local government, the Nanjing Gold Thread and Gold Foil General Factory plans to strengthen the protection of senior craftsmen and cultivate a new generation that will support traditional foil-making, so that this precious skill can be protected and inherited (as shown in Picture 6.88).

6.3.9 Mud-Mold and Stone-Mold Casting

Among the three casting techniques in ancient China (mud mold, lost-wax casting, and iron mold), mud-mold casting was used the longest, had the widest application, and was the most important. This traditional technique still exists in many areas in China today.

Picture 6.88 Buddha statue decorated with gold foil

6.3.9.1 Multi-Mold Casting Pot

Small and medium castings produced in batches often adopt reusable multi-molds (or semi-permanent mud molds), and the most representative one is used for casting iron pots, more specifically, cauldrons.

Smelting and Casting, Chap. 9 of *Heavenly Creations* contains the following on this casting technology: "A casting mold for a cauldron is made up of inner and outer layers. The inner mold is made first. After it has dried in the sun for several days, an outer layer mold is made to cover the inner one according to the size of the cauldron. The artisans making outer molds should be extremely careful with the operation, or the molds will be useless if they have even a tiny defect. Once the mold has dried, the mud is kneaded and smelted in the furnace, such as the kettle, which is subjected to pig iron. The back of the furnace is ventilated through pipes, and the surface of the furnace is kneaded and tapped. Molten iron produced from one smelter can cast ten to twenty cauldrons. After being melted, the iron is taken from the smelter mouth by an iron ladle padded with earth. A cauldron is molded with about one ladle of molten iron. This is poured into the hole at the bottom of the mold. The mold cover should be removed before the iron cools down to see if there are any cracks or defects. Then the cauldron, glowing red and not yet turned to black, should be mended with a little molten iron where cracks appear. Then smooth it over with wet grass leaves to leave no trace" (as shown in Picture 6.89). *Code of Great Ming Dynasty* records that materials used for cast iron pot include white charcoal, fried pieces, magnetic powder, green *gantu* soil, bamboo sieve, horsetail silk, and hemp, with a similarly described practice as mentioned above.

Picture 6.89 Casting cauldrons

This traditional handicraft is still widely used. For example, Wangyuanji workshop in Wuxi, Jiangsu province, founded in the 17th year of the Daoguang period (1837) in the Qing dynasty, is famous for its casting of thin-walled iron pots. Most of the modeling materials come from local areas, including rice awns, bran ash, magnetic powder, coal ash, and pine soot ash. Long-grained rice bran planted in mountainous areas is the best bran ash, because its shell is thick and firm, and it is made into powder after being baked into ash. The mud mold has high fire resistance, high strength, and durability. Bran ash must be screened, and the properties of coarse ash and fine ash are different. It must be kept dry for later use. Magnetic powder has great viscosity and toughness, and is rich in organic colloid. It must be stored in a water tank for later use. Pine soot ash is used as coating.

There are four kinds of molding materials. The back material is made of yellow mud and rice awns with a ratio of 10:1, mixed in water. The yellow mud should be put into the pool, soaked and stirred into slurry, mixed with rice awns, and repeatedly stirred with a harrow until it is uniform. Then, the mud is repeatedly trampled by the craftsmen until soft according to the traditional practice, and kept for one day and night. The intermediate material is used for connecting the workblank and the mixed mud, and is made by mixing magnetic powder, coal ash, and water. The surface material is made of magnetic powder slurry, coarse bran ash, fine bran ash, and water. After repeated trampling, the surface of the mud becomes shiny and tough.

The lower mold is made of a workblank and yellow mud, which is placed on the ground. Then the back material is stacked at the mouth edge of the mold and pushed to the bottom of the mold. The mud material should be 40–60 mm thick. After the surface of the mold is decorated and leveled, a hole is dug out at the center (where the scraper is placed). It takes about 10 days to dry the mud mold, during which time it is beaten firmly with a rope bolt every other day, trimmed with a scraper, and dried in the sun, which qualifies it as a lower mold.

The upper mold is made based on the lower one. The lower mold is laid flat on the ground, piled with yellow mud, and smeared into a flat-topped cone. A gate is opened in the center of the top, and air holes are punched around the top. After drying, the upper mold is turned over with the lower one, scraped flat with a scraper, and then the inner surface is polished, thus forming an upper mold (as shown in Picture 6.90).

Picture 6.90 Mud molds for casting pots, cited from *China at Work* by Rudolf P. Hommel

After the mud molds have dried, they must be baked at a high temperature of 500–600 °C, until the surface is red, and the mold surface should be coated with bran ash and trimmed with fine sandpaper. The gate mouth is ground with a cutter to ensure that the cavity thickness is uniform and the gate wall is smooth. Before casting, the surface of the mud molds is brushed with paint made of pine soot ash and water, and dried for later use. After assembling the upper and lower molds, insert bamboo splits in the hole and rotate along the cavity to test whether the thickness is uniform, and fine-tune it where necessary.

A small blending furnace is used for iron melting. As mentioned in *Heavenly Creations*, the furnace is shaped like a kettle, and the air duct is inserted directly into the center of the furnace from behind it, so that charcoal can be fully burned, and the furnace temperature can reach as high as 1,450 °C. The temperature of molten iron is about 1,400 °C. Quick pouring is required, and then the excess molten iron in the hole is dug out to flatten the navel of the pot.

The biggest advantage of the multi-mold casting pot is that a pair of mud molds can be used dozens to hundreds of times, and the production cycle is short. Even though it is a purely manual operation, it can be produced in large quantities with high efficiency. The thinnest part of this iron pot is only 0.5–0.8 mm. Although it is made of gray cast iron, it is tough. A pot with a diameter of 40 cm can be flattened to 29 cm without cracking. Because of its high quality and light body, it is ideal for cooking, which makes it well received by the people for a long time.

6.3.9.2 Stack Casting in Foshan

According to folklore, stack casting in Foshan, Guangdong province began halfway through the Ming dynasty, giving it a history of more than 500 years. At first, it was mostly used to cast small daily necessities and artwork. Since the 1920s, it has been used in industrial production to make components, locks, gears, and racks for textile machinery and sewing machines, with an accuracy level of 5–7 and a smoothness level of 3–6.

Stack-casting molds are made of clay and sawdust. Clay is sourced locally, which is light yellow and is a type of thin clay. The clay must be dried and crushed into powder. The coarse sawdust, used as back material, is mostly made of Chinese fir quality, which easily carbonizes when fired. The fine sawdust, used as surface material, is mostly made of hard miscellaneous wood that do not easily carbonize when fired, resulting in a bright and clean profile.

The mold for stack casting is composed of a metal mold and a template, and a mold frame is added when molding (as shown in Picture 6.91). After the mold and core are made, they are dried in the shade or sun, or baked at a low temperature (as shown in Picture 6.92). During casting, the mold and iron are bound with iron wire and painted with mud. The roasting temperature ranges from 800 to 900 °C. The traditional method is to use firewood as fuel and look at the color of the flames using experience as judgement. Due to the high mold temperature, good thermal insulation, and slow cooling rate, iron castings with a wall thickness of 0.5 mm can be cast without the color becoming white, and their mechanical properties are superior to those of ordinary sand castings (as shown in Pictures 6.93 and 6.94).

Modern stack-casting technology is developed with the rise of large-scale machine production to meet the needs of batch castings. It is mostly used for casting small objects such as piston rings, and its costs are lower than that of other casting methods. Some of its technological basis and measures are consistent with the ancient stack-casting techniques. Based on the historical development of casting technology, we can see that the traditional stack-casting process is the predecessor of modern stack-casting and shell-mold casting, which explains that foundries in Foshan and Guangzhou, Guangdong province, call it thin-shell mud-mold casting. In the 1990s, traditional stack casting died out in Foshan and many craftsmen and equipment were lost. Fortunately, the Foshan Foundry and Guangzhou South China Sewing Machine Factory kept detailed records, which enable us to know the outline of this technology.

Picture 6.91 Stack-casting boxes

Picture 6.92 Stack-casting molds

6.3.10 Collecting-and-Throwing (Shoupaohuo) Work and Tibetan Copper Forging Technology

The traditional copper forging process was called "*sa*" in ancient times, but is more commonly known as *shoupaohuo* (collecting-and-throwing work), which still exists in many areas today, especially in Tibet.

6.3.10.1 Collecting-and-Throwing Work

Mr. Wen Tingkuan made a field investigation on the collecting-and-throwing work in Beijing in the 1950s. The representative works from the Ming and Qing dynasties include the large bronze statue at the Lama Temple and the door and window ornaments at the Palace Museum.

Taking the copper head at Shijiazhuang Martyrs Cemetery as an example, the following is a brief introduction to this technique:

(1) Raw materials, tools, and equipment
   Red copper plate: The thickness of the bronze head at the Martyrs Cemetery is 1.5 mm, while the copper plate must be 2 mm thick.

   Welding flux: Round carving work should be hammered and welded in blocks. The flux is self-made. Copper and zinc are half-melted in a crucible, poured into a fine sand pile, and condensed into crushed slag, then mashed with

Picture 6.93 Demonstration of master Huang Wenxing, photographed at the *Ancient Chinese Science and Technology Exhibition* held in Canada

a hammer until it's the size of small grains of rice, and mixed with 25% borax powder to form the welding flux.

Iron rod anvil: Forged with square iron bars, bent at both ends and ground flat at the top. When in use, the iron rod is pierced into the branch and obliquely supported on the ground.

Square iron anvil: When copper plates are beaten, it is padded below and can be inserted into the ground or wooden frame.

Fine sandbag: A white cloth bag is filled with fine sand. The copper details are beaten and padded under it.

Glue making: Heat, melt, and mix rosin, chalk, and vegetable oil (with a weight ratio of 4: 9: 1), and condense it for later use. It is softened by heating it just before use, as a pad for chiseling.

Steak hammer: A steel hammer with a flat and square end and a sharp and round end.

Picture 6.94 Stack casting

Throwing hammer: An iron hammer with round and flat ends.
All kinds of chisels: Chisels with pointed heads, round heads, and flat heads.
(2) Manufacturing process:

1) The copper plate is cut and the pattern traces chiseled out.
2) Heating and beating.

Annealing and beating, which is a kind of cold forging, must be repeated. A copper plate is called a "raw blank" when it is not beaten. After the pattern is chiseled out, it is heated and annealed to increase the ductility of copper, which is called a "cooked blank". The cooked blank becomes brittle after it is beaten, and can be beaten again after annealing. It takes many rounds of annealing to finish the collecting-and-throwing work.

There are five beating methods (as shown in Picture 6.95).

Throwing: It is often used in large convex places, such as the nose and eyebrows of the head sculpture. Place the copper plate on the square iron anvil and beat it on the back with a throwing hammer to make the copper plate extend and protrude. When hammering, it mainly depends on the downward force, which is called "virtual hammering".

Moving: When there is a higher bulge on one part of the copper plate, but there is not enough copper in this part for the extension, then it is necessary to "push" the copper bulge around with force. This method is called "moving" (a way of moving

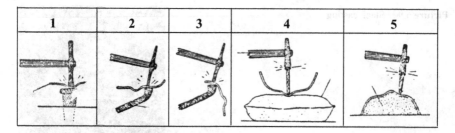

Picture 6.95 Methods of beating: 1. throwing 2. moving 3. collecting 4. pointing 5. chiseling

copper from one place to another). You can put this part on the iron anvil, hammer it with the flat head of the steak hammer, move the copper plate by hand while hammering, and move the copper around to the throwing place.

Collecting: Move the copper plate around to the back (opposite to throwing), such as the forehead, cheek, chin, etc., all of which need to be retracted back. Collecting also uses a steak hammer to virtually hammer on the iron rod anvil. When hammering, the copper plate should be kept rotating and repeated several times before it can be collected.

Pointing: For example, when the convex area of a copper plate is small (such as a nose tip) or sharp edges and corners (such as temples), since it is difficult to work accurately using an iron rod anvil, fine sandbags are put under the copper mold and the tip of the steak hammer is used to hammer gently. This method is called pointing.

Chiseling: The copper surface is not neat enough and uneven, so chiseling is needed. The copper mold of the gasket is filled on the back with special soft glue pressed by fingers, and then the worker uses a hammer to hit the chisel on the copper surface carefully to show the details of the copper wares.

Hammering the copper mold requires a combination of throwing, moving, collecting, and finally pointing and chiseling. The front half of the copper head is shown in Picture 6.96 and the back half in Picture 6.97.

(3) Filing, aligning, and welding

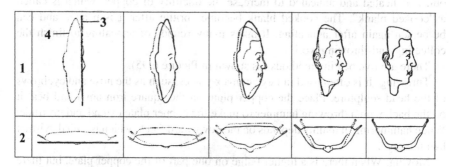

Picture 6.96 Forming of the front half of the copper head: 1. vertical section 2. cross section 3. copper plate 4. shape after hammering

Picture 6.97 Forming of the back half of the copperhead: 1. profile 2. top

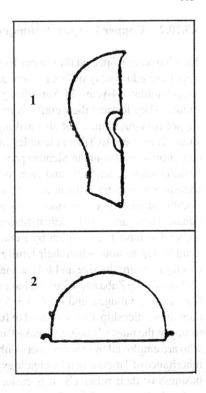

Round carving usually requires two or more copper plates, which means they need to be welded together. Before welding, the copper edge is draw-filed and tied with iron wire. Take the welding flux and sprinkle it on the joint, and fire it to bond the welding flux until it fills the joint. Then heat the copper molds, so that the copper-zinc alloy will melt and penetrate into the joint to weld the copper molds firmly together.

(4) Trimming and coloring

After the copper joint is draw-filed, the product is finished once it has been colored or gold-plated.

In the 1960s, this author visited Beijing Metal Crafts Factory to investigate the collection-and-throwing work, and the general situation, as observed, is the same as that described by Mr. Wen Tingkuan in his article as mentioned above. Yixian County, Hebei Province is a famous producing area for collecting-and-throwing work in North China, and there are still some manufacturers and workshops that use this technique.

6.3.10.2 Copper Forging Techniques in Kham, Tibet

Tang Xuxiang inspected the copper forging techniques in Naye village, Gama township, Changdu county in Tibet. There are ten families in the village, six of which are coppersmiths. 44-year-old Yongzhong Zaba and his brother Wengzha are coppersmiths. They learned their crafts from their father since childhood, mainly forging copper Buddha statues, but also making exquisite silver ornaments and daily necessities. They went to Tibetan temples in Shiqu county, Baiyu county, Ganzi Tibetan autonomous prefecture in Sichuan province, and Yushu in Qinghai province, to make Buddha statues, artifacts, and architectural ornaments. The forging skills of their ancestors were very influential in Tibet's Dongkang (Eastern Kham) area, and the Buddha statues they cast once enjoyed the privilege of not being consecrated by lamas. There are two Buddhist Statue Measurement Sutras in the family, which are said to have been written by Renqin Daji, the Great Living Buddha of Karma Temple. Up to now, when their family make Buddha statues for temples, they still strictly measure the size and follow the steps according to the ancient method.

Yongzhong Zaba and Wengzha have six apprentices, all of whom are young people from nearby villages, and it takes 3–5 years to finish their training. After finishing their apprenticeship, they still need to further improve their skills, and some continue to stay at the master's house to make Buddha statues; there are also some apprentices who are employed by other coppersmiths to start earning a living. The most common inheritance of Tibetan metal technology is still the apprentice who inherits his father's business or their relative's. It is easier for them to see the true story behind their crafts. In the manual family environment, the sons are exposed to their parents' working process every day, which makes them more knowledgeable than others in the professional field, since it's a unique growth process. Yongzhong Zaba and Onza have three sons and one daughter altogether. Tenzin Luobu, the eldest son, is 25 years old. After graduating from primary school, he started learning his craft from his father for 10 years. Ji Rao Luobu, the second son, is 20 years old. He also studied crafts with his father after graduating from primary school and has been doing crafts for 5 years. Chai Rangduoding, the third son, is 15 years old, graduated from primary school and has been an apprentice for 1 year. They also have an apprentice named Nisi Quben, 19 years old, who studied crafts for 3 years and was a relative from Yoba Town. In this way, this family can not only pass on the technology to the younger generation in an all-round way, but also complete the orders from temples. For the apprentices, they can not only learn the basic skills of crafts with their parents, but they can also observe how their parents deal with the outside world, learn to process information, seize opportunities, and gradually become mature.

The Yongzhong Zaba family are now making Buddha statues for the temples in Naqu. Its modeling features belong to the Karma school, the features of which lie in that Buddha statues have different sitting positions and handprints, but they all wear erect caps with gold edges. The Buddha statues of local Kagyu sect made by coppersmiths in Karma township all wear this type of hat ornament. As the Karma Temple burned down a few years ago, the main hall was almost completely reduced to ashes. Since most of the early statues inside were clay sculptures, the fire caused

Picture 6.98 Great copper Buddha

serious damage. Every coppersmith in this area has undertaken a certain amount of copper-making projects, with powerful families making copper Buddha statues and small families making metal fittings. The statue of Karma pa made by the Yongzhong Zaba family is about 3 m high (as shown in Picture 6.98). The raw materials are provided by the temples while the production is undertaken by the coppersmiths for free. Karma pa is the founder of Kagyu sect, a branch of Tibetan Buddhism, as well as the founder of Karma Temple, which has a very lofty status.

The Yongzhong Zaba family is in charge of making the "*gengtuo*" of Karma pa statue, which is the flammule of the Buddha statue. Yongzhong Zaba thinks that the flammule of Karma pa, Sakyamuni, Tsongkhapa, and Guanyin is all the same. There are six kinds of sacred objects in to decorate the flammule. "*Jiaqiong*" (Jīvajīvaka, a kind of bird) is placed on the top, because it flies higher than all the other birds, symbolizing that the Buddha Dharma is as high as heaven. In the eyes of Tibetans, the sun, moon, and stars are the most sacred auspicious symbols, and the *jiaqiong* is decorated with the sun, moon, and stars, which shows that the Buddha and Dharma are as sacred as the sun, moon, and stars. "*Sengeng*" is a lion, the king of beasts, symbolizing the supremacy of Buddha Dharma. "*Langbuti*" is a white elephant with infinite strength, symbolizing the boundless Buddha Dharma. "*Qiuxie*" is a capricorn, which represents a firm belief in Buddha. Yongzhong Zaba says that the "*qiuxie*" will not let go when it bites something. "*Lubomo*" is a dragon girl, with a human body at the top and a snake at the bottom, symbolizing wealth. "*Buqiong*" is a little boy riding on the back of a deer, symbolizing the kindness of Buddha. This is how an ordinary Tibetan coppersmith understands his traditional culture and his own work. Based on this understanding, the daily work of coppersmiths is related to religious beliefs and emotions, which contributes to the uniqueness of their work. Although this kind of work is also related to economy and life, from the religious and emotional aspects, it transcends the general work behavior and is full of intelligent creativity. This is essentially different from Han craftsmen and Bai craftsmen who regard making Buddha statues as economic endeavors. It is natural for Tibetan craftsmen that religious awareness rises to religious emotion throughout the whole process of making Buddha statues.

The Yongzhong Zaba family's measurement methods for making Buddha statues are very strict. Take a Buddha statue with a height of 2 m as an example:

(1) Divide 2 m into 8 parts to get the first cardinality: 25 cm. This number is very important, which is the length of the Buddha statue's head, and it is also related to the measurement of other parts.
(2) Divide the first cardinality into 12 parts, and get the second cardinality: 2.08 cm, which is the only parameter of many detailed measurements. The two numbers of 25 cm and 2.08 cm are the basis for measuring the size of the Buddha statues. As mentioned earlier, the two measurement sutras, written by Renqin Daji, the Great Living Buddha of Karma Temple, were passed on to the brothers by their father.
(3) The total height of the Buddha statue's abdomen to shoulder is 3×25 cm. Below the abdomen is 0.8×25 cm, the chest width is 2×25 cm, the hand length is 25 cm, the lower arm length is 16×2.08 cm, the upper arm length is 20×2.08 cm, the waist width is 18×2.08 cm, and the abdomen width is 15×2.08 cm.

The foot length is 25 cm + 5 cm, and the cross-legged width of the Buddha statue is $(4 \times 25$ cm$) + (4 \times 2.08$ cm$)$. The lotus pedestal is divided into two layers, with the height of the upper layer 10×2.08 cm and the height of the lower layer 25 cm.

Through calculation, the dimensions of each part are obtained, thus forming the total height of the Buddha statue. The calculation method of the size of the large copper Buddha statues in Tibetan Buddhism is a stylized formula summed up according to the experience of shaping and creating Buddha statues. This modeling pattern ensures the individual characteristics of Tibetan Buddhist art. Even under the impact of modern social changes, especially modern culture, Tibetan Buddhism still maintains the influence of religious spirit in their crafting culture. For thousands of years, the spiritual life and economic life of coppersmiths have changed little. Both old and young coppersmiths work in a relatively closed environment, receiving the same religious influence and inheriting the craftsmanship left by their ancestors in the same way (as shown in Picture 6.99). This craft of building Buddha statues is so laden and imbued with Tibetan tradition that it is of immense social value to Tibetan culture.

Another important factor is the unity of the style and structure of tools used in Tibetan metal processing. Tibetan coppersmiths usually use a combined-sleeve steel anvil when forging. This kind of anvil is very convenient to use. An iron bar is supported by a carpenter's tripod, and the upper end of the iron bar is forged into a flat square cone, which can cover all the anvil for use. The front end of the steel anvil is made into different shapes with different sizes, inclinations and radians, and changeable shapes according to what is needed; the rear end is made into a tube shape. The coppersmith can change the steel anvil at any time, and can forge various

Picture 6.99 Lifting and pressing

shapes. The steel hammer used by Tibetan coppersmiths is also very characteristic, with a length of 15–20 cm. The hole for the wooden handle is a round hole in the middle of the hammer. The hammer's surface is made into round head or square head according to the smith's needs. Its surface varies greatly from plane to inclined plane to circular arc surface. The longer the coppersmith works, the more hammers he will have. A good coppersmith has high requirements for tools, since good tools produce good work. These anvils and hammers are unique to Tibetan areas. Coppersmiths in Changdu and Shannan and Tibetan coppersmiths in Yunnan, Sichuan, Qinghai, and Gansu, all use the same anvils and hammers. The unity of Tibetan coppersmith's tools makes us have to ask questions from a historical perspective: In the process of social and cultural evolution, how are they able to consistently maintain the unity and invariability of these techniques? Is it because of the national character of religious belief that the unity of Tibetan technology culture is promoted?

The above field investigation data and intuitive feelings are only one aspect of the Tibetan metal technology' history. These seemingly backward handicrafts once played a great role in the development of a nation's history and culture, and still have important value today.

6.3.11 Gold, Silver, and Fine Gold Technology

6.3.11.1 Evolution

Gold, silver, and fine gold handicrafts are traditional techniques of making furnishings and decorations with precious metals such as gold, platinum, and silver. The beginning of this technique can be traced back to the Shang dynasty, using hammer techniques to extend gold, which was mostly used for vehicle ornaments. During the Spring and Autumn period and the Warring States period, gold tableware and hooks appeared, which were made by casting and engraving. In the Han dynasty, gold seals were still made mainly by casting and engraving. The expansion of territory and foreign exchanges made gold products more abundant, and the techniques such as wire pinching, filigreeing, welding, and inlaying appeared, with the style showing clear influences of Roman culture. During this period, nomadic people's skills advanced quickly both in technology and variety because of their frequent migration, which required articles for daily use to be portable, and to meet their unique taste. They created a variety of ornaments, earrings, and headdresses and were proficient in a variety of techniques, such as casting, hammering, embossing, wire drawing, and inlaying.

In the Tang dynasty, the system of mining and craftsmen in service was reformed, and the scale of the market and industry of gold, silver, and fine gold craft were expanded. Craftsmen learned from the technologies of South Asia, West Asia, and ethnic minorities, and came up with a variety of processes such as gold melting, gold pasting, gold plating, gold grinding, gold coating, gold painting, golden engraving, gold twisting, gold thread engraving, gold thread couching, gold foil painting, gold inlaying, and gold wrapping. The diversified development of gold and silver products was supported by the time's level of technology, and some processes are still in use today. Gold and silver products in the Tang dynasty include eating utensils, containers, jewelry, medical utensils, and other miscellaneous utensils. Their shapes and patterns were changeable and beautiful, and clearly influenced by Central Asia and West Asia.

The Song dynasty advocated Neo-Confucianism, which opposed extravagance and selfish desires. Therefore, the shapes of gold and silver products became mostly high reliefs with an artistic style that tended to be delicate, which was handled meticulously from the whole to the local. Among the minority regimes coexisting with the Song dynasty, Liao's gold, silver, and fine gold crafts were relatively developed, with most being harnesses, wine utensils and burial utensils, and more extensive shapes and patterns than Han products.

In the Yuan dynasty, the technology of gold, silver, and fine gold in the south was more developed than that in the north. Because of its sophisticated technology and multi-purpose melting and casting methods, their appearance was more vivid and fine. *Yin Cha* (as shown in Picture 6.100), a masterpiece by famous artisan Zhu Bishan, is a representative work that marked the artistic attainments and technical

Picture 6.100 Yin Cha

level of this era, and reflected the influence of the literati's way of life, which avoided politics, pursued emptiness, and leisure, on gold, silver, and fine gold technology.

Preciseness, slenderness, and hollowing out are the characteristics of gold, silver, and fine gold techniques in the Ming dynasty. This slenderness aesthetic tendency made the filigree craft incisive and vivid. The gold, silver, and fine gold craft in the Qing dynasty inherited the complicated style of the Ming dynasty, which was often used in combination with inlaying. It was full of decoration, rarely left blank, and contained more carvings. The ornamentation was three dimensional and heavy, and the filigree production paid attention to technical difficulty and visual magnificence. All kinds of products strove for luxury, which tended to be over-decorated and formatted.

6.3.11.2 Raw Materials, Tools, and Machinery

The metals used are gold, platinum, silver, and copper. Jewelry includes pearls, precious stones, jadeite, and jade. The welding fluxes include gold welding flux and silver welding flux. There are also silver blue glaze and other chemical materials and bonding materials.

According to the required hardness, color, and specifications of the finished products, raw materials are prepared and primary processing implemented, such as smelting, plate casting, strip casting, sheets rolling, and wire drawing. There are 43 types of sheets and wires according to their thickness and diameter.

Gold is divided into pure gold and K gold. Copper or silver must be added to pure gold to improve its hardness. K gold is the abbreviation of Karatage of gold, which includes 9 K, 10 K, 12 K, 14 K, 18 K, 20 K, and 22 K, and pure gold is 24 K.

Commonly used tools and machinery are hammers, pliers, tweezers, files, chisels, wire pinching plates, wire-rubbing wood, wire-rubbing plates, wire sticks, cones, iron sand plates, and iron pliers, as well as modern machinery, such as wire-drawing machines, sheet-rolling machines, chain-making machines, wire-rubbing machines, engraving machines, punching machines, and rolling machines.

6.3.11.3 Filigree Technology

(1) Making filigrees

A filigree is a metal strip with one or three metal wires wound or combined according to certain requirements. Its basic shapes include arch filigree, standard filigree, bamboo node filigree, auspicious filigree, vine filigree, wheat ear filigree, small pine filigree, phoenix eye filigree, twist filigree, and pigtail filigree (as shown in Picture 6.101).

They are used on the surface of large products which need decoration. Various kinds of filigrees can be changed into many continuous patterns through different arrangements and combinations or weaving techniques, such as palindromes, woven mat patterns, and lantern holes. There are also patterns mostly used as shading lines, such as turtle back lines, *fangsheng* lines, and ancient coin lines. In addition to geometric patterns, there are also some realistic patterns, such as

Picture 6.101 Filigrees

flowers (peony, chrysanthemum, sophora flower), animals (dragons, phoenixes, butterflies, bats), landscapes, clouds, and flames (as shown in Picture 6.102).

(2) Production techniques

Heaping: After blending carbon powder and hyacinth orchid with water, the basic shape is molded. Stick the filigrees to the shape with hyacinth orchid and weld them firmly. This process is suitable for hollow works, and the welding must be successful from the first try, which requires high technical proficiency.

Filigree: Two or more layers of filigree patterns are superimposed by bonding and welding.

Weaving: Various textures are woven with different techniques, such as braids, mats, and lanterns.

Knitting: The surface's ornamental texture is made by interspersing tidying warps and wefts.

Pinching, also known as "wire pinching", is suitable for plane patterns expressed by lines. Repeated patterns can be made simultaneously: Wrap a certain number of filigrees or plain filigrees evenly and symmetrically side by side on a stick and bond them together with pigskin fat. After drying, cut off the filigrees according to the required length, then bend the filigrees into the specified pattern with tweezers, burn it, melt off the fat, and attach the separated filigree patterns at the required position one by one.

Filling: The figure profile is made by inlaying thick plain filigrees, and then the pattern made with a flattened single filigree or plain filigree is filled in the profile, and the welding flux is applied to fix it.

Combining: Gold, silver, and fine gold handicrafts are delicate and complicated, which means they often cannot be completed at once. They need to be made separately and then combined, which is commonly called "combining".

Welding: Welding is often interspersed in the whole manufacturing process and is an important means of working with gold, silver, and fine gold.

(3) Inlaying techniques

Chiseling: It refers to chisel engraving, and the chiseled metal sheet needs to be padded on a flexible backing. The pad glue is prepared with a precise amount of carclazyte, rosin, and peanut oil. There are four kinds of chisel engraving techniques: *yin* chiseling, *yang* chiseling, flat chiseling, and hollowing. The flat chiseling means removing the pattern lines on the metal surface; hollowing is to chisel off the ground after chiseling, to form a hollowing effect.

Real inlaying: It refers to inlay, including mosaic, bezel inlay, prong inlay, burnished work, and gold foil painting. Mosaic must be made by chiseling out patterns on the surface of the body, filling the grooves with metal wires, and then hammering the inlaid wires and polishing the result. Bezel inlay means wrapping the inlay with the edging which is made in accordance with the shape. Prong inlay is mostly used for ornaments such as rings. The inlaid pearls or precious stones are attached by using the upright "prongs" sticking up from the base bracket. Burnished work means cutting gold foil or silver foil into patterns, pasting them on the surface of utensils with glue paint, brushing them several times, drying them, and finally burnish them until the gold and silver foil are

Picture 6.102 Peach blossom patterns

exposed. Gold foil painting means covering the pattern with gold and silver foil and hammering it to make it fit onto the surface of the wares.

(4) Surface treatment

The most commonly used method is gold plating. Before electroplating appeared, traditional gold plating technique was the most commonly used method. This method means putting the cut gold foil into a crucible and heating it to about 600 °C, and then adding mercury to form a gold amalgam, which is poured into cold water to become mushy "gold mud". This gold mud is brushed evenly on the surface of the object, and baked with a smokeless charcoal fire to let the mercury evaporate, so that the gold attaches itself to the object's surface.

6.3.11.4 Historical Masterpieces

A prime example is the eagle-shaped golden crown top and crown belt, which was discovered at the Hun Tomb in Aluchaideng, Hangjin Banner, Inner Mongolia, in 1972, and are the relics of the Warring States period. It is now kept in the Inner Mongolia Museum. The top of the hemispherical crown is decorated with four groups of wolves and sheep lying opposite to each other. The eagle on the top is a hollow chiseled gold piece. Its head and neck consist of two pieces of turquoise, which are connected with the eagle body by gold wire and can swing left and right. The crown belt consists of three semicircular golden flower belts, decorated with tigers, horses,

6 Metallurgy and Metalworking

Picture 6.103 Eagle-shaped golden crown top and crown belt, the Warring States period

and sheep. The whole piece of work integrates chisel engraving, embossing, gem inlaying, wire drawing, and other techniques, which reflects the superb level of gold, silver, and fine gold techniques of the grassland nationalities at that time (as shown in Picture 6.103).

Another example is the gold ornament of the Northern Qi dynasty, which was discovered in the Lou Rui Tomb in Taiyuan, Shanxi province, in 1981, and is now stored at the Shanxi Institute of Archaeology. The decorative piece adopts embossing, chisel engraving, and hollowing techniques to complete the bottom support, and then contains inlays of pearls, agate, sapphire, turquoise, shell, and glass, resulting in the beauty and magnificence as shown in Picture 6.104.

A golden hairpin decorated with buildings and figurines, unearthed from the tomb of Zhu Houye, Yizhuang Prince of the Ming dynasty, in Nancheng county, Jiangxi province, in 1958, and is now kept in the National Museum of China. This golden hairpin shows the fineness of lines and is completely transparent, which shows that the focus of technology at that time shifted from chisel engraving to filigrees.

6.3.11.5 Tongzhou Filigree Inlay Factory and Shanghai Laofengxiang Silver House

Since the late Qing dynasty, Beijing, Shanghai, Guangzhou, and Suzhou have become industry leaders in gold, silver, and fine gold technology, especially Tongzhou Filigree Inlay Factory and Shanghai Laofengxiang Silver House.

Picture 6.104 Gold ornament, Northern Qi dynasty

Tongzhou Filigree Inlay Factory gathered all the masters in Beijing, who created works that are elegant and magnificent, incorporating the characteristics of court art, yet continuously innovating along with the times. For example, the *Fengming* Bell, designed by Bai Jingyi and made by Wang Zaigao, used gold, platinum pinching, combinating, filigreeing, heaping, and welding of gold and platinum, as well as white diamond inlaying techniques, and won the Excellent Design Award of South East Asia Diamond Design Competition in 1983. Due to mismanagement, the factory went out of business in the 1990s, but some artisans still continue their craft and passing on their skills.

Shanghai Laofengxiang Silver House is a famous time-honored brand. It opened as Yang Qinghe Silver House in 1773, but was renamed Fengxiang Silver House in the 28th year of the Daoguang period (1848), when it had 100 employees (as shown in Picture 6.105). In 1952, it was renamed the Shanghai Gold and Silver Jewelry Store, and in 1985, it was given the name it has today. In 1996, it was established as a joint-stock company with nearly 200 senior artisans and technicians, and nearly 1,000 outlets in China, with annual sales reaching 4 billion yuan, quickly becoming the most powerful enterprise in this industry.

Lao Fengxiang's techniques are comprehensive, including filigreeing, hollowing, welding, polishing, hammering, mold casting, chisel engraving, rotary cutting, weaving, staggered inlay, bead frying, and gilding. It is located in the 10-mile-long foreign market (referring to old Shang and now a bustling, cosmopolitan city), which not only enables it to absorb the essence of traditional crafts, but also to keep up with the times. Its designs and production methods are imbued with the flavor of the times. Its fine products have won numerous awards at home and abroad, and some of them are displayed in museums (as shown in Picture 6.106). The company has a history of 234 years. From Fei Ruming, it has gone through six generations, first being passed down to Fei Zushou (the second generation), and then Fei Chengchang (the

Picture 6.105 Photo of Laofengxiang Silver House during the Republic of China

third generation), Tao Liangbao and Bian Bingsen (the fourth generation), Zhang Xinyi (the fifth generation), and Shen Guoxing and Wu Beiqing (the sixth generation) at present. Zhang Xinyi began to study the craft in 1972 and studied under Tao Liangbao and Bian Bingsen. The fine works produced by Zhang Xinyi reflect the level of contemporary gold, silver, and fine gold techniques. With the improvement of the national economy and people's living standards, the gold, silver, and fine gold manufacturing techniques represented by Lao Fengxiang will have an even broader development space.

6.3.12 Miao Silver Ornaments

The Miao ethnic minority has a long history of making and wearing unique silver ornaments. Ethnical beliefs and emotions make the Miao attach great importance to the distinctive personality and cultural identity of silver ornaments. Through the wisdom of Miao silversmiths, the beauty of silver ornaments is put on full display, making them durable as well as beautiful (as shown in Picture 6.107).

There are many branches of the Miao ethnic minority, and different Miao areas choose different patterns and shapes for their silver ornaments, while still following the same general rules. Miao silver ornaments are famous for their beauty and complexity. Women wear as many as 30 or 40 pieces of silver ornaments, weighing more than 10 or 20 pounds, ranging from silver clothes, crownpieces and necklaces to bracelets, earrings and rings. It often takes 1 year to process such a set of excellent silver ornaments. Skilled silversmiths can fully support themselves solely by doing silver work. The processing fees earned can go halves with silver materials.

Silver jewelry metalwork needs to be done with fire and the constant beating and trimming of the silver. The manufacturing process can be summarized as chiseling,

Picture 6.106 Crafts made by Lao Fengxiang 1. Longhua tower 2. Bugatti car model

filigrees, welding, and knitting. These techniques are not separated from each other, but need to be interspersed and flexibly used depending on each different shape.

6.3.12.1 Chiseling Technology

The phoenix patterns, dragon patterns, horse patterns, and beast patterns on the Miao's "silver clothes" are all made mainly by relief chiseling, but, at the same time, they are interspersed with special expression techniques such as tin mold turning, hollowing, and texturing.

The chisel is the most important tool. It is made by the craftsman himself to fit his needs. It is best to use steel, since it has good durability and does not easily deform, but copper can also be used (as shown in Picture 6.108).

6 Metallurgy and Metalworking

Picture 6.107 Miao women's holiday costumes

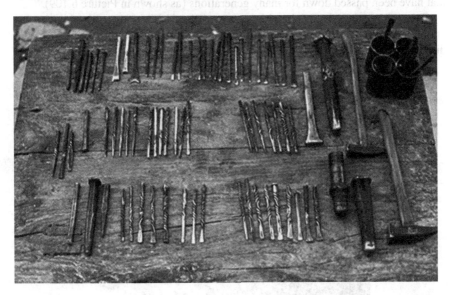

Picture 6.108 Chisels

Each silversmith has hundreds of chisels, which are divided into different categories, such as linear-shaped chisels, sparrow eye chisels, point chisels, surface chisels, and texture chisels. There are many models of the same type of chisel. For example, a linear-shaped chisel can vary in thickness and width, with the thicker one used to carve the outer contour of a pattern, and a thin, sharp one used to carve decorative lines on patterns. Thick and dull chisels carve lines with chaotic and subtle visual effects, while thin and sharp chisels can carve clear and bright effects, which are perfect for carving flying hairs of animals, clouds, and human faces.

The kind of chisel used is based on the silversmith's experience and understanding of the desired pattern. Different silversmiths will have different ways of thinking and shaping the same silver pattern. When carving silver clothes, they understand the principle of plane composition and the reasonable distribution of points, lines, and surfaces, and know the decisiveness and borrowing force when hammering, and can properly express different patterns through the use of different chisels. This way, the sculpture's meaning and the aesthetic appeal of the pattern are enhanced, and people are given the privilege of an exquisite view. Long-term practice makes it possible for craftsmen to not simply copy existing patterns, but to pursue more personalized content as well.

The basic patterns of Miao silver clothes still follow the patterns of copper molds that have been passed down for many generations (as shown in Picture 6.109).

Chiseling pattern is a job with strong subjective initiative, in which many emotional factors are added. Silversmiths adopt different expressive techniques for displaying animal bodies. For example, some have fine flying hairs arranged in a

Picture 6.109 Copper mold

fan shape, some have them arranged with curly grass patterns, and some have them arranged with fish scale patterns. The same kind of silver garment decorated by different craftsmen will embody different understandings and decorative techniques, either being more patterned or more realistic with clear lines. The chiseled patterns should be finely treated on the surface of the pattern in combination with the change of relief position. The outline should be clear but not rigid, and the detailed texture should be neat and flexible. When the outline is chiseled, it is often necessary to use a bit more force, similar to how indentation is a result of writing on paper with a bit more force. When working with chisels, you should raise the wrist, and use strong or weak force according to the ups and downs of the pattern. At the end of a line, the chisel finally naturally empties it into the background, and the art of combining reality and emptiness creates the flexible beauty of chiseling technology.

6.3.12.2 Filigree Technology

The Miao's silver crownpieces, silver horns, silver combs, and silver hairpins all have exquisite craftsmanship, which are shaped and decorated with silver filigrees that are as fine as hair. To make filigrees, fine silver wires with a high purity and a diameter of about 0.3 mm should be used. When creating silver wire, it should be as long and uniform as possible, and then it should be pressed flat. After annealing and softening, make a few small twists on the one end with pliers. Then put them on the self-made filigree wood and rub them into twist-shaped filigrees (generally, they need to be annealed and rubbed three times in a row). The tighter and smoother the twist, the more excellent it becomes. Attach the twisted filigrees in rows on the nails of a wooden board, stick them together with ox glue, remove them after the glue has dried (as shown in Picture 6.110), and then pinch the filigrees of different patterns.

Finally, the most difficult and crucial process is to weld the filigrees onto the silver sheet. If the fire is not hot enough and the welding flux is too old, the filigrees will curl up. If the fire is too hot, the filigrees will melt. If welded too many times, it will cause burn pits (also known as "eliminated materials"), which gravely affects the overall aesthetic. Welding is an art of time control, and the coordination of the craftsman's eyesight and body movement will be necessary in this process.

Filigree techniques have greater autonomy in artistry, while pursuing the matching of texture and pattern changes of filigrees on plain silver sheets. The labor cost of making high-quality filigree jewelry is much higher than that of common jewelry. Therefore, fine filigree jewelry is rare (as shown in Picture 6.111).

6.3.12.3 Knitting Technology

Knitting is one of the key elements of the Miao people's silver metalworking techniques. Take the swelling necklace (as shown in Picture 6.112) as an example: Firstly, it is forged into four square silver bars with a uniform thickness, a diameter of about 0.7 cm, and length of about 150 cm. After annealing, two of the silver bars are tied,

Picture 6.110 Filigrees

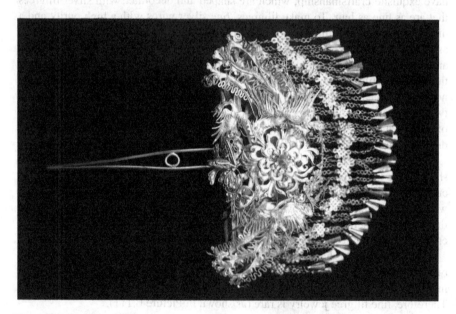

Picture 6.111 A fine filigree hairpin

in parallel, to a round wooden stick with a length of 2 m. The middle section of the silver bars is bound firmly with ropes, and after being twisted into two groups of winding shapes, the two groups of silver bars are wound together, and the two ends of the twist are hammered together firmly with a hammer. The shape is then made

Picture 6.112 Swelling necklaces

with wooden tweezers and the silver tube wrapped around the twist. The knitting is finished once the two are welded together.

The processing method of the millet flower bracelet is also unique: Wood is used as the skeleton of the bracelet. First, 12 filigrees with a length of about 70 cm are attached to the center of the wood, then the knitting rule of "one build two, two build one" is used as a mantra to knit one end before the other. Then the result is annealed in a charcoal fire and the wood is burned away. Then the filigrees are trimmed with a mallet, the bracelet knocked into a spindle shape, and while bending it slowly, the two ends are wound and welded with filigrees, until finally a movable connecting structure is added as the finishing touch (as shown in Picture 6.113).

The charm of knitting is that you must see the whole process with your own eyes before you can understand how the final effect is achieved. Knitting is seemingly colorful and complicated, but it is actually just the coherence of a series of simple actions. The shaping is very important in knitting, and it is necessary to rely on simple tools to change the shape of the soft silver bars, so as to present a perfect and full contour from multiple angles.

The above-mentioned three technological types, as well as the mold turning and welding processes interspersed among them, generally reflect the overall character of Miao silverwork techniques. Miao silversmiths have been making silver ornaments for a long time. Due to its uninterrupted history, the techniques and patterns have come together perfectly, making silver ornaments an unequivocal part of the Miao culture to this day. The interpretation of the Miao totem of the *Butterfly Mother* (myth in the Miao ethnic minority) gave birth to a new technological form, which made this cultural symbol sublimate and deepen with the help of silver ornaments.

No matter which regions they are from, the silverwork techniques and culture of the Miao people all have some things in common since the silversmiths of each region

Picture 6.113 Millet flower bracelet

have always interacted with each other. *Casting the Sun and Moon* in *Transporting Gold and Silver*, the ancient Miao song sings: "Silver is made by the Han people. We use it to melt gold and silver and make ornaments for our girls". The communication between Han and Miao silversmiths has been enriching the shape and craft of Miao silver ornaments.

People are the key elements in developing techniques. While patiently observing the creativity and step-by-step methods of Miao silversmiths, it is not difficult to get a sense of freedom, ease, and calm. Metalworking masters in China and even in the world have one characteristic in common: They are very good at finding metal materials and discovering simple processing methods solely from their environment. So are the Miao silversmiths. For example, the Miao wire-drawing machine is actually a specially treated bench. A hole is dug in the middle of the bench to make the silver wire pass through in case of the disturbance of the force when drawing the wire; with the aid of the bench-supporting structure, a wooden stick with holes at both ends is fixed and silver wires are twisted in the middle part of the wooden stick. When in use, the wooden stick is rotated with tools, and the pulled silver wire can be smooth and flat as long as the force is uniform. Such simple tools are easy to control and manage, and all processes can be completed quickly and perfectly by hand while simply sitting there.

Miao craftsmen may be scattered across mountain villages and have no chance of getting in touch with modern tools and machinery, but they have a thorough understanding of their techniques and tools through their experience. Most silversmiths are also the designers and manufacturers of their own tools. They make tools that fit their hands according to their needs. This independent spirit is precisely what the operators of modern large industries lack.

6.4 Conclusion: Present and Past of Traditional Metal Technology

In China, both bronze and iron smelting were invented later than in the West. The reason why it was able to catch up was that they had a series of major original inventions with their own characteristics, such as shaft wood structure supports, copper sulfide smelting, split casting, and pig iron smelting (as shown in Pictures 6.114, 6.115, 6.116, and 6.117).

The splendid Chinese metal civilization has played an extremely important historical role in the formation and development of the Chinese nation, and contributed to both individual and national wealth. In China, there are profound social reasons for the unsuccessful transformation of traditional metal technology into modern metal technology as a whole. Looking at the 5,000-year history of Chinese metals, most

Picture 6.114 Ancient Egyptian bronze casting

Picture 6.115 Bronze head wearing gold mask in Guanghan city, Sichuan province

Picture 6.116 Iron forging depicted on an ancient Greek pottery bottle

of the major inventions and creations appeared before the Song and Yuan dynasties. After that, with the exception of the copper smelting with bluestone and zinc smelting, the development of the mining and metallurgy industry manifested more in the demand for larger quantities driven by the increase of population and the needs of daily life rather than better quality. The reason lies in the fact that the old political and economic system dominated by the autocracy and landlord economy declined since the Song dynasty, and frequent regime changes brought heavier oppression

Picture 6.117 Han stone relief of cattle pulling a plow

and harsher taxation for the people. When the whole society fell into disorder or even disaster, how could we liberate our minds and nurture our creativity? Therefore, the emergence of new technologies could only wait for the advent of a new era. Nevertheless, there are still many traditional metal technology widely used today in production and daily life because of their own vitality, such as cutting tools, bell metal instruments, gold foil, and silver ornaments. Some of them cannot be replaced by modern technology. For example, lost-wax casting is valued by scholars because it contains precious technical qualities. A large number of reasons support the fact that traditional metal technology not only has great historical value, but also important modern value (as shown in Picture 6.118). Through the joint efforts of communities, enterprises, artisans, and experts, the protection, inheritance, and development of traditional metal technology looks promising. This book contains a brief introduction to 12 existing traditional metal technology, while others, such as the *baisha* method, cloisonné enamel, mottled copper, black copper inserted silver, and iron painting, are not included due to page limitations. If readers are interested, please refer to *Chinese Handicrafts: Metallurgy and Metalworking* and other related works.

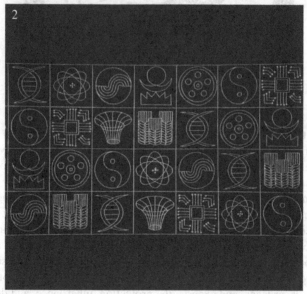

Picture 6.118 1. China Millennium Bell, made in 1999 to welcome the new millennium, incorporating nine traditional handicrafts. 2. Bell beam decorated with golden patterns, including modern ones such as atoms, integrated circuits, fairing, and DNA

Chapter 7
Sculpture

Luo Xingbo

"Sculpture" is the common plastic art of carving and modeling. It refers to the creation of artistic images that reflect social life as well as pure representations of the sculptor's artistic expression by carving hard or plastic materials. The emergence and development of sculpting are closely related to human production activities, and is influenced by religion, philosophy, and other ideologies.

Carving refers to the use of free-standing carving, relief carving, openwork carving, inlaying, and refining of materials such as jade, colored stone, wood, ivory, and shells. This complete artistic modeling process can be seen as "subtraction". Modeling refers to the use of soft materials such as clay, dough, gypsum, and yak butter, which can be seen as "addition". Sculpting techniques are among the most colorful categories of traditional crafting techniques in China. The development of sculpting techniques contains the amalgamation of the wisdom and skills of generations of craftsmen and authors.

7.1 History

The origin of sculpting can be traced back to primitive society. Liang Sicheng says in *The History of Chinese Sculpture*: "When art first began, sculpture was the start. When our ancestors lived in the wilderness, they first had to chisel stones into tools to survive; after that, there were living quarters for them to create things in. Hence, when the practical art of sculpting began in the Stone Age, so did its artistic form." The stone tools and bone tools that appeared in the prehistoric era can be regarded as the beginning of sculpting in China. Back then, the sculptures were simple, decorative, and symbolic, rich with the flavor of primitive religion and witchcraft.

L. Xingbo (✉)
School of Humanities, University of Chinese Academy of Sciences, Beijing, China
e-mail: luoxb@ihns.ac.cn

© Elephant Press Co., Ltd 2022
H. Jueming et al. (eds.), *Chinese Handicrafts*,
https://doi.org/10.1007/978-981-19-5379-8_7

In the early stages of human civilization, the raw materials of sculptures were mainly natural stone and animal bones. After entering the Neolithic Age, human beings started creatively using clay. Pottery itself was simple and practical, with uncomplicated and distinct patterns. The three elements of carving, modeling, and engraving were fully applied. The hair and mouth of the painted pottery bottle with the head-shaped mouth, unearthed in Qin'an, Gansu province, from the Yangshao culture were carved, and the nose, forehead and face were modeled, which showed a distinct level of technical skill (as shown in Picture 7.1). The clay statue of a goddess unearthed along the Niuliang River, which belonged to the Hongshan culture around the Liaohe River Basin, is one of the few surviving sculptures of the Neolithic Age. It is a semi-free-standing carving and is shaped as a mask. Inlaying techniques were used to create it, but the eyes were left empty, and were inlaid with turquoise jade pieces after modeling, which is proof of a high technical skill (as shown in Picture 7.2). In addition to pottery, there were carved ivory tubes unearthed at Dawenkou cultural site and jade carving ornaments from the Hongshan culture dating back to the same era (as shown in Picture 7.3).

Sculpting techniques in the Shang and Zhou dynasties were prominently reflected in the bronze wares and jade wares which changed from being practical to being accessories and ritual vessels. A large number of jade wares and turquoise ivory cups inlaid with animal faces, unearthed from the Fuhao Tomb at the Yin Ruins, Anyang, Henan province showed that sculpture techniques reached a fairly high level at that time.

During the Spring and Autumn period and the Warring States period, sculpture techniques began to contain more aesthetic techniques. Bronze ware production did not only progress in casting, modeling, and ornamentation, but the lost-wax casting method also appeared, which made it possible to create complex and exquisite devices, and greatly improved the ornamental and artistic quality of bronze wares. For example, the techniques for creating the penetrating *Ding* with coiled chi-dragon patterns of the Spring and Autumn period (as shown in Picture 7.4) required the wax mold to be pasted on the inner mold, then engraved, and then placed into the outer mold. This was a typical utensil made by the early lost-wax casting method. During the Spring and Autumn period and the Warring States period, due to the development of urban architecture, brick carving technology began to flourish. Patterned bricks, railing bricks, and especially tile bricks, all had quite an exquisite ornamentation, which was common for sculptures in this era.

The Terracotta Warriors of the Qin dynasty are the most representative work of large-scale sculpture projects (as shown in Pictures 7.5 and 7.6). They were made from clay, and both the people and horses were life-sized. All statues have vivid expressions, different shapes, and different costumes. The Terracotta Warriors, representing the advanced sculpting techniques at that time, are both magnificent and mighty.

Stone reliefs are representative of sculptures in the Han dynasty, which mainly consisted of flat carving, using line carving and line relief modeling, with simple lines and vivid images. Stone reliefs often show certain plots and contents, so they are called "epics on stones". During the Han dynasty, various sculpting techniques were

7 Sculpture

Picture 7.1 Painted pottery bottle with the head-shaped mouth in Qin'an, Gansu province

perfected, and the types of sculptures greatly expanded. Many outstanding pottery figurines, bronze figurines and jade carvings were made at that time. The storytelling figurines unearthed in Chengdu, Sichuan province (as shown in Picture 7.7), are rough and concise in shape, vivid in expression, and show strong dynamism. The Galloping Horse Treading on a Flying Swallow unearthed in Wuwei, Gansu province (as shown in Picture 7.8), is full of spirit, with three legs up in the air and one hoof standing on the back of a dragon finch. Its shape is novel, which not only shows the lightness and swiftness of the horse, but also the balance of the center of gravity of the whole sculpture, reflecting the scientific mechanical principle. In the Han dynasty, jade carving techniques developed further, coming up with openwork carving and relief carving, as well as using decorative techniques such as carving lines and millet patterns.

With the prosperity of Buddhism, the theme of sculptures changed during the Wei, Jin and Northern and Southern dynasties. Buddha statues, grottoes and other large sculptures were mostly produced. The Yungang Grottoes in Datong, Shanxi province

Picture 7.2 Clay statue of a goddess along the Niuliang River

Picture 7.3 Jade dragon from the Hongshan culture

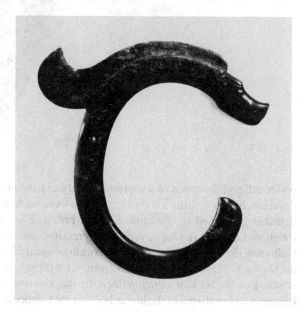

(as shown in Picture 7.9), the Longmen Grottoes in Luoyang, Henan province, and the Mogao Grottoes in Dunhuang, Gansu province, are all world-famous large-scale religious stone carvings. Metal carving and paint carving developed to a certain extent. Jewelry coiled with gold and silver filaments and bronze mirrors decorated

7 Sculpture

Picture 7.4 The penetrating *Ding* with coiled *chi*-dragon patterns

Picture 7.5 The terracotta warriors of the Qin dynasty

by combining line carving with relief carving all flourished in this period. Lacquering techniques also began to gradually expand to paintings and sculptures.

The Sui and Tang dynasties' national strength was strong, and the foreign trade and cultural exchanges were frequent, which ensured brilliant achievements in sculpting. In addition to a large number of religious statues, *sancai* (Tang Tri-color glazed

Picture 7.6 Details of the terracotta warriors of the Qin dynasty

Picture 7.7 Storytelling figurine

7 Sculpture

Picture 7.8 Galloping horse treading on a flying swallow of the Eastern Han dynasty

Picture 7.9 Sculptures at the Yungang Grottoes

ceramics) is the epitome of sculptures in the Tang dynasty (as shown in Picture 7.10). Figures, birds, animals, camels, etc. were all depicted as plump, round, vivid, and life-like. Their craftsmanship and artistry displayed the height of sculpture techniques in the Tang dynasty.

Sculptures of the Song and Yuan dynasties paid more attention to practicality and aesthetics, and the division of labor became more and more standardized but also diversified. However, religion was still an important theme of sculpting in this period. Jade carvings were known for their charming color (also known as "pretty color") (as shown in Picture 7.11) and were carefully and skillfully brought forth using the material's natural form, color, and texture, and reflected the creativity of the artists. For example, the Jade Cats of the Song dynasty had a female cat meticulously carved from a large piece of white jade, while the six kittens were carved in different colors:

Picture 7.10 Figurine of a camel carrying musicians

Picture 7.11 Pretty color jade turtle

yellow, black, and tortoiseshell. During the Song and Yuan dynasties, painting and calligraphic works flourished, as well as inkstone carving. There were many excellent works carved on the *Duan* inkstone and *She* inkstone.

During the Ming and Qing dynasties, jade carvings, ivory carvings, bamboo carvings, wood carvings, inkstone carvings, root carvings, brick carvings, ceramics carvings, clay sculpting, dough modeling, mask art, and shadow play carvings made great progress. For example, the large-scale jade carving King Yu Taming the Flood (as shown in Picture 7.12) not only showed skillful relief, engraving, and line carving techniques, but also reflected a high level of sculpture modeling. Carving and modeling techniques matured, forming a precise routine from materials to techniques, and the works became more refined and patterned. This can be seen from the complexity and piling of architectural decoration in the Ming and Qing dynasties, which made the focus on creating new sculptures more limited. Although the level of the materials and scale were vastly superior, they lacked innovation.

In the first half of the twentieth century, great changes took place in Chinese society. Western sculpting methods were introduced to China and collided with traditional sculpting methods, resulting in modern Chinese sculpting methods. During the Republic of China, traditional folk sculptures coexisted with modern sculptures. The former contained works such as *Nirenzhang* (Clay Figurine Zhang), the Huishan clay figurine, Mianrentang (Dough Figurine Tang), wood carvings, and brick carvings while the latter were mainly works that fell under western sculptures that were created by sculptors who had returned from studying abroad and had been trained by art schools. Traditional sculpting began to decline, and some people thought that when the Western sculpture methods were introduced to China, the national sculpture traditions were interrupted. In fact, throughout the twentieth century, our nation did not produce any sculpting methods nor concepts, expressions or forms of any global significance. In the past hundred years, we have created sculptures according to the sculpture rules and standards formulated by Westerners.

Picture 7.12 Jade Mountain: Yu the Great taming the flood

As traditional sculpting declined, western sculpting spread, rooted and bloomed in China. In 1920, the Shanghai University Fine Arts College set up a sculpture department. Like other schools established later, its teaching system mainly comes from realistic sculpting in the Beaux-Arts de Paris and the Soviet Union. Li Jinfa, Liu Kaiqu, Zheng Ke, Zhou Qingding, Jiang Xiaohe, Zhang Chongren, Hua Tianyou, Wang Linyi, Zhang Jingyuan, and Pan Yuliang were all outstanding representatives of modern Chinese sculpting in the twentieth century. They not only created excellent sculptures, but also explored and changed the art together with scholars. They have brought China a sculpture art system based on western realistic sculpting techniques and have a full understanding and meticulous grasp of it.

In 1958, Liu Kaiqu, Fu Tianchou, Zhang Songhe, Xiao Chuanjiu, Hua Tianyou, Ceng Zhushao, Wang Bingzhao, and Wang Linyi created large white marble reliefs of the Monument to the People's Heroes, including Destruction of Opium at Humen, Jintian Uprising, Wuchang Uprising, May Fourth Movement, May Thirtieth Movement, Nanchang Uprising, Anti-Japanese Guerrilla War, and Successfully Crossing

Picture 7.13 Relief group of monument to the people's heroes (partial)

the Yangtze River (as shown in Picture 7.13). This relief group, as a masterpiece of Chinese sculptures in the twentieth century, has had a profound impact on later sculptures.

From a global perspective, Chinese sculpting is different from the West in aesthetic theory and creative practice. As far as the world is concerned, sculptures can be roughly divided into memorial, religious, decorative, and artistic creations. Among them, memorial and religious sculptures are most represented throughout history and have the most influence. Most of these two kinds of sculptures look realistic. Western sculptures, created by Western artists, are based on stone carving art accumulated over thousands of years, and are a combination of natural realism and natural beauty, and are thus extremely realistic and expressive. Chinese sculptures, on the other hand, go beyond realism and beauty, and combine the realistic beauty of human behavioral thinking with illusions and imagination. The interpretation of realism in Chinese history is different from that in the West. Influenced by philosophy, Buddhism and literary tradition, Chinese aesthetic concepts pay attention to meaning first, then spirit, and then form. The shape of the sculpture is not the starting point of Chinese traditional sculptures, but rather artistic meaning. The creation of the sculpture is guided by the artist's spirit, thus resulting in a piece that is not modeled after an imitation of beauty, but after imagination and idealized beauty.

In ancient China, the social status of sculptors was very low, and their lives were rarely recorded in literature. Although generations of sculptors have left countless exquisite works, we know very little about the sculptors themselves. Only a few outstanding craftsmen's names have been handed down, such as Sima Da, who passed on Chinese Buddhist carving art to Japan, Yang Huizhi, whose representative work was the Guanyin with Thousands of Hands and Eyes during the Kaiyuan period (713–741) of the Tang dynasty, He Chaozong, a porcelain sculptor from Dehua in the Ming dynasty, and Shi Dabin, a purple clay artist.

Throughout history, we can see the development of Chinese sculpting techniques. Each kind of technique was passed down from generation to generation through oral transmission. After the techniques matured, they often formed a paradigm, which was the basic criterion of this technique. For example, the Buddha statues of Tibetan Buddhism have a whole set of calculation methods for making dimensions, which are summed up through the experience of shaping and making traditional Buddha statues. This paradigm not only ensures the shape and style of bronze Buddha statues, but also shows the maturity of this technique.

7.2 Sculpting Techniques

7.2.1 Carving Techniques

7.2.1.1 Free-Standing Carving

One of the most common carving techniques is to use knives and chisels to remove unnecessary parts from hard materials to complete the work. They generally require omni-directional and three-dimensional carving.

There were free-standing carving works dating back to the Shang dynasty that showed evidence of more advanced techniques. The stone carvings unearthed at the Fuhao Tomb in Anyang, Henan province, mainly used free-standing carving techniques except for the line carving of the eyes. There are 16 stone carvings from Huo Qubing's tomb in Maoling, Xingping county, Shaanxi province (as shown in Picture 7.14), which are free-standing carvings made of granite according to the natural form of stone.

7.2.1.2 Relief Carving

Relief carving is a carving technique with which images are flattened through carving and engraving. According to the thickness of the surface, it can be divided into high relief, bas-relief, and thin relief. The compression degree of a high relief is small, the spatial structure is close to that of a free-standing carving, and even the local treatment completely adopts free-standing carving. Bas relief has a low position, large compression, and strong sense of plane, which is similar to painting. Thin relief mainly uses lines and faces, which are thin yet three-dimensional. Relief carving is generally attached to the plane, so it is mostly used for architectural decoration. The

Picture 7.14 Stone carving at Huo Qubing's tomb

contents, forms and materials of relief carving are rich and colorful, and the materials used are stone, wood, and ivory.

During the Han dynasty, the popularity of lavish burials led to the prevalence of building luxurious ancestral halls and tombs, and artworks carved and narrated on building materials came into existence. Some carved pictures directly on stone, while others carved pictures on molds first, and then pressed them into brick work blanks and fired them. At this time, relief techniques developed so rapidly that it was already quite advanced during the Jin and the Northern and Southern dynasties. *The Picture of Buddha-worshipping* at the Gongyi Grottoes (as shown in Picture 7.15) is full of composition, exquisite carving and has a strong sense of volume, which is an outstanding representative of relief carving at that time.

7.2.1.3 Openwork Carving

Openwork carving, also known as hollow carving, is a sculpture technique with transparent space in free-standing carving or relief carving. It refers to hollowing and digging out, but it also includes the technique of forming perforated space by systematically eliminating material from the outside in. It can adapt to the requirements of two-sided viewing, so that two adjacent spaces are both separated and connected (as

Picture 7.15 Picture of Buddha-worshipping from the Northern Wei dynasty

shown in Picture 7.16). There are two kinds of openwork carving: single-sided and double-sided, which are often used as decoration in building components, such as doors and windows.

7.2.1.4 Inlaying and Embedding

The process of inserting one material into another or several materials is called inlaying or embedding. Inlaying is to stick one material on the surface of another raw material, while embedding means to lodge materials between raw materials.

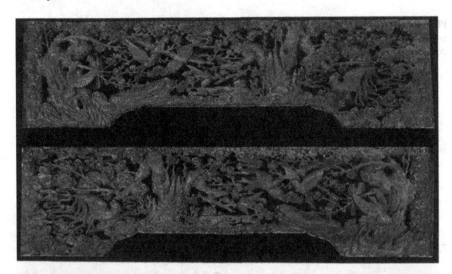

Picture 7.16 The double-sided wood openwork carving

The early inlaying and embedding technology in China is mainly decorative. For example, the ivory cup unearthed from the Fuhao Tomb at the Yin Ruins was inlaid with turquoise in the shape of eyebrows, eyes, and a nose. During the Spring and Autumn period and the Warring States period, bronze inlaying and embedding techniques were developed. Shallow grooves were carved on the surface of bronze wares according to desired patterns, then other materials were inlaid, and finally, the surface was ground flat. During the Warring States period, the technology of inlaying gold and silver (as shown in Picture 7.17) was quite mature.

Turquoise, precious stone, red copper, gold, and silver are common materials used in the inlaying process. In the Tang dynasty, the mother-of-pearl inlay was widely used in the decoration of lacquer ware, wood ware and bronze mirrors. In the Ming dynasty, another kind of inlay, called *baibao* inlay, used gold, silver, precious stones, mother-of-pearl, ivory, beeswax, and agarwood as inlay materials.

Picture 7.17 Inlayed gold and silver beast-head-shaped decoration

7.2.1.5 Carving and Polishing

This refers specifically to the carving techniques of jade wares. After the jade stone is carved, it is ground with water and sand to make it smooth and more refined.

In the northern jade-carving system represented by the Hongshan culture and the southern jade-carving system represented by the Liangzhu culture, the technology of carving and polishing was fully applied.

7.2.1.6 Deep and Shallow Carving

Deep carving is common in lacquer carving and bamboo carving, while shallow carving is common in bone carving, jade carving, seals, bronze wares, architecture, and stone carving.

In the Tang dynasty, lacquer carving mainly used deep carving technology, in which they first coated thick vermilion lacquer with lacquer cement, and then carved on the lacquer layer after it dried slightly. Jiading bamboo carving in the Ming dynasty used this contrasting technique to show bamboo carving's artistic charm. The bamboo carving of the Jinling School (as shown in Picture 7.18) mainly used line carving, done with a shallow knife, which was a simple knife method that could create vivid and interesting works.

7.2.1.7 *Yin* and *Yang* Engraving

*Yin*graving refers to removing the parts other than the contour lines of images and characters, so that the images and characters clearly stand out. *Yin* engraving is to engrave on a plane to represent images or characters.

Yang engraving has techniques such as ground flattening, ground patterning and engraving on the flattened ground. There are two main methods of creating *yin*

Picture 7.18 Bamboo carving of the Jinling School in the Ming dynasty

engravings: One is to cut concave lines or block surfaces directly on bone and stone; the other is to remove the part other than the outline of the object image first, and then engrave the details in the protruding part, which is called *yin* engraving on the flattened ground.

7.2.1.8 Through-Carved Work and Micro-Carving

Through-carved work refers to digging into and engraving the inside of materials and removing the parts that do not need to be used. It is often used in ivory carving and bamboo engraving. To show the complex internal structure, long carving tools are often needed.

Micro-carving (as shown in Picture 7.19) refers to carving images, characters, or three-dimensional images on tiny materials such as fruit stones, rice grains, and hair using shallow carving. Micro-carving techniques came into being quite late, and modern micro-carving often relies on microscopes, which require more advanced engraving tools.

Picture 7.19 A fruit stone carving from the Qing dynasty

7.2.2 Modeling Techniques

7.2.2.1 Round Modeling

Round modeling technology is the most common modeling technique, which refers to the use of piling, kneading, and molding to shape soft materials such as mud and clay into finished products.

Painted pottery heads unearthed from the Hemudu site in Yuyao, Zhejiang province, are the earliest known round modeling works in China, dating back about 6,700 years. The pottery sculpture of a naked pregnant woman found in Dongshanzui, Kazuo county, Liaoning province, was a work from the Hongshan culture, which showed that as early as the Neolithic Age, both southern and northern China had mature round modeling techniques. After the rise of Buddhism, a large number of modeling products with religious themes were created. Buddha statues in the Mogao Grottoes in Dunhuang and the Kizil Grottoes in Xinjiang are made of clay. When making large clay sculptures, rocks are skillfully used as inner bases to make the Buddha statues more durable.

7.2.2.2 Floating Modeling

Floating modeling refers to the modeling technique of creating three-dimensional plastic art protruding on an even plane by stacking. Floating modeling can be categorized as high and shallow. Yang Huizhi, a famous sculptor in the Tang dynasty, who is also called the "Sculpture Master", made floating sculptures on the wall to compete with Wu Daozi, the "Painting Master". Floating models are generally made of grass clay, but some are made of wood or stone.

Picture 7.20 Da Afu

7.2.2.3 Colored Modeling

Colored modeling is the technique of applying paint to the shapes made of mud and clay, and the process is complicated, including dozens of processes such as binding skeletons, applying large quantities of mud, strengthening, calendering, preventing cracks, backing powder, painting, gilding, polishing, inlaying eyes, and installing accessories.

Most of the clay sculptures in ancient China are colored models. *Nirenzhang* (Clay Figurine Zhang) and the Huishan clay figurines are two well-known modern colored models. The former was produced by the Zhang family in Tianjin in the late Qing dynasty, who were good at making colored models for several generations, hence the name. The latter is the general name of the Huishan clay sculptures in Wuxi, Jiangsu province, and Da Afu (as shown in Picture 7.20) is the most distinctive work.

7.2.2.4 Molding

Molding refers to the process of making molds to create sculptures, which is common in the production of bronze wares, tiles, and portrait bricks.

Shang and Zhou bronzewares needed to be molded into clay molds first, then dried and carved with decorative patterns, and finally turned into block models.

In the Han dynasty, both the government and commoners' buildings required a large number of tiles and decorative bricks, and the trend of lavish burials also prevailed, which greatly increased the demand for portrait bricks. The emergence of molding techniques made large-scale production possible. The wooden models of portrait bricks have their own characteristics. For example, the wooden models in Henan province are divided into *yin* carving and *yang* carving, and those in Sichuan are divided into bas-reliefs and line carving.

7.2.2.5 Bodiless Lacquering

It is also known as dry ramee-lacquer, and was first produced in the Warring States period. The specific method required pasting hemp cloth on utensils or molds, repeatedly painting the cloth, and taking out the internal mold after drying to obtain the body. The shape made by bodiless lacquering is light in weight and easy to move, which was often used in the production of Buddha statues.

In the Eastern Jin dynasty, Dai Kui was the first to apply the traditional bodiless lacquering technique to create Buddha statues and made a five-body five-generation Buddha ramee-lacquer statue for the Nanjing Waguan Temple. Jian Zhen, a famous monk in the Tang dynasty, passed away in Japan. His disciples preserved Jian Zhen's appearance and posture in good condition by using the bodiless lacquering method. The statue of Jianzhen is currently preserved in Japan, where it is regarded as a national treasure (as shown in Picture 7.21).

Picture 7.21 Sitting statue of Jian Zhen

7.3 Types of Sculptures

7.3.1 Carving

7.3.1.1 Stone Carving

Igneous rocks, sedimentary rocks, and metamorphic rocks can all be used as raw materials for stone carving. Igneous rock is hard and does not weather easily, so it is often used as architectural decorative stone carving, but it is difficult to process, so granite is used more commonly. Limestone and sandstone are sedimentary rocks, limestone is brittle, cracks easily and is perishable, but it is convenient to process; sandstone is hard, but its structure is loose and easily wears down, so it is often used to carve large-scale works with rugged simplicity. Marble is a type of metamorphic rock. It is characterized by its fine grains, dense quality, moderate hardness, and it is easy to process, which makes it suitable for carving relatively small works. However, its main component is calcium carbonate, which is easily damaged and weathered. Stone carvers should fully consider the characteristics of various stones and carefully select them. For example, the statue of Lao Zi at Qingyuan Mountain, Quanzhou city, Fujian province, was carefully carved with local stone materials that did not weather easily, and it is still quite intact after hundreds of years of being exposed to wind and rain (as shown in Picture 7.22).

Tools of stone carving include hammers, chisels, axes (used for splitting materials), chopping axes (used for cutting out finer surfaces on coarser stone surfaces), saws, grinding wheels, and sandpaper.

Stone carving can be divided into three steps: Clearing, depicting, and polishing.

Picture 7.22 1. Stone statue of Lao Zi at Qingyuan Mountain, Quanzhou city. 2. Stone Lion in Quyang, Hebei Province

Clearing refers to removing waste materials, showing a rough outline, and using sledgehammers and chisels to chisel from all directions to obtain a regular outline.

Depicting refers to the detailed processing of the rough outline of clearing, first hitting the high part, then hitting the low part and concave holes. If there is a plaster draft of the same size, a point line machine (a kind of carving tool) can be used for detailed depictions to obtain the accurate shape (as shown in Picture 7.23).

Polishing is also called grinding. A relatively flat large body can be ground with a grinding wheel, file and sandpaper first, and then polished with polishing paste and cloth wheel.

The most famous stone carvings in China are Qingtian, Quyang, Shoushan, and Hui'an stone carving, Tibetan Gesar painted stone engraving, and Huizhou's three types of engraving.

Qingtian stone carving is famous for the stone that is used. Most Qingtian stones are pyrophyllite, and have a delicate texture, moderate hardness and strong carving ability. *Dengguangdong, Fengmenqing,* and *Huangjinyao* (these are different kinds of stones for carving) are the best qualities among those stones. The works carved from Qingtian stones are colorful, shiny, and unique (as shown in Picture 7.24).

There are two types of chisels used in Qingtian stone carving: round mouth chisels and square mouth chisels, with different widths, ranging from half a *fen*, one *fen*, two *fen* to one *cun* (1 *fen* is about 0.33 cm, and 1 *cun* is about 3.3 cm), which are used for beating, poking, hollowing, leveling, and polishing (as shown in Picture 7.25). Carving knives include flat knives, round knives, and oblique knives. Thorn strips are simple and important tools and come in various shapes, such as round, flat, oval, oblate, and flat square, which are used for fine carving.

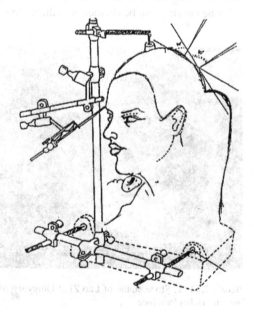

Picture 7.23 Carving stone with a point line machine

Picture 7.24 *Yandang Chunnong* (The flying geese constitutes the spring), Qingtian stone carving by Du Zhengqing

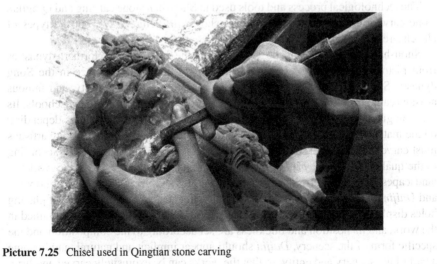

Picture 7.25 Chisel used in Qingtian stone carving

The technological process of Qingtian stone carving can be divided into six steps: Material selection and layout, blanking and poking, hole carving, fine carving and finishing, base matching and installing, polishing and waxing. The spread of skills depends on word-of-mouth teaching. In the long-term practice, people have used their own mnemonic chants to memorize the skill. Taking the skill of proportion of characters in stone carvings as an example: "The length of the whole body is equal to that of seven heads, and the upper body and legs are three times longer than the feet. The width of the shoulder is twice as wide as the heads, and 1.5 longer than that of the hips. The position of the eyes is in the middle of the face, and their length is two fifths of the face. The lower jaw is parallel with eyebrow respectively, the length between the eyebrow and nose is equal to that of the ear."

Another example of a mnemonic chant for remembering how to carve the character's expression: "If you want the figure to laugh, bend its eyes and turn its mouth up. If you want the figure to worry, bend its mouth and wrinkle its eyebrows. If you want the figure to be kind, carve a *Guanyin* face. If you want the figure to be crafty, carve triangular eyes. If you want the figure to be evil, squeeze the eyebrows, eyes, nose, and mouth together."

These mnemonic chants are the crystallization of the wisdom of generations of artists. Lin Rukui, Zhou Baiqi, Ni Dongfang, and Zhang Aiting are outstanding representatives of the contemporary Qingtian stone carving art.

Shoushan stone carving uses pyrophyllite from Shoushan village, Fuzhou city, Fujian province, as source material, which is warm, clear, colorful, flexible, and easy to carve. There are three kinds: field pit, water pit, and mountain pit. Field pit stones are extremely rare, with the best stone quality, slightly transparent or translucent, with hidden radish texture and sometimes a red checkered pattern (as shown in Picture 7.26). Water pit stones are full of luster, including such types as *Shuijingdong* and *Yu'naodong*. *Gaoshanshi*, *Shanbodong* are examples of mountain pit stones.

The technological process and tools used in Shoushan stone carving and Qingtian stone carving are basically the same (as shown in Picture 7.27), but the types of chisels are slightly different.

Shoushan stone carving can be traced back to the Southern and Northern dynasties from historical records and became a tribute to the Imperial Palace in the Song dynasty. Shoushan stone seals were all the rage in the Ming dynasty, and famous artisans came forth in large numbers in the Qing dynasty, forming many schools. Its technological characteristics are the application of different techniques depending on the materials, fineness of the carving, and the levels of richness. Carving artisans must carve the shape based on the original shape of the stone, lay it out according to the quality, and display its beauty according to its natural color. In the works of landscapes, flowers and fruits, the main techniques are placing holes and hollowing, and *Daijin*. Placing holes is the basis of hollowing, and different methods of placing holes display different sceneries and levels. With *Daijin* the joint body is retained in the work, and its position and thickness are set according to the composition and the specific form of the scenery. *Daijin* should appear implicit and natural, and become a part of the scenery and entity, so that the works can be exquisitely carved and firm.

Picture 7.26 *Elegant Gathering in the Western Garden*, Shoushan *Tianhuang* (a yellow type of Shoushan stone) stone carving by Lin Fei

Picture 7.27 Traditional tools used in Shoushan stone carving

There are many famous stone carvers in Shoushan, including Guo Gongsen, Lin Hengyun, Feng Jiuhe, and Guo Maojie.

In 2006, Shoushan's stone carving skills were added to *List of the First Batch of National Intangible Cultural Heritage*. In 2007, Feng Jiuhe and Lin Hengyun were recognized as the representative inheritors of the project.

Quyang stone carving in Hebei province is most known for its white marble carving. The reserves of white marble produced in China are large, white, and crystal-clear, which makes it suitable for meticulous carvings. Buddhism prevailed in the Northern Wei dynasty, which made Quyang stone carving develop rapidly. During the Tang and Song dynasties, Quyang stone carvings spread all over the country and were widely used in the construction industry. Kublai Khan of Yuan built Dadu (Beijing), and the schematics of *Huangcheng Zhouqiao* (bridge around the Imperial

Palace), designed by Quyang stone carving artisan Yang Qiong, were selected and later used as the blueprint for the Jinshui bridge. After the founding of the People's Republic of China, Quyang stone carving artisans participated in the construction of the Monument to the People's Heroes, the Great Hall of the People, and the History Museum. Contemporary inheritors of the techniques include Lu Jinqiao and Zhen Yancang.

The representative of the stone carving art in the south is Hui'an stone carving. In Fujian province, there is a saying that "Quyang represents the north and Hui'an the south".

The stone carvings on the tomb of Tang military governor Wang Chao are the earliest Hui'an stone carvings, dating back more than 1,100 years. The Wan'an bridge in Luoyang township, Quanzhou, Fujian province, dates back to the Song dynasty and is a famous historical bridge in China. It is also the largest stone carving project in the history of Hui'an carving, with Buddha statues carved on piers and stone towers (as shown in Picture 7.28). There are also inscriptions of the *Record of Wan'an Bridge* written by Cai Xiang, who presided over the bridge construction.

Picture 7.28 Stone carving of Wan'an bridge in Luoyang township

Hui'an stone carving uses mainly granite, among which the best quality of granite is called *"Fengbai"*. Most of the works are practical utensils, steles, architectural ornaments, and garden sculptures.

The techniques of traditional Hui'an stone carving are commonly known as *"daqiao"* (carving skillfully). It mainly includes four processes: kneading, engraving, picking, and carving. "Kneading" means to get a workblank sample, *"engraving"* is to dig out the useless stone after the workblank sample is kneaded, "picking" is to remove the redundant stone outside the carving piece, and "carving" is to cut, chop and shape. After the above steps are completed, there are two more processes: trimming and configuring the base. "Trimming" refers to surface polishing, and "configuring the base" refers to using various bases to stabilize the works and play a foil role, so that the main body is prominent and eye-catching.

Hui'an stone carving masters include Li Zhou, from the Kangxi to Qianlong period in the Qing dynasty, who was respected as a master by southern Fujian artisans. Many of his works have been passed down, such as Fuzhou Wanshou bridge stone lion. The prosperity of Hui'an stone carving in the Qing dynasty was closely linked to the training of a large number of apprentices by Li Zhou. Jiang Renwen was another famous artisan. He rebuilt the Sun Yat-sen Mausoleum in 1924. He led more than 30 craftsmen to Nanjing and successfully completed the task, which made Hui'an stone carving craft famous.

7.3.1.2 Stone Engraving

It refers to the engraving art of expressing pictures and texts engraved from the stone surface to its inside, which is characterized by a plane display, also known as engraving stones.

The earliest stone engravings in China are mainly rock paintings, such as those in the Langya area of the Yinshan mountains, Inner Mongolia, in the Heishan area in Jiayuguan, Gansu, in Huocheng at the northern foot of Helan Mountain, northern Xinjiang, and at the Kunlun Mountain pass in Pishan county, southern Xinjiang. There are three types of stone reliefs in the Han dynasty: immortals, historical stories and daily life. Stone seals also became popular at that time.

Stone engraving in the Tang dynasty mainly used the technique of *yin* engraving lines, which not only incorporated the conciseness of stone engravings of the Han and Wei dynasties, but also foreign techniques. The characters were beautiful and plump, and the lines were slender and delicate. Fifteen paintings of ladies were engraved on the stone coffin unearthed from Princess Yongtai's tomb in the Tang dynasty (as shown in Picture 7.29). The beauty of their shapes is very rare. The whole work has smooth and symmetrical lines and a mellow and soft sense of rhythm, which makes it the most representative work of stone engraving line paintings of that time.

Before starting a rock painting stone engraving, the dark stone skin needs to be removed first. Then lines are chiseled on the stone surface, often using the *yin* engraving method. Grinding-and-engraving techniques were used first, and the engraving method appeared later, but the skill of knocking-and-engraving was used

Picture 7.29 Stone line painting of Princess Yongtai's tomb in the Tang dynasty

all the time, which was the main method for making rock paintings, which was usually chiseled using flint tools. Stone engraving techniques are also commonly used for inscription and seal cutting.

Stone engraving includes *yang* engraving, *yin* engraving, deep engraving, and shallow engraving. *Yang* engraving has three techniques:

Ground flattening: A flat knife or shovel is used to cut the vacant surface to make the image part bulge.

Ground patterning: There are sand ground and pattern ground to show the patterns.

Engraving on flattened ground: The ground is flattened around an image or text so that the patterns remain raised and, thus, are clearly displayed.

Yin engraving, also known as "sunken engraving", has two main production methods:

Chiseling concave lines or blocks directly on stone.

Yin engraving on flattened ground: First, remove the parts other than the outline of the image, then engrave the details with lines on the protruding outline.

Deep and shallow engraving are distinguished by looking at the depth from the relief's surface to the starting line of the engravings. Deep engraving is often used for chiseling stone tablets, stone reliefs and rock paintings, while shallow engraving is often used for stone line paintings. They can both be done using *yin* engraving or yang engraving.

Tibetan Gesar painted stone engravings are the most representative of the minority stone engravings. They occupy an important position in Tibetan art history and are also the carrier of Gesar's cultural inheritance. It mainly depicts the heroic epic *Biography of King Gesar* (as shown in Picture 7.30) and combines stone engraving skills with painting in a unique artistic style, which is mostly found in the Seda, Shiqu and Danba counties of Ganzi Tibetan autonomous prefecture, Sichuan province.

Generally, locally produced shale is used, on which the outline is engraved with a chisel first, then drawn with a pen, and then engraved by means of erect engraving and scraping. After the engraving is completed, the white pigment should be brushed on the picture as the primary color, and then colored after drying. There are six kinds of color pigments: red, yellow, blue, white, black, and green. After coloring, a clear lacquer or *gonglishui* (a kind of material) is needed to coat to protect the color of the painting.

In 2006, Gesar painted stone engraving was approved by the State Council to be added to *List of the First Batch of National Intangible Cultural Heritage*. The skills are inherited by people like Niqiu, Juere, Zhaluo and Qieqiong.

Picture 7.30 Gesar painted stone engraving

"Huizhou's three types of engraving" refers to the popular wood engraving, stone engravings and brick engravings that use the Huizhou style. With their long history and exquisite techniques, they enjoy a high reputation at home and abroad.

Brick engraving was initiated by Bao Si, a Huizhou kiln craftsman in the Ming dynasty. Its techniques include flat engraving, relief engraving and vertical engraving. The themes include feathers and flowers, dragons, tigers and lions, forest gardens and landscapes, and drama figures, which resonate strongly with folk customs. Brick engravings in the Ming dynasty were rough and simple. In the late Ming and early Qing dynasties, the style gradually became more delicate and complicated, paying attention to story and composition, and sharpening the level of openwork engraving.

Wood engraving is mainly used for the decoration of buildings and furniture, using free-standing engraving, relief engraving, openwork engraving, and other techniques. The themes include figures, landscapes, flowers, animals, insects and fish, palindromes, antiquities, characters, and various auspicious patterns. Free-standing carving of Hu's ancestral hall in Longchuan village, Jixi county, Anhui province is famous for its wood engravings. The 20 partition doors in the main hall are pictured with groups of lotuses (harmony). The painting was made by Xu Wenchang, a great writer and painter in the Ming dynasty, and was beautifully engraved. It is a wood engraving of the best quality that symbolizes the harmony between the home and the outside, and common prosperity, which embodies the core concept of Chinese traditional culture and has been well preserved up to now (as shown in Picture 7.31).

Picture 7.31 Wood engravings of Hu's ancestral hall in Longchuan, Jixi county

Stone engraving is mainly used for the decoration of colonnades, gate walls, archways and tombs, and their themes are mainly animals and plants, antique patterns, and calligraphic works.

Huizhou's three types of engraving are closely coordinated with the whole building, using reasonable layouts, ingenious structures, exquisite decorations, and rich symbolism. They reflect the influence of the Huizhou culture such as the Xin'an painting school, Huizhou printmaking, seal cutting, ink carving, and ink model carving.

7.3.1.3 Jade Carving

As early as 8,000–7,000 years ago, the Chahai-Xinglong culture around western Liaoning and southeast Inner Mongolia had well-made dagger-shaped jade wares. The Zhaobaogou culture from 6,000 years ago also possessed quite advanced jade wares. By the late Neolithic Age (as shown in Picture 7.32), the production of jade wares was widely distributed, came in large quantities and categories, showed fine craftsmanship, and was separated from tools to become decorations and ceremonial sacrificial utensils. The 755 pieces of jade unearthed from Fuhao Tomb show that it was fashionable in the Shang dynasty to use and wear jade. In the Zhou dynasty, jade wares widely used different materials, including sapphire, white jade, serpentine jade, turquoise, and crystal. Jade wares were included in *Rites of Zhou* as a standard, which reflected a strict hierarchy. During the Spring and Autumn period and the Warring States period, the moral, political and religious significance of jade wares promoted the development of jade carving techniques.

The trend of lavish burials in the Han dynasty elevated funeral jade even more. Because of the trade with Xiyu, or the Western Regions, Hetian jade became very popular on the mainland. Jade articles at this time were mainly used as ritual vessels,

Picture 7.32 Jade *cong* (vessel) of the Liangzhu culture

funerary wares, and seals. From the Wei and Jin dynasties to the Sui and Tang dynasties, jade ware techniques were at a low point and its production was not great.

In the Song dynasty, the government set up a special institution for making imperial jade articles, which was called Jade School, and the hollowing technology developed rapidly. During the Yuan, Ming and Qing dynasties, jade carving techniques reached their peak. In the Yuan dynasty, specialized agencies were set up in Beijing and Hangzhou to manage jade production, which lasted until the Qing dynasty. *Ancient Jade Catalogue* was published in the first year of the Zhengyuan period (1341) in the Yuan dynasty, which was the first catalogue of jade articles. *Dushan Jade Jar*, which was created in 1265, weighed about 3,500 kg and was carved showing 13 kinds of beasts of the sea, symbolizing the tolerance and courage of the rulers of the Yuan dynasty.

Jade carvings were beautifully made in the early Ming dynasty. In its late period, jade carving was divided into two schools: the north school and the south school. The north school was rich in shape and rough in production, while the south school was exquisite in both materials and carving. The main characteristic of that time was openwork carving done in three layers. Song Yingxing's *Heavenly Creations* (as shown in Picture 7.33) records: "As the cutting begins, a round iron disc is made and mounted on a frame connected with pedals underneath and a basin of sand is placed beside it. The disc is turned by pedals, at the same time sand is sprinkled on it so as to cut through the jade. Finally, the jade will be cut open as sand is gradually added into the basin." The jade carving process at that time was very similar to that of modern times. To a certain extent, the popularity of jade carving technology was promoted by Ming dynasty literati, who loved jade very much; the works of famous artisans received a warm welcome.

During the Qianlong and Jiaqing periods (1736–1820) in the Qing dynasty, jade carving reached its peak, while the government was in charge of the mining and use of jade materials, prohibiting private mining. Jade carvings with animals deliberately pursued the meticulous painting method of the Song and Yuan dynasties, and the development of landscape painting was reflected in jade carvings, so the expression form of *shanzi* (as shown in Picture 7.34) appeared, which took landscapes as their main body, but dotted it with figures, buildings, boats, and vehicles, ground and cut deeply, but finely and magnificently.

Jade is hard, so it must be ground and cut with mound machines using water-accompanied emery, which has higher hardness than jade. Mound machines have different shapes and can sand jade into different lengths and shapes. With the continuous improvement of metal materials, mound machines were gradually diversified, and work efficiency improved. Mound machines are composed of brackets, rotating shafts, and transmission belts, which have not changed greatly since the Ming dynasty.

A "pretty color" is the main characteristic of jade, which takes center stage through carving. Jade has different colors and shapes. In order to make full use of jade materials, jade carvers creatively use the method of displaying the pretty colors, by designing and carving objects according to the natural colors and textures of the jade, resulting in wonderful works.

Picture 7.33 "Cutting Jade" from *Heavenly Creations*

Picture 7.34 Details of *shanzi* carving

Using jade, loving jade, and collecting jade are closely related to Chinese traditional culture. As early as the pre-Qin period, jade was closely related to people's beliefs, etiquette, and daily life. Confucianism says that a gentleman is better than jade, and Xu Shen's *Discussing Writing and Explaining Characters* says that jade has five virtues, all of which prove the Chinese people's love of jade.

Jade carving tools (as shown in Picture 7.35) are divided into cutting, carving, drilling, polishing and abrasive tools.

Cutting tools, such as circular saws, band saws and wire saws, are mainly used to cut larger jade materials.

Carving tools are divided into two categories: Iron mounding tools and diamond tools. All kinds of iron mounding tools correspond to the stages of creating coarse work blanks, fine work blanks and fine carving. They include:

Picture 7.35 Common tools for jade carving

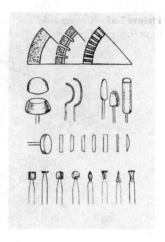

Lateral mounding tool is a round sheet disc, and is used to grind and cut a work blank surface during cogging.

Chisel mounding tool is a small cutting tool used to grind and cut small sections when producing fine work blanks and can also be used for sticking and laying.

Hook mounding tool is similar to the chisel one, and the thickness of the plate varies along the mouth. It has the characteristics of being small, flexible, and strongly adaptable, which can be used to hook and cut detail textures and can also be used for top grinding and cutting on the plane.

Drill mounding tool is a ring-shaped metal tool, used specifically for drilling and grinding large planes.

Rolling mounding tool is a rod-shaped, spherical, jujube-shaped, or pointed grinding tool that is suitable for grinding along the rises and falls of the jade surface.

Boring mounding tool is a spherical iron grinding tool for grinding the bore of open utensils.

Curved mounding tool is a grinding tool bent with iron wire, and used to cut out the bore of a small vessel.

Nail mounding tool is a grinding tool in the shape of a nail, commonly known as "nail" or "bell mouth". It comes in many sizes and shapes, and has many functions. It is small and flexible and is the main tool for grinding details.

Diamond tools are basically the same as iron mounding tools in shape, specification, and function, except that a layer of diamond powder is plated on the working face to combine the iron mounding tool with an abrasive so that diamond mortar is not needed for processing.

Drilling tools are used for drilling holes. To save materials, it is often necessary to acquire coring materials when drilling large holes.

Polishing tools include a rubber mill, wooden mounding tool, leather mounding tool, felt wheel and cloth wheel.

The hardness of the abrasive is higher than that of jade. Using an abrasive and water, jade can be worn down. Natural abrasives are commonly used, such as quartz sand, garnet sand, as well as synthetic abrasives, such as silicon carbide and artificial emery, and polishing powders, such as iron oxide powder and alumina powder.

At present, the most famous jade carvings in China can be found in Yangzhou, Jiangsu province, in Xiuyan Manchu autonomous county, Liaoning province, and in Fuxin city in Liaoning.

Yangzhou jade carving: The Han, Tang and Qing dynasties were the peak periods of Yangzhou jade carvings. In the middle of the Qing dynasty, Yangzhou became the center of jade carvings in China, famous for being good at carving large jade wares. During the Qianlong period, the salt official of Lianghuai (area along the Huaihe River) set up a jade bureau at Jianlong Temple in Yangzhou, and paid tribute to the imperial court according to the annual tribute requirements. Most of the jade mountain carvings, which weighed more than 500 kg, in the Qing Palace, were made by jade workers in Yangzhou. Among them, King Yu Taming the Flood, which weighs more than 5,000 kg and is called the king of jade, was made of green and white jade in Xinjiang Uygur autonomous region. With a height of 224 cm, this rare treasure took 150,000 workers six years to complete and cost more than 15,000 taels of silver.

Yangzhou jade ware is divided into six categories: censers and vases, figures, flowers and birds, animals, antiques, and *shanzi* carving. *Shanzi* carving and *lianzihuo* (jade wares with chains) have distinct local characteristics and are the best jade carvings in Yangzhou. This skill has been approved by the State Council to be included in *List of the First Batch of National Intangible Cultural Heritage*. It is inherited by Jiang Chunyuan and Gu Yongjun.

Xiuyan jade carving belongs to the north school (as shown in Picture 7.36). It rose in prominence from the Qing dynasty's Qianlong to Daoguang period. It combines the characteristics of wood carving, stone carving, clay sculpting, embroidery, paper cutting, shadow figures, and painting, and forms a jade carving system with three-dimensional free-standing engraving and relief carving as the main techniques, supplemented by line carving, engraving and openwork carving, as well as techniques such as the skillful use of pretty colors and threads, resulting in a unique artistic style. Traditional Xiuyan jade carving has five categories: *suhuo* (displaying utensils), figures, flowers and birds, animals, and flowers. Wang Yunxiu, a famous contemporary artisan, was already carving jade at the age of fourteen, and is the representative inheritor of Xiuyan jade *suhuo* technology.

Fuxin agate carving: Fuxin city, Liaoning province is famous for producing agate, which has excellent texture and abundant reserves. The main component of agate is silicon dioxide, which is a stone pith with different colors distributed in a ring. Fuxin agate has a magnificent texture, many varieties and seven degrees of hardness, equivalent to jadeite.

Since the Liao dynasty, the people of Fuxin have mined, excavated, and processed agate, which became a tribute in the Qing dynasty.

Fuxin agate carving is divided into five categories: figures, birds and animals, flowers, *suhuo* (displaying utensils), and water drops. Li Hongbin, the inheritor of this national intangible cultural heritage, is famous for his *suhuo* (displaying utensils) technique.

Picture 7.36 Xiuyan jade carving

7.3.1.4 Ivory Carving

It refers to a carving with ivory as raw material, The earliest sign of ivory carving dates to the Neolithic Age. More than 20 carved ivory handicrafts were unearthed at the Hemudu site in Yuyao, Zhejiang province, among which there were many fine works. At the Dawenkou site in Baotou, Ningyang, Shandong province, carved ivory barrels, and combs were found.

Ivory carving techniques were also applied to carving horns and bones in the early days, as bones and ivory were similar in texture. For example, bone carvings of eagle heads were found at the Xingkai Lake area in Heilongjiang province.

The ivory carving process in the Shang and Zhou dynasties were quite meticulous, imitating bronze patterns and often inlaid with turquoise. For example, the cylindrical ivory carving unearthed from the Fuhao Tomb has fine patterns all over the surface, and the body of the *kuipan* cup is inlaid with turquoise in *taotie* (a mythical ferocious animal) and *kui*-dragon patterns (as shown in Picture 7.37). Bone carving was very common in early China, but the size of the bones was small and the texture rough. Later sculptors gradually gave up on using bone as carving materials.

Picture 7.37 *Kuipan* ivory cup

During the Song and Yuan dynasties, multi-layer carving of ivory balls was created, and each layer could be rotated freely and carved in its own pattern. Now it can reach up to 45 layers.

The painting art of the Ming dynasty, especially the realism in portrait painting, influenced the ivory carving techniques, and the lines of the works became more delicate and exquisite. Rhino horn carvings were also quite famous, and the most representative artisan was Bao Tiancheng in Suzhou city. His Immortal Riding a Raft Cup (as shown in Picture 7.38) used the natural form of the rhino horn to carve rocks, peach trees and old men.

The Qing dynasty was the peak period of ivory carving. Under the direct intervention of the emperor, many fine products were produced. There are twelve volumes in *Yueman Qingyou Book*. The ivory carvings in the book, inlaid with colored stone, jasper, agate, and hawksbill, with strong three-dimensional sense and rich colors, depict the entertainment activities of maids in the palace from the first month of the lunar calendar to the twelfth. It is known as the representative work of ivory carving in the Qing dynasty. Miniature ivory carving was also very popular. From the Yongzheng to Qianlong period, miniature hollow ivory boxes were often used.

Ivory carvings were mainly produced in Beijing, Guangzhou, Shanghai, Nanjing, Tianjin, and Fuzhou. Ivory carvings in Beijing, dominated by flowers, birds, and ladies in ancient costumes, are finely carved, pure and simple, and some of them are

Picture 7.38 Rhino horn carving Immortal Riding a Raft Cup by Bao Tiancheng

rendered in color. Guangzhou was famous for its exquisitely carved ivory balls (as shown in Picture 7.39).

Ivory is the fangs of wild elephants, which has the structure of circular line patterns, an outer layer, inner layer and core, a delicate and compact texture, being

Picture 7.39 Ivory ball

hard but not brittle, and having a certain toughness. The structure and texture of ivory determine the technological characteristics of ivory carving. Early ivory carving only relied on grinding, resulting in rough and simple products. After the emergence of metal crafting, ivory carving was made by tools such as chisels, scuffle hoes and files. Modern ivory carving tools are more diverse. The former process mainly uses sawing, milling, and chiseling, while the latter process mainly uses scuffle hoes, brazing, scraping, filing, and grinding.

It is often said in the tooth carving industry that: Craftsmanship accounts for 30% and tools for 70% of the carving process. This means that the improvement of tools made ivory carving techniques richer and more efficient. At present, there are sawing machines, work blank discharging machines, brazing machines, flexible shaft machines, polishing machines, prototype machines, lathes, and various specifications of knives, files, milling, drilling thallium, sawing, and polishing tools.

The technological process of ivory carving can be divided into six steps:

Process design: It is necessary to give full play to the fineness and cleanliness of materials, so that the beautiful dentin and texture can be reflected.

Cutting: Cut the whole piece of ivory according to the draft of the design, and use the ivory to the fullest, both as economically and completely as possible.

Work blank discharging: It is divided into two processes: coarse blanks and fine blanks. The former is a carved blank, which requires the removal of the ivory's skin, cracks, dirt, decay, and residue. Then, the main line of the work is determined, the base is leveled, the surplus materials removed, and the procedures followed from top to bottom, from front to back, first master to second, and from whole to local. The latter means accurately carving the details of the work on the basis of the rough blank and using small knives to drill with a mounding tool and grind it down.

Brazing: On the basis of the work blank's depiction, make it more detailed and accurate and remove any traces that are left to achieve a smooth surface.

Burnishing: After being soaked in water, it is ground with sandpaper, grass knots and bamboo shoot shells to remove any subtle processing traces.

Polishing: First, use colorant to "cut" the whole body with the sheepskin mounding tool rotating at high speed, or use polishing paste to "cut" the whole body, then wash it with soap and polish and wax it after it is dry.

During the Ming and Qing dynasties, ivory carvings were mostly produced by imperial workshops, and artisans mainly came from Beijing, Guangdong, and Yangzhou. In the late Qing dynasty, artisans returned to common workshops one after the other, and their skills spread far and wide.

Nowadays, the ivory carving techniques of Beijing and Guangzhou are the most distinctive. The northern ivory carving school, which is characterized by its magnificence and exquisiteness, is good at carving ladies, figures and flowers, fully showing the delicate texture of ivory, and combining free-standing carving, relief carving and hollow carving.

Guangzhou ivory carving belongs to the southern school. The production focuses on carving skills, paying attention to the bleaching and colored decoration of the ivory. The techniques include carving, inlaying, and weaving. Most of them are *yin* engraved, protruded, engraved, and dyed. Because the climate is warm and humid, ivory does not crack easily. This makes it suitable to make works that require drilling engraving and transparent openwork carving, represented by ivory balls, flower boats and micro-engraved paintings, and calligraphic works. Ivory pieces can be as thin as paper and translucent. Dyeing is also very delicate and charming, absorbing the western light-and-shade method, which greatly enhances the three-dimensional sense.

7.3.1.5 Brick Carving

It is also known as brick engraving, colloquially known as *huahuo*, which refers to a craftsman's fine and delicate work. These carvings were commonly used in palaces, Buddhist temples, tombs, and residential buildings.

There are many similarities between brick carving and pottery carving. At the Yanxiadu site, rudimentary jujube-pattern thin bricks have been unearthed that date back to the Warring States period. At Xianyang No.1 Palace site, hollow bricks with dragon and phoenix patterns with vigorous lines and meticulous depictions have been unearthed, dating back to the Qin dynasty.

The Han dynasty was the heyday of ancient brick carving, which mainly focused on the production of brick reliefs. Among them, the brick reliefs in Sichuan showed true mastery of the skills, resulting in wide distribution, consistent production, rich variety, and both vivid and diverse techniques. Back then, brick reliefs were mainly made with *yang* lines that were often painted, supplemented by *yin* lines, and the tiles themselves gilded to enhance the effect.

After the Han dynasty, the use of brick reliefs gradually decreased. By the Northern Song dynasty, the realistic style of brick carving gradually took shape, with fine carving techniques, and well-proportioned and vivid images. The Jin dynasty tombs in the south of Shanxi province are characterized by their plain brick carvings, showing rigorous and meticulous techniques (as shown in Picture 7.40).

During the Ming and Qing dynasties, architectural brick carving flourished like never before, and an architrave, *queti* (a support used to bear weight), door cover, gatehouse, cornices, roof ridge, spirit screen, and archway all became the carriers of brick carving techniques such as line carving, openwork carving, bas-relief, and high relief. The gold bricks fired in the imperial kiln in Lumu town, a northern suburb of Suzhou, was thick and heavy, which made them suitable for carving complex and exquisite patterns. They were widely used in gatehouses, spirit screens, and architraves. Guangdong brick carvings are mostly used in gable walls, gatehouses, and in both the sides and eaves of gates.

The raw bricks used in brick carving include hollow bricks, square bricks, glazed bricks, blue bricks, and stone bricks. Pig blood and brick powder are often used in carving. Pig blood is mixed with brick powder to be used as an adhesive to repair brick surface defects, holes, and cracks.

Linxia, Gansu province, is a famous hometown of brick carving, which was called Hezhou in ancient times, so Linxia brick carving is also called Hezhou brick carving. As early as in the Song dynasty, Linxia brick carving was already quite advanced.

Picture 7.40 Brick relief

It was also widely used in various buildings in the Yuan and Ming dynasties. The screenwall of Prosperity Brought by the Dragon and the Phoenix in front of Bafang Hall North Temple built in the late Ming and early Qing dynasties is a masterpiece of Linxia brick carving.

Linxia brick carving is characterized by kneading and engraving. Kneading is to shape clay by hand or mold, and then burn it into bricks in the kiln, while engraving is to engrave relief patterns on blue bricks. Carving techniques include relief, openwork carving, high relief, flat carving, and hollow carving. Works are usually composed of several layers of patterns with a strong sense of hierarchy (as shown in Picture 7.41).

Picture 7.41 Linxia brick carving

7.3.1.6 Wood Carving

The history of wood carving techniques in China can be divided into three periods, namely, the early, middle, and late periods. In the early period, from the Spring and Autumn period and the Warring States period to the end of the Eastern Han dynasty, figurines of people and animals were common. Architectural wood carvings were also applied. Due to the introduction of Buddhism, a large number of temples were built in the Southern and Northern dynasties, and thus, in the middle period, from the Wei and Jin dynasties to the Song and Yuan dynasties, architectural decoration led to the progress of wood carving skills. In the Tang and Song dynasties, besides architectural wood carving, other wood carving crafts also appeared in large quantities, such as

Picture 7.42 Wood carving of three generations, cited from *Chinese Local Handicrafts*

stationery and furniture. Wood carving techniques are diverse, but many use the same high-quality wood such as rosewood and boxwood. The late period, from the Ming and Qing dynasties to the Republic of China, was the heyday of traditional wood carving (as shown in Picture 7.42).

The raw materials used for wood carving then were softwood and hardwood. Loose softwood such as basswood, ginkgo wood, camphor wood, and pine wood, are suitable for carving works with simple modeling and general images. Hardwood, such as mahogany, boxwood, rosewood, and almond wood, with its rigidity, fine texture, and bright color, is a first-class material for carving. It does not easily break or deform and is suitable for carving works with a complex and fine structure.

Wood carving tools include hand chisels, round chisels, upturned chisels, butterfly chisels, carving knives, and triangular chisels. Among them, the triangular chisel is a drill-bar type, which is dexterous and convenient to use, but has poor stress. The other five types are divided into two types: the chisel hoop type and the drill-bar type. The chisel hoop type is more solid and is used for beating rough work blanks and stripping the ground.

The production process of wood carving is divided into nine steps, including brushing samples, stripping contours, stripping the ground, dividing layers, dividing blocks, carving fine blanks, trimming, sandpapering, and fine carving. Brushing samples means copying artwork on a wooden board with paper, which is called a pattern board. The carver hits the contour line according to the pattern and then takes off the wood pad. He holds the chisel with his left hand, and the chisel's edge leans back onto the contour line. At first, it forms an angle of 45 degrees. His right hand beats the chisel handle with a small axe, gradually chipping away the open space. The angle of the chisel knife gradually decreases to zero with the deepening of taking off the wood pad. In layered artwork, it is necessary to start carving from the bottom layer, with the pattern board from low to high and the artwork from far to near, which is just the opposite of taking off the wood pad from near to far, and from high to low. The main purpose of dividing blocks is to make the layered pattern board three-dimensional. After completion, the fine blank carving is carried out, from bottom to top, starting with depicting the secondary, then primary, and ending with the details. Fine blank carving is mainly done by knife carving, including shovel carving, off-the-ground carving, fine carving, regular carving, engraving, reverse carving, straight carving, cutting, and picking. Trimming is the supplement of fine blank carving, starting from the unused space, repairing it from the bottom to the top with a flat knife, and getting rid of scars and warped corners. Sandpapering is a necessary step in the finishing process. Fine carving is the last step of wood carving, which should result in a completed and beautiful piece, with smooth lines, uniform density, and full colors.

China's wood carving skills are widely distributed, with many categories and schools. The most representative ones are Dongyang wood carving in Zhejiang province, boxwood carving in Yueqing, Zhejiang, Chaozhou wood carving in Guangdong province, cinnabar-gold lacquer wood carving in Ningbo, Zhejiang, puppet head carving in Zhangzhou, Fujian province, and *Nuo* mask carving in Xiangdong, Pingxiang, Jiangxi province.

Dongyang wood carving, named after its place of origin, Dongyang city, Zhejiang province, is the first of the four major wood carving schools in China. High-grade wood with a fine and tough texture such as camphor wood, rosewood, *nanmu*, teak wood, and sandalwood are used as source materials. They are dried and straightened through sawing and fumigating and can be carefully carved and preserved for a long time.

The Dongyang woodcarving school (as shown in Picture 7.43) can be divided into architectural decorations, furniture decorations, ornaments, and religious articles. With the characteristics of scattered perspective composition and using a grand layout, Dongyang wood carving often uses auspicious animals (dragons, phoenixes, cranes, deer, bats, etc.), flowers and trees (twin lotus flower, peony, pomegranate, and ganoderma lucidum), characters (mythical people, historical celebrities, four beauties, and literati), folk customs, calligraphic works, and abstract patterns (cloud, water, grass, rocks, flowers, and birds).

Most of Dongyang wood carvings are plane carvings, mainly with relief. Multilayer relief, scattered perspective compositions, and plane decoration result in its unique carving techniques and artistic features. The surface of the wood is not treated nor painted, to preserve the elegant wood texture and color. In recent years, based on their inherited traditions, Dongyang wood carving artisans have continuously innovated and developed multi-layer carving, bark carving and letter-style wood carving. Lu Guangzheng and Feng Wentu are the main inheritors of these skills.

Boxwood carving in Yueqing, Zhejiang province, is named after the boxwood it uses. This kind of wood is tough, yet smooth in texture, delicate in grain and the color is yellow and delightful, which is the reason for being suitable for fine carving.

Picture 7.43 Dongyang wood carving Three Heroes Fighting against Lü Bu by Lu Guangzheng

Traditional boxwood carving in Yueqing is mainly based on mythological stories and historical figures, with a distinct personality and proper handling of creative methods and block ratio, delicate knife techniques, and life-like figures.

Model clay drafting is a typical process of Yueqing boxwood carving. Unlike wood carving, it can be processed directly on the basis of the draft, so it can often create a perfect image (as shown in Picture 7.44). The next step is to carve and treat the clay draft as the prototype. Wang Duchun and Gao Gongbo are the contemporary inheritors of boxwood carving in Yueqing.

Chaozhou wood carving originated from areas around Chao'an, Jieyang, Chaoyang, Puning, Raoping and Chenghai in the east of Guangdong province, and has distinct local characteristics. Chinese fir is often used for architectural decoration, while camphorwood is mostly used for furniture carving, since it prevents moisture and damage done by insects, and does not rot or deform for a long time. After carving, it is usually lacquered with gold to make the finished product brilliant in color. Rich in subject matter such as flowers, birds, insects, fish, fruit of all seasons, rare birds and beasts, and folk myths, it uses almost anything. The most typical technique of this carving is literally called "exquisite carving". The most famous form is with gold lacquer. The carving is decorated with gold foil, which is called "black with gold" since it is set off by black lacquer or multicolored lacquer, which is then

Picture 7.44 Yueqing boxwood carving Hide and Seek by Wang Duchun

called "multicolored with gold". In 2006, this skill was added to *List of the First Batch of National Intangible Cultural Heritage*. Li Denong and Chen Peichen are the representative inheritors of this technique.

Ningbo cinnabar-gold lacquer wood carving is characterized by its lacquering rather than its carving, that is, the carving accounts for 30% and the lacquering accounts for 70%. The selected wood is mainly fragrant, moisture-proof camphorwood, with cinnabar as the primary color and gold foil as the decoration. The lacquering process includes the rough lacquering of wood base, filling slots, pasting grass cloth, lacquering and decorating with gold foil (as shown in Picture 7.45).

Zhangzhou puppet head carving is made from camphorwood in the areas around Zhangzhou, Xiamen, and Quanzhou, Fujian province. The heads are finely carved, exquisitely painted, and have a rigorous shape and an exaggerated expression (as shown in Picture 7.46). It preserves the painting styles of the Tang and Song dynasties and pays special attention to the depiction of the characters. Artisans pay special attention to a face's distinct shapes (eyes, mouth, nose, eyebrows, and ears) and five bone structures (frontal bone, nose bone, cheekbones, upper and lower jawbone) when carving. Depending on the character, identity, experience, and temperament of the character, they use modeling, lines, and colors to express its charm.

Nuo mask carving in eastern Hunan province is an important part of *Nuo* culture, which are used for *Nuo* instruments, *Nuo* dances and *Nuo* operas. *Nuo* masks are made of camphorwood. They use ancient figures as prototypes, and often exaggerate their depictions on the masks to give it an extra level of vivacity. After carving, craftsmen paint and lacquer them to fully display the solemn and gorgeous artistic image.

They also use different facial expressions and decorations to display the different roles, such as the mighty, the loyal, the ferocious, and the treacherous. The expressions and crown decorations of *Nuo* masks have a specific cultural connotation and significance, a simple expression, and a clear regional style.

7 Sculpture

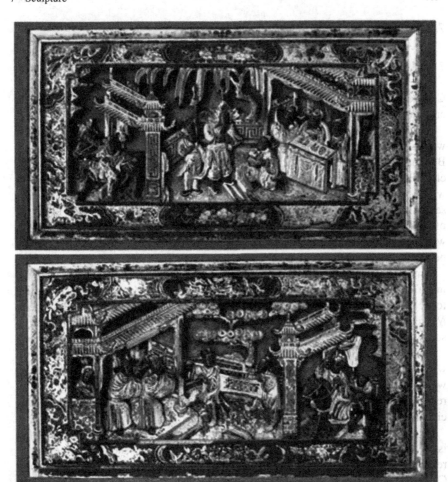

Picture 7.45 Flower boards with cinnabar-gold lacquer wood carving from the Qing dynasty

Picture 7.46 Puppet head carving in Zhangzhou

7.3.2 Modeling

7.3.2.1 Clay Sculpting

Clay sculptures are made of clay and are mostly made by hand. Most of the objects are shaped like people and animals.

Clay sculpture art began in the Neolithic Age. After the Han dynasty, Buddhism was introduced, temple buildings were widely built, and thus clay sculptures of Buddha became essential, which greatly expanded the field. In the Tang dynasty, clay sculpting developed a lot, and Yang Huizhi, known as the master of sculpture, was the most outstanding representative. He apprenticed with Zhang Sengyou at the same time as Wu Daozi, who became a famous painter, while specialized in the art of sculpting and became a famous artisan. There was even a saying that went, "Daozi was famous for painting, and Huizhi for sculptures, who inherited Sengyou's magic pen."

Small clay sculpture toys became popular in the Song dynasty, and artisans who make a living by making clay sculptures appeared. From the Yuan and Ming dynasties to the Qing dynasty, clay sculpting rose to new heights, forming two schools, namely, *Nirenzhang* (Clay Figurine Zhang) in the north and Huishan Clay Figurines in the south, each with their own characteristics and still deeply loved by people today.

Nirenzhang (Clay Figurine Zhang) is named after the founder Zhang Changlin, whose color sculpture skills have been passed down the family line to this day. His works are full of natural emotional appeal, striving to express historical figures and real figures, daring to face life, and conform to the aesthetic standards of modern citizens (as shown in Picture 7.47).

Nirenzhang's colored sculptures are made of clay and pigment. The best clay is the clay with a strong viscosity, delicate texture, and no sand. It is divided into red, yellow, gray, and black colors caused by the different areas it is sourced from. Clay should be screened, its impurities removed, then dried in the sun, smashed, and soaked in containers. Then the soaked clay is placed on a slate and a little cotton

Picture 7.47 *Nirenzhang*'s work Everything Goes Well by Zhang Yuting

wool is added. It is repeatedly beaten with a mallet, then to be put into a jar. After having been stored for one week, it can be used.

Traditionally, the pigments are mainly mineral pigments, of which the brewing method is complicated and inconvenient. Now many different pigments can be supplied and used, such as traditional Chinese painting colors, gouache, acrylic colors, and oil painting colors.

The pressing tools used for *Nirenzhang*'s colored sculpture include clappers, presses, molding tables or clay plates, iron molding knives, and mallets. Brushes are used to paint the sculpture. Kneading techniques include conceiving small drafts and kneading large shapes, details, heads, and hands. Painting techniques include trimming blanks, background colors, main colors, and head coloring.

Huishan clay figurine was made on a certain scale in the Ming dynasty, and black mud found at the foot of Huishan Mountain was used. That kind of mud is dark, delicate, and moist. It is not necessary to be beaten and rubbed extensively, so it can be used rather quickly after being dug out. After the clay figurine is kneaded, it is allowed to dry naturally, without an oven or open flame. The resulting dried clay work blank is very hard. The finished products are roughly divided into two categories, i.e. colored clay figurines and hand-kneaded clay figurines. The former can be produced with molds (as shown in Picture 7.48), copied in large quantities and spread far and wide. The latter is all made by hand, and most of them are based on Chinese opera, also giving it the name "hand-kneaded opera".

Colored clay figurines have unique freehand brushwork, and their main colors are distorted and exaggerated. *Da Afu* is the most representative of this kind of clay figurines. It vividly portrays an amiable, lovely, and chubby child. It sits cross-legged and is holding a green lion. Its shape is steady, with a full, smiling face. The eyebrows are curved, the nose is straight, and there is a diamond bun on top of its head. The figure shows a mighty quiet spirit and slightly simple, honest modesty (as shown in Picture 7.20).

The themes of hand-kneaded clay figurines include opera, street life, mythological stories, folklore, and auspicious birds and beasts. Among them, the characters in Kunqu Opera and Beijing Opera are the best.

Huishan pays attention to the materials used for kneading clay figurines by hand. Before kneading the mold, the clay should be thoroughly kneaded, and the dryness, wetness, softness, and hardness of the clay should be adjusted to the optimal state. There are two methods: (1) kneading-and-inlaying, and (2) printing-and-inlaying. The steps of kneading-and-inlaying are printing, kneading, wrapping, inlaying, and then pressing. The head and figure of printing-and-inlaying method are printed with molds, and only the hands are kneaded into inlays. After that, there was an important adjustment process, called *nieshizi* (kneading the shape), that is, revising the shape of clay figurines.

The painting process and method of colored clay figurines and hand-kneaded clay figurines are roughly the same. The dried clay blanks need to be trimmed with sand blanks, regular blanks and burnt shards to make the work blanks smooth and firm.

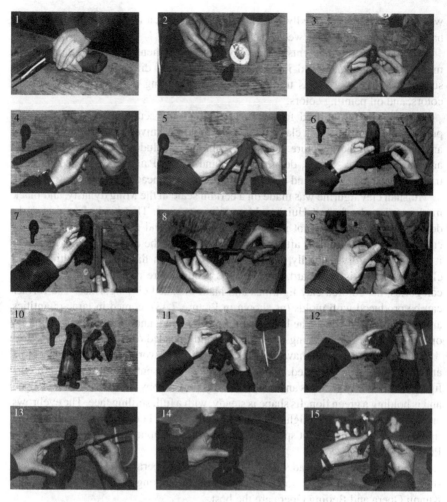

Picture 7.48 Process of hand-kneading a clay figurine 1. kneading mud 2. mold sealing 3. trimming 4. kneading feet 5. splicing body 6. dressing the lower body 7. patting the body 8. kneading and cutting hand shapes 9. dressing the upper body 10. body before insertion 11. inserting hands 12. inserting the head 13. pressing patterns of clothes 14. placing on a base 15. *Pinching shizi*

The painting of clay figurines is very important. As the saying goes, "work blanks account for 30% and color accounts for 70%", and "painting is more important than the work blanks", which means that if there are some defects in the blank, it can be fixed with paint.

7.3.2.2 Dough Modeling

It refers to the technique of sculpting with doughs. Most of the sculptures are characters and opera stories. Tang Zibo and his son Tang Suguo are outstanding contemporary dough modeling masters. Tang Zibo (1882–1971) was good at painting. He incorporated folk art and literature into the creation of dough sculptures. He made an outstanding collection with footprints all over the world. In 1956, the Central Academy of Fine Arts set up the Tang Zibo Studio for him. Qian Shaowu, a famous sculptor, praised Tang's masterpiece One Hundred and Eight Stars of the Water Margin as having a "unique charm" and "outstanding admirable structure". Tang Suguo is gifted, studied hard with his father, and went to the Central Academy of Fine Arts for further study. He graduated from the Sculpture Department of the Academy in 1967, and inherited the family's skills in 1978, establishing Tang Suguo Studio. He has a profound technical foundation and cultural accomplishments and is determined to seek innovation. He integrates Chinese and foreign sculpture art and has created many fascinating masterpieces (as shown in Pictures 7.49, 7.50 and 7.51), which have earned him a high international reputation. In 1996, UNESCO awarded him the title of "Master of Chinese Folk Arts and Crafts". In 1997, Tang's dough sculptures exhibition and seminar were held at the United Nations headquarters in New York. It is said that Tang Zibo led folk dough sculptures to the Art Hall, and Tang Suguo has inherited his father's business and brought Chinese dough sculptures onto the world stage by integrating the strengths of China and the West based on tradition. The experience of the Tang father and son shows the direction that traditional handicrafts should follow to keep pace with the times.

7.3.2.3 Butter Sculpting

This usually refers to yak butter sculptures. It is a unique Tibetan Buddhist art that uses unique materials, techniques, and content (as shown in Picture 7.52).

Yak butter sculptures means they literally make sculptures out of yak butter. The yak butter made with yak milk is white and delicate, with good viscosity, toughness, and plasticity, but it easily deteriorates, changes color, and melts. The techniques for making yak butter sculptures have been explored and summarized by Tibetans through long-term practice.

Tang visited the masterpieces of many masters of art in Europe, and appreciated them, so he aspired to make innovative works such as Rodin and Picasso.

Yak butter sculptures are categorized as painted sculptures, but its color is not painted on the surface. Pigment is mixed in the yak butter according to set quantities to make various blanks. These blanks are then used to complete the sculpture. The strength and support that yak butter can offer is small. The beauty of the sculpture is mainly reflected by the sculpture itself, which is small and exquisite, fully showing the characteristics of precision, complexity, ingenuity, and exquisiteness. For large butter sculptures, wood is used as the skeleton, plant ash is added to the butter to increase its toughness and strength, and then a colored work blank is used for coating.

Picture 7.49 The statue of Huang Binhong, a master of traditional Chinese painting. This work shows the effect of freehand ink blooming on dough sculptures, which echoes the style of Huang's landscape works and reflects the lofty realm of the grandmaster

The yak butter sculpture of the Ta'er Lamasery in Qinghai province is the most representative of its kind. It must be made in a cold place. A pot of ice water should be placed beside the sculpture rack to keep the hands of the sculptor cold at any time and prevent the butter from melting (as shown in Picture 7.53). The production process is divided into four processes: Make the skeleton, make rough blanks, apply coating, and load the plates.

It is a monk's main task and duty to learn this skill in the lamasery. Therefore, monks are the masters of yak butter sculptures. They mainly study *gongqiaoming* (craftsmanship) in Tibetan Buddhism, including such books as *Statue Measurement Sutra, Proportionality Theory, Colorimetry, Axis Painting, Painting History of Wise*

Picture 7.50 Statue of Liang Shuming, a master of Chinese studies. Mr. Liang's education and personality are admired by the Chinese people. Tang's works are famous for being vivid. This can be seen in the focus on depicting Liang's eyes and lips to show the character's resolute and unyielding character

Men, and *Object Map and Proportionality*. They usually sculpt various Tibetan-style border decorations and the eight treasures of Tibetan Buddhism, practice basic skills, and make yak butter sculptures, and learn sculpting in winter.

Sugar sculptures, commonly known as sugar-blowing, is a kind of manual art where sugar with different food coloring is heated and softened, and then with the help of scissors, small combs, small bamboo knives, and other tools and bamboo chips, springs, gypsum powder, and other auxiliary materials, artisans use blowing, kneading, rubbing, pulling, pinching, pressing, picking, cutting and other techniques to make sculptures.

Sugar sculptures are vivid in shape and rich in color, and their manual techniques are mainly divided into two categories, namely blowing and molding, which are called *paohuo* and *touzihuo* in the industry. Blowing pays attention to fast and accurate placement because the sugar film solidifies rapidly at low temperatures, and the

Picture 7.51 Statue of Rodin, Master of sculpture

lower the temperature, the faster it solidifies, which requires artisans to have quick eyes, an even breath, and skilled movements. Otherwise, it is difficult to complete the sculptures. Blowing is often used in sugar sculptures to create the shapes of animals and plants, while molding pays more attention to the use of structure and color and is common in sugar sculptures of characters.

Sugar sculpting skills are widely distributed in China, and artisans usually carry wooden cases laden with materials to make and sell sugar sculptures on the spot (as shown in Picture 7.54). If the temperature is too high, the sugar sculpture cannot be molded. The temperatures between September and March are more suitable. This means that artisans only have a limited time to take to the streets to make and sell sugar sculptures every year.

Picture 7.52 People admiring yak butter sculptures

Picture 7.53 The sculptor using ice to keep his hands cold while making yak butter sculptures

A wide range of sculpture techniques have played an important role in the development of art, culture, economy, and society throughout China's history, and are still deeply beloved by the people. Since 2006, many sculpture crafts have been added to the *List of National Intangible Cultural Heritage Protection*, and their inheritance and development are expected to be more effective.

Picture 7.54 Sugar sculpture artisan on the street of Beijing

Chapter 8
Weaving and Tying

Wang Lianhai

Weaving is based on the stems, leaves, skins, and cores of plants, which are interwoven and formed in a certain weaving form to make products, daily necessities or handicrafts. The materials used are most common in rattan, willow, bamboo and grass (as shown in Picture 8.1).

8.1 The History of Weaving

In the Paleolithic Age, there was a primitive weaving technique where the fibre of plants was used to weave string bags which could be used to sling stones at prey. In the Neolithic Age, woven products were used all the time. About 7,000 years ago, more than 100 pieces of reed mat fragments were unearthed from the Hemudu site in Zhejiang province. The larger ones are more than one-meter long. Although most of them have decayed, their compact structure can still be observed. Several reed strips were unearthed woven together: the vertical and horizontal reeds being woven into diagonal lines, better known as a "herringbone pattern". These reed mats (as shown in Picture 8.2) have a clear and smooth texture with a uniform herringbone pattern, which indicates that it was formerly used as bedding. More than 200 woven bamboo objects (including baskets) were discovered at the Qianshanyang site in Xingwu county, Zhejiang province. These woven objects can be traced back to about 4700 years ago. The bamboo strips had all been carefully scraped and ground to become smooth and fine, and woven in a variety of patterns such as herringbone, cross, diamond, and plum blossom. At that time, when they finished firing their pottery, they put the pottery workblank on the weavings to dry in the sun. This

W. Lianhai (✉)
Academy of Arts and Design, Tsinghua University, Beijing, China
e-mail: wanglianhai327@163.com

© Elephant Press Co., Ltd 2022
H. Jueming et al. (eds.), *Chinese Handicrafts*,
https://doi.org/10.1007/978-981-19-5379-8_8

Picture 8.1 1. The weaving market in the past 2. A corner of the contemporary weaving market

caused the patterns of the weavings to imprint onto the surface of the workblanks, thus preserving them up to the present day. There are more than 100 pottery pieces with imprinted weaving patterns unearthed at the Banpo Cultural Site in Xi'an. On the pottery pieces, four knitting methods can be discerned: (1) longitudinal lines, (2) diagonal lines, (3) entanglement knitting, and (4) loop knitting.

8 Weaving and Tying

Picture 8.2 A piece of a reed mat unearthed at the Hemudu site

Written records of weaving craftsmanship during the pre-Qin period are abound, for example, *The Classic of Rites* records the usage of bamboo mats and baskets. *The Pattern of the Family* in *Classic of Rites* says that after washing in the morning, pillows and sleeping mats must be stored and are forbidden to be shown to others. *Funeral Records* says that when shrouding the deceased for burial, bamboo mats were used for monarchs, cattail mats for officials, and reed mats for scholars. This illustrates not only the rich variety of materials and mats at that time, but also the hierarchical division of rank.

The exquisite materials, superb workmanship and extensive usage of weaving and tying in the Spring and Autumn period and the Warring States period were so good, that their products are on par with modern woven products. More than 30 pieces of bamboo pillows, fans, mats, baskets, roofs, and other woven products were unearthed from Chu tomb No. 1 in Mashan, Jiangling, Hubei province. The bamboo fan, for example, is made out of three warps and one weft, in which the warps were red bamboo strips and the wefts black bamboo strips. The fact that the width of the strips is only 0.1 cm is especially astounding. In the city of Changsha, Hunan province and the city of Xinyang, Henan province, many woven products dating back to the Warring States period were unearthed, such as mats, fans, containers, curtains, and baskets, made with a variety of patterns, such as herringbone, cross, square, octagonal, hexagonal hollow, rectangular, and coiled spirals.

Picture 8.3 Bamboo tubes unearthed from Mawangdui No. 1 Han tomb

In the Han dynasty, people put mats on their beds. The most commonly used were rush mats, scirpus mats, thick bamboo mats, and fine bamboo mats. Thick bamboo mats were used by the common people while fine bamboo mats called "*dian*" were for those in higher positions. *Dian* with character patterns were unearthed at both the Mawangdui tomb No. 1 in Changcha, Hunan province and the Western Han Tomb in Pingjibao, Yinchuan, Ningxia Hui autonomous region. At the Mawangdui tomb No. 1, they discovered bamboo frames wrapped in a fine silk, used for incensing clothes and quilts, as well as 48 bamboo baskets, each embroidered with the Chinese characters of what they once contained, such as clothes (衣笥), silk fabrics (缯笥), beef (牛脯笥) and eggs (卵笥) (as shown in Picture 8.3).

In the Tang dynasty, cattail stems were used to weave clothes and sails, and places like Cangzhou produced wickerwork boxes. From the wicker suitcase that was unearthed from Tang tomb No. 105 in Astana North District, Xinjiang, it can be seen that the history of weaving is at least 1,000 years old. At the beginning of the nineteenth century, a string of five straw *zongzi* (sticky rice dumplings) from the Tang dynasty was unearthed in Turpan, Xinjiang. The five *zongzi* were made by cutting open straw and then "holding the corners and nesting" the strands. This is the oldest straw toy unearthed up to now (as shown in Picture 8.4).

In the Song dynasty, bamboo weaving in Dongyang, Zhejiang province included dragon lanterns, flower lanterns, lanterns with rotating shadow figures, incense baskets, and flower baskets. The workmanship was exquisite: it was fine and compact, with 120 bamboo strips per square inch, and a variety of patterns. The Sichuan Dazu Stonecarvings have a famous carving called "chicken girl", where the chicken coop uses a hexagonal woven pattern.

The most famous straw weaving from the Ming dynasty is a cattail hassock, which has evolved from a monk's meditation seat to an everyday seat (as shown in

8 Weaving and Tying

Picture 8.4 String of straw *zongzi* dating back to the Tang dynasty, unearthed in Xinjiang

Picture 8.5). The "*shu* mats" produced in Shucheng, Anhui province are exquisite due to the material used. It only uses the top-layer of bamboo as strips. The strips are thin and soft, with exquisite patterns, which made them highly valued by the imperial palace.

The stools, chairs and couches in the existing Ming dynasty furniture often use these "top-layer mats". The structure of the mat on the dalbergia odorifera chair from the Ming dynasty stored at Tsinghua University is exactly the same as that of today's

Picture 8.5 Modern cattail hassock

Picture 8.6 Dalbergia odorifera chair from the Ming dynasty, Stored at Tsinghua University

straw mats. The mat incorporated in the "beech nanguan hat chair" is woven from rattan, which is very dense and firm and has not withered, even after 500 years (as shown in Pictures 8.6 and 8.7).

The bamboo weaving tools used in the Ming dynasty can be seen in the picture of *"the winnow"* in *Heavenly Creations* (as shown in Picture 8.8), which shows woven baskets being carried and placed underneath to catch the grain being threshed.

The most extensive weaving project came from the early Qing dynasty and was called the "willow palisade" in English, also known as the *"Shengjing border wall"*, or more simply referred to as the "willow wall" or "border of woven strips". It was an extremely long fence built to strengthen the national borders. It was started during the Qing dynasty's Shunzhi period (1644–1661) and was completed during the Qing's Kangxi period (1662–1722). The woven walls started from Fengcheng, Liaoning province in the south and reached the north of Jilin city in the north. It contained 21 border gates, where officers and soldiers checked in and out, and residents were prohibited from crossing the fence.

During the Qing dynasty's Kangxi period, the water bamboo summer sleeping mat from Yiyang, Hunan province was reputed to be "as thin as paper, as bright as jade, as flat as water and as soft as silk". Later, bamboo and rattan baskets produced in Suzhou were called "tiger hill baskets", named after Suzhou's nearby hill of the same name. During the Qing dynasty's Jiaqing period (1796–1820), weaving and braiding hats using wheat straw rose in popularity in Henan and Shandong provinces and lasted for more than 100 years. The straw braided hats from Xingyang, Henan province, were named "bridge braid hats" because they first emerged near the Huiji bridge.

Picture 8.7 Rattan-woven mat seat

The straw braid hats from Lucheng, Shanxi province are called "Lu braid hats". Shandong is most famous for the straw hats produced in Yexian county. Farmers in these areas wove a large number of straw hats, which were famous for their bright color and uniform width. They also exported a large number of straw hats during the beginning of the Republic of China. Since winning the Panama Gold Medal, the output has increased day by day and reached its peak. In 1911, Liu Xisan of Yexian county turned straw hats collected from the countryside into wide-brimmed straw hats and put them on the market to compete with foreign goods. In 1929, Shengxifu's straw hats and straw braided hats won the first prize at the Philippine Expo, which was the start of the well-known Shengxifu hat industry.

During the Qing dynasty's Daoguang and Tongzhi period (1821–1875), Chengdu's porcelain (also known as "china") and woven bamboo wares became famous. The one who made this happen was Zhang Guozheng from Chongzhou, Sichuan province. Their main characteristic lies in the bamboo covers wrapped around the fired porcelain vases, pots, cylinders, and boxes, which enhanced the preservation and look of the porcelain. In 1916, this type of porcelain and woven bamboo wares won a silver medal at the Panama Expo.

Picture 8.8 Picture of *"the winnow"* in *Heavenly Creations*

8.2 Bamboo Weaving

China is rich in bamboo, including more than 250 kinds of bamboo, such as moso bamboo, sulfur bamboo, clumping bamboo, large-leaved bamboo, henon bamboo, and black bamboo. Bamboo is most densely found in southern regions such as Zhejiang, Fujian, Sichuan, Guangdong, Jiangsu, Guangxi, and Hainan provinces.

8.2.1 Varieties of Bamboo Weaving

Bamboo weaving can be divided into daily necessities, tools, furniture, and ornamental products.

8.2.1.1 Daily Necessities

Daily necessities woven from bamboo include mats, baskets, needle baskets, boxes, fans, pillows, lamps, hats, dustpans, strainers, conical hats, and bases for flower arrangements. The varieties of bamboo mats are the most numerous, but they are mainly used as bedding in summer. Bamboo mats are most famously and most widely distributed in Dongyang and Shengxian in Zhejiang, Shucheng in Anhui, Quanzhou and Putian in Fujian, Chengdu in Sichuan, Yiyang in Hunan, and Taiwan. The bamboo mats produced in these places all have their own characteristics (as shown in Picture 8.9).

"*Shu* mats" are made of local fishscale bamboo. Top-grade *shu* mats, with a warp and weft density of 32 strips per *cun* (about 3.3 cm), are the densest bamboo mats in China. It is made of fishscale bamboo, which is thin, dense and soft. It can be broken into soft bamboo strips. This woven mat can be rolled up to a length of 5–6 *cun* (about 17–20 cm), which makes it very convenient to carry and store. Fengdu in Sichuan province is abound with mats made of thin bamboo strips. There are more types of mats such as Wenzhou straw mats, Taiwanese rattan mats, Qianshan bamboo mats, and many more.

Picture 8.9 Bamboo mat made in Lishui, Zhejiang province

There are many kinds of bamboo baskets, such as carrying baskets, hanging baskets, baskets with cover, net baskets, and banneton baskets. The rectangular *yuanbao* basket, which is a basket in the shape of shoe-shaped gold ingot with semicircular upturned ends, is also called a "waist basket" for shopping and visiting relatives (as shown in Picture 8.10).

People in Hangzhou weave flower baskets, lotus leaf baskets and multi-angle baskets with henon bamboo. The multi-angle ones can be further divided into hollow baskets and fish-eye baskets depending on how they were woven. The banneton basket is used as a heat source or for baking. The basket holds an earthen bowl, on which a charcoal fire is lit (as shown in Picture 8.11).

The Asian conical hat, also known as farmer's hat or Asian rice hat, is used to shade from the sun and rain. There are single-layer and double-layer hats. One kind of conical hat is lined with oil paper or treated with coconut root-bark between the layers to enhance its ability to protect against the rain (as shown in Picture 8.12).

In modern times, bamboo safety helmets are used as protection during labor, thanks to their good elasticity and ability to greatly cushion blows.

Picture 8.10 Bamboo-woven *yuanbao* basket

Picture 8.11 Banneton basket with charcoal fire for heating

8.2.1.2 Bamboo-Woven Tools and Utensils

Bamboo-woven tools and utensils include winnowing fans, pack baskets, bamboo cages, and storage baskets. Winnowing fans are farm tools, which are used for milling rice, winnowing, filling, and washing rice and vegetables. Bamboo baskets include fish baskets, back baskets and wastebaskets. Back baskets are the main means of transportation in underdeveloped areas. They are constantly being used, be it for working in the fields, going up the mountain to pick herbs, going to market, or visiting relatives and friends. If you carry a big back basket, it will stick out two *chi* (about 67 cm) above your head. In Guizhou, Sichuan and Yunnan provinces, special "back baskets" are used to carry children (as shown in Picture 8.13). There are two

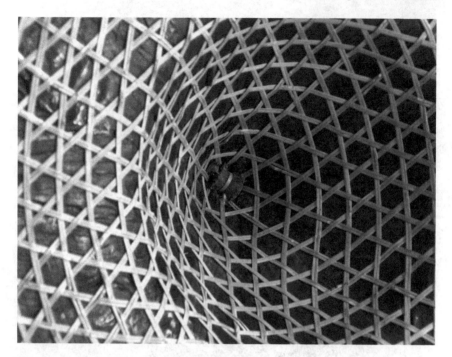

Picture 8.12 The inside of a conical hat lined with oil paper

types of bamboo cages: One type is a cage for containing poultry or insects, such as birdcages and cricket cages, while the other is a support cage or as a cage that covers utensils or appliances, such as the banneton cage and fumigating cage.

Bamboo fishing cages are often used by the Li people in Zhuhai, Guangdong province, and Hainan province, and the Drung people in Yunnan province to catch fish. These fishing baskets are long and narrow, without a lid but with reverse bamboo strips inside the opening. In ancient times, this kind of fish trap was called a *gou* (as shown in Picture 8.14).

8.2.1.3 Bamboo-Woven Furniture

The structure of bamboo-woven furniture mostly imitates wooden furniture, such as beds, tables, chairs, stools, cabinets, bookshelves, screens, and curtains. However, some furniture is also made by fully embracing the characteristics of bamboo. For example, the structure of a bamboo chair is different from that of a wooden chair. Its front and rear legs are made of a single piece of bamboo and an inner circular gap is cut out at the bend to wrap the horizontal rim on the side. The horizontal rim is also bent on all four sides by a single piece of bamboo that wraps the four legs, making it stronger yet using the least amount of material (as shown in Picture 8.15).

8 Weaving and Tying

Picture 8.13 Bamboo-woven back basket for carrying children in Sichuan province

In ancient China, a *ji* was a book case woven with bamboo. People often used them when they were going out to study or when they went to Beijing to take the imperial examinations. This is where the phrase *fu ji* comes from, which literally means "carrying a case of books and leaving home to study". This book case was woven with wide and thick bamboo strips (as shown in Picture 8.16). Its warp strips are several times wider than the weft strips, and the number of warp strips is several times less than that of the weft strips, which makes it much lighter than a wooden box of the same size.

Picture 8.14 1. Weaving fishing cages in Doumen district, Zhuhai city, Guangdong province. 2. Setting up the fishing cages in Doumen district, Zhuhai city, Guangdong province

8 Weaving and Tying

Picture 8.15 Bamboo-woven furniture in Anji city, Zhejiang province

Bamboo curtains are usually woven with cotton thread or hemp thread as warps and bamboo strips as wefts, while some are woven with bamboo strips as warps and wefts. For both, the outer frame is made of bamboo poles and the inner frame is woven with bamboo strips with different widths but a similar thickness to the woven patterns used in the curtain (as shown in Picture 8.17).

8.2.1.4 Bamboo-Woven Artwares

The "bamboo over porcelain" in Chengdu, Sichuan province uses porcelain as the body and covers it in a sheath made of fine bamboo. This cover is "shaped according to the shape of the porcelain and tightly attached to it". There can be no gap between the porcelain and the woven bamboo nor between the joints of the bamboo strips. It is astounding, though it does beg the question of who started incorporating porcelain into bamboo weaving and how.

Bamboo weaving in Dongyang, Zhejiang province, focusses on recreating the three-dimensional images of people, utensils, buildings, and animals. In 1915, Ma Fujin, a bamboo weaving artisan in Dongyang, Zhejiang province, won an award at the Panama Pacific International Exposition for his work "Kuixing Cedou". In 1981, a large incense burner pavilion with a height of 1.5 m was made in Dongyang. It incorporated more than 30 different kinds of weaving patterns, praised as the "essence of contemporary Dongyang bamboo weaving".

Picture 8.16 A bamboo-woven book case in Shandong province

Bamboo weaving paintings were famous in Nanjing, Jiangsu province, and Lishui, Zhejiang province. The width of bamboo silk used in these woven paintings is only 0.5 mm. It interweaves warps and wefts to form patterns. There is also a kind of bamboo-woven lithophane, in which warp and weft yarns are not dyed, but the textures are perpendicular to each other to form patterns that become visible when reflected by light.

Liangping, Sichuan province, houses a unique bamboo curtain that is known as "the best curtain in the world". According to *Annals of Liangshan County*, during the reign of Emperor Taizong (976–997) of the Northern Song dynasty, a craftsman named Yan Hongshun, was able to process bamboo to weave curtains as thin as silk. This bamboo silk was used to make paper as well as door curtains and window curtains, and then given as tributes to the imperial court. During the Qing dynasty's Guangxu period (1875–1908), Fang Bingnan, a famous painter and artisan, created bamboo curtain paintings, which spread to Wusheng and Nanguang. The warps of the painting are filaments with a diameter of only 0.18–0.25 mm, which are obtained by reeling off raw silk from coco on; the wefts are bamboo silk. They are woven on a special curtain loom, brushed with tung oil, and then hung level until the curtain holds its shape (as shown in Picture 8.18).

Picture 8.17 Bamboo-woven curtain with the Chinese character "福", which means blessing in Zhejiang province

8.2.2 The Craft of Bamboo Weaving

Breaking bamboo refers to splitting bamboo lengthwise using a knife for the initial cut by hitting the back of the knife with a mallet. If you are to split the shorter bamboo, first lay the knife stuck inside the bamboo back down on the workbench, then hold the bamboo in one hand and the knife in the other, and finally hit both on the table together to help the blade back on its way through the bamboo. The grain of bamboo is straight and long, and it will naturally expand, crack and give way after being cut into. However, sometimes the knife's blade will shift. When this happens, the blade should be turned towards the narrowest side. Bamboo must be split in one go, so that the width of the two lengths of bamboo are the same (as shown in Picture 8.19).

Cutting thin strips of bamboo is also called "starting the strips" (*qimie*). One way is to start cutting along the diameter of the bamboo tube and each bamboo strip will contain both the green outer skin, fibres and yellow inner skin. Another way is to cut along the circumference of the bamboo. Strips from the outermost layer of green bamboo are called "head bamboo strips" and from the second layer are called "secondary strips", while the innermost layer is called "yellow inner skin". The toughness and glossiness of the head strip are the best, while every subsequent strip decreases in quality. When splitting thin bamboo strips, hold the knife in your right hand and then use your left hand to push the bamboo strips onto the blade.

Picture 8.18 Making a bamboo curtain in Liangping, Sichuan province. 1. Spinning. 2. Weaving. 3. Hanging

Picture 8.18 (continued)

After the blade goes in, the bamboo strips can be pulled apart into two strips. When splitting long bamboo strips, step on one half with your foot, hold the other half with your left hand, and continue to cut. Some craftsmen like to hold one half in their mouth and pull the other half with their left hand (as shown in Picture 8.20).

Scraping strips means processing split bamboo strips into strips of the same width and thickness. For this, two beveled knives are placed opposite each other, the distance between the knives adjusted, and the tips of the knives nailed onto the worktable. The bamboo strips are inserted between the two knives and pulled through.

Weaving. There are two main weaving methods: One is plane weaving, used for woven products such as mats, curtains and bamboo paintings. Some of these weaves

Picture 8.19 Breaking bamboo in Fenghuang county, Hunan province

only use bamboo as warps and wefts, some use bamboo as wefts and other fabrics as warps. Other warps and wefts are a mix of bamboo and grass or bamboo and rattan. The other method is three-dimensional weaving, used to create products such as storage baskets, baskets and cages, which consist of a flat bottom and four sides. When you weave, you should start from the bottom, then bend the warp strips and stand upright to weave the sides, then slightly fold the opening of the basket to install a hoop handle. Some baskets also require a lid to be woven (as shown in Picture 8.21).

Let's take a look at how to weave a round basket. To start, you cross several bamboo strips and lay them out into a circle. These are the warp strips (as shown in Picture 8.22).

8 Weaving and Tying

Picture 8.20 Holding one half of the bamboo strip with your mouth when splitting it

Then you take another bamboo strip and weave it through the warps. You do this by pulling it over one warp and then under the second warp until you have gone through all the warps and come full circle. These are the weft strips that are added circle by circle. After weaving the bottom, you bend the warp strips and position them upright. Then continue to weave the weft strips through them. This will create the sides of the basket. After reaching the desired height, you can begin to slightly fold the opening of the basket and finally install the handle.

There are many different ways of arranging and structuring the warps and wefts of bamboo wares, such as flat weaving, twill weaving, heterogeneous weaving, twisted weaving, and multi-warp and multi-weft weaving. The width of the warps and wefts can also be changed, such as wide warps and narrow wefts, narrow warps and wide

Picture 8.21 Weaving a bamboo basket in Kunshan city, Jiangsu province

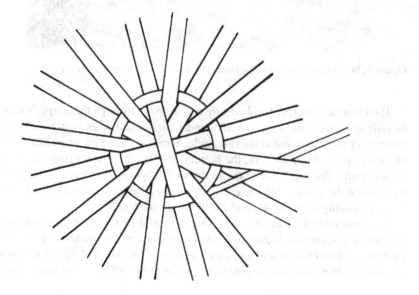

Picture 8.22 Schematic diagram showing the first step for weaving a bamboo basket

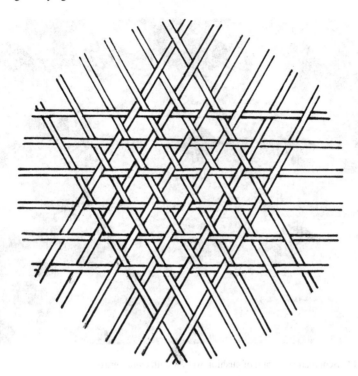

Picture 8.23 Schematic diagram showing the pattern of a bamboo cage

wefts, and alternating widths between warps and wefts. It is precise because there are so many different ways to weave bamboo wares that such a vast variety of patterns and colors is possible (as shown in Picture 8.23).

8.3 Straw Weaving

The materials used for straw weaving are wild grasses and by-products of crops, resulting in many different types of straw weaving, such as cattail weaving, rush weaving, reed weaving, wula sedge weaving, yellow straw weaving, wheat straw weaving, palm fibre weaving, corn husk weaving, rice straw weaving, sorghum stalk weaving, and Indian aster weaving. The most famous of them are wheat straw weaving from Shandong, Hebei and Henan provinces, yellow straw weaving from Jiading district, Shanghai, and from Gaoyao district and Dongguan city, Guangdong province, golden silk straw weaving from Zhejiang province, Chinese alpine rush weaving from Hunan province, palm fibre weaving from Xinfan town, Sichuan province, corn husk weaving from Shandong province, and wula sedge weaving from northeast China.

Picture 8.24 Schematic diagram of cushion woven with cattail grass

8.3.1 Cattail Weaving

The scientific name of cattail grass is "typha", a perennial herb, which grows at the waterside or pond, and is most common in North and Northeast China. The leaves of cattail are thick, soft and tough, containing cottony mesophyll. The main varieties of products woven from cattail are cattail mats and bags. The former (as shown in Picture 8.24) are made of flax or cotton thread as warp and cattail leaves as weft. The warp and weft are twisted into a mat, which is soft and light. The latter has the ventilates well and filters water. The large cattail bags can hold fresh fish and vegetables, the medium ones are used to hold snacks and grains, and the small ones, which are only 5–8 cm long, are used for making dried *tofu*.

8.3.2 Reed Weaving

Reed weaving uses reed as raw material and is mostly produced around lakes and rivers. The area in the north where mainly reed mats and reed foils are woven is Baiyangdian Lake in Anxin county, Hebei province. Reed mats are also called *lu* mats. The main procedures for weaving them include breaking, soaking, rolling, and weaving. Reed stalks are round tubes with joints. Before weaving, they should be "broken" or split. The split reeds are then laid flat on the ground, repeatedly rolled over four or five times with stone rollers to make them flat and soft, and finally woven into reed mats following a herringbone pattern. Weaving reeds into a strip of one and a half feet wide is called "folding", which is used to block grain stores.

The screen of reeds is woven with whole reeds, in which the warp is made of flax thread or string. As each reed strip is woven, with the warp spaced at one foot or more, the double strings are twisted at 180° to fasten them (as shown in Picture 8.25).

8.3.3 Wheat Straw Weaving

Wheat straw weaving is widely distributed. Wheat is planted in large areas in Heilongjiang, Jilin, Liaoning, Inner Mongolia, Shandong, Shaanxi, Shanxi, Hebei, and Henan provinces. Wheat straw is abundant as a resource. Wheat is often woven into daily necessities and toys, mainly straw hats, handbags, shoulder bags, coasters, and fans.

The procedure for making straw hats include selecting straw, pulling its core, soaking and weaving. Different weaving methods can create different weaving structures. The simplest and most basic method is flat braiding, which uses three, five, seven, or eleven strands. The so-called "strand" refers to the number of wheat straws that are braided together. The more strands, the wider the braid. There are hundreds of kinds of wheat braids in Laizhou, Shandong province, but the most common varieties are the "big centipede", "small centipede", "folding" and "orchid folding" (as shown in Picture 8.26).

The braids used for straw hats are flat strips which are sewn together into pieces and can actually be sewn into different articles as well.

A grasshopper cage is the most typical example of a wheat straw woven toy and is woven in a spiral shape. Once the grasshopper is put inside the cage, the cage is then knotted and sealed. In Fengxiang, Shaanxi province, wheat straw is dyed green, red, purple, and blue, and woven together with colorless wheat straw into colorful pendants.

8.3.4 Palm Fibre Weaving

Palm fibre weaving uses palm leaves. Palm trees are evergreen trees that grow in the southern provinces. Palm fibre weaving can be divided into two types: (1) palm leaf weaving and (2) palm fibre weaving.

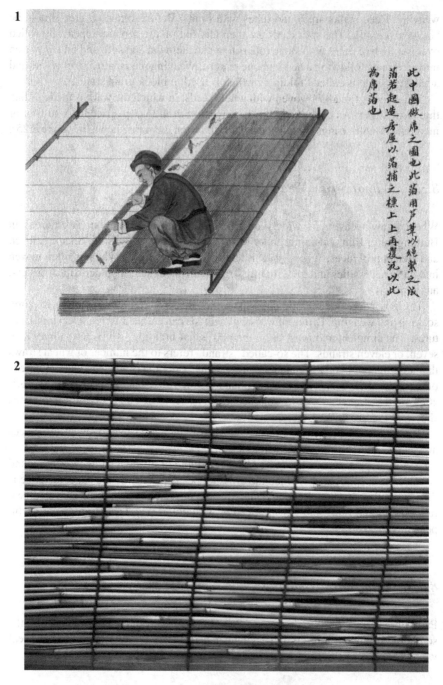

Picture 8.25 1. Weaving screen. 2. The screen of reeds in Hebei province

Picture 8.26 Braids used for straw hats in Laizhou, Shandong province

8.3.4.1 Palm Leaf Weaving

The palm leaves are divided into narrow strips and boiled to enhance their toughness. The weaving techniques mainly consist of folding, passing, covering, pulling, and buckling, with some weaving techniques being given a specific name, such as "single belly", "double belly" and "bamboo mat weaving". Palm leaf weaving is mainly used for toys. Many artists use local materials to make and sell them while walking through the streets. Traditional varieties include a mantis, dragonfly, frog and cicada (as shown in Picture 8.27).

8.3.4.2 Palm Fibre Weaving

Palm fibre weaving uses palm fibres, which are made by splitting the palm leaves into filaments with a cutting needle. The split palm fibre should be soaked, whitened (or bleached) and dyed. Xinfan town in Xindu county, Sichuan province is famous for its unique palm fibre weaving, which is known as "Xinfan palm fibre weaving".

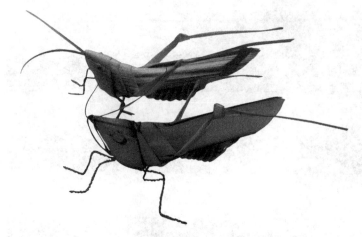

Picture 8.27 Palm leaf weaving grasshoppers made by Wang Wending

This process began in the Qing dynasty's Jiaqing period (1796–1820), when they mainly produced shoes, hats, bags, and toys. These toys can come in the shape of elephants, fish, sheep, birds, and more. Instead of a frame or skeleton, these toys are supported by the elasticity of the palm fibre. During the weaving process, palm fibre is twined on the molds (including wood, mud or paper molds) to form warp threads and then woven with weft threads. The molds are taken out after weaving to a certain extent and then transferred to another mold to continue weaving. Finally, the mold is taken out and the edge is woven to close it off.

8.3.5 Corn Husk Weaving

Each corn husk consists of nearly 10 leaves. The outermost two layers are dry and thick, but the inner layers can be used for weaving. Firstly, the corn husk needs to be separated into leaves and dried. Then the leaves are bundled and soaked in water to soften them before weaving. Once done, they are split into thin strips by hand and woven into single-strand ropes, double-strand ropes or multiple-strand ropes to be used as raw materials. Wooden molds are needed for weaving corn husks. Take a bag as an example. First, the bottom of the bag is woven using a wooden mold and struck with a wooden hammer to make it smooth and compact. Then, the peripheries of the bag are woven, after which it can finally be taken off the wooden mold to insert the edge and install the hoop handle (as shown in Picture 8.28).

Picture 8.28 Corn husk weaving cradle in Henan province

8.4 Rattan and Wicker Weaving

8.4.1 Rattan Weaving

Rattan generally refers to herbs and woody plants with stolons or climbing vines. Ivy, gray vines, mixed black rattan, podagrica rattan, common rattan, and white rattan are used for weaving. Among them, common rattan and white rattan are the most frequently used ones. Rattan plants are mainly cultivated in Guangxi, Guangdong, Fujian, and Yunnan provinces, among which Guangdong rattan is the most famous.

The materials used in rattan weaving are divided into the rattan strips, the rattan skin and the rattan core. The cut rattan can be used for weaving after it has been beaten, picked, dried, stretched, peeled, and bleached. Peeling refers to the cutting off or peeling of the outer shell of the rattan, which can only be done by splitting the rattan as well. Thick rattan can yield four pieces of rattan skins, one from each of the four sides.

Thick rattan can be used for weaving the core structure or frame, while the rattan skin and the rattan core can be used to weave other parts. Most rattan wares use bamboo, wood and wicker to weave the frame and use rattan skin to cover the surface.

Picture 8.29 Rattan chair kept in Tsinghua University

Rattan products can be divided into three categories: furniture, household equipment and toys.

Rattan furniture includes desks, chairs, beds, cases, cabinets, bookshelves, tea tables, flower racks, and folding screens. They are much lighter than their wooden counterparts. The mesh fabric in rattan furniture ventilates well, which makes it suitable for summer (as shown in Picture 8.29).

Rattan household equipment includes mats, plates, baskets, boxes, flowerpot covers, lampshades, coasters, and cup covers. For convenience, rattan skin can be woven into a long piece of material, called a "rattan mat", which can be cut at the desired size when used.

Rattan baby strollers were popular in the middle and late twentieth century. The frame of the stroller was made of wooden sticks and bamboo, and wrapped with rattan skin. There is a rattan seat and table in the stroller, which is flexible and detachable. It shows that it is reasonably designed and pragmatic (as shown in Picture 8.30).

Picture 8.30 Rattan baby stroller in Beijing

8.4.2 Wicker Weaving

Wicker weaving is a general term that includes a variety of woven materials, such as the twigs of the red wattle, the yellow wattle, false indigo, apple tree, and willow. Among them, wicker, twigs of a willow, are the most commonly used. Various types of willows can be used in wickerwork, such as weeping willows, dryland willows, dappled willows, barren-ground willows, and sandbar willows, which are mainly found along the Yellow River Basin and the Huaihe River as well as in Liaoning province and Inner Mongolia autonomous region.

Wicker used for weaving needs to be peeled, which exposes its white core called "white wicker".

Wickerwork also requires a "base" or a "starting point" first. The most common method is the "米"-shaped warp-setting method, also known as the "米"-staked warp-setting method. To do this, you take 16 pieces of wicker in groups of 4, in which two groups cross vertically. Then, you take one group of wicker and cross them obliquely from the upper left to lower right, while crossing the other group obliquely from the upper right to lower left. At this point, there are four layers of wicker in the center (as shown in Picture 8.31). After the base has been woven, the weft strips are pressed from the outer edge of the "米" shape to the center of it. This method is called "pressing and extracting", which involves pressing one weft strip on one warp strip, followed by another warp strip on the next weft strip, etc. After two or

Picture 8.31 Shallow basket woven using the "米"-shaped warp-setting method in Shandong province

three turns, the warp can be divided. This means that the warp in the "米" shape, as well as the space between the warps, are evenly distributed. Then, the weft strips can be woven into the "米" shape. The finished product contains an equal amount of warps and wefts. This is called a "flat pattern" or "flat weaving". There is also a "cross"-shaped warp-setting method, also known as the "cross"-staked warp-setting method (as shown in Picture 8.32). In this method, 8 pieces of wicker are taken in groups of 4. The first group of 4 is punctured with a knife, then the second group of 4 is pulled through the gaps so that the two groups of wicker are crossed vertically and form a "cross" shape.

It still uses the weaving method of "pressing and extracting". After two turns, the warp is divided, which means each two warps form an angle of 45°. The warp must be divided while weaving the warp strips; otherwise, the warp strips will pull themselves straight under the pressure of its own elasticity. In addition, there are other warp-setting methods, such as the "Feng" shape, which means it's shaped like the Chinese character "丰", the sunflower shape, the twist shape, and the turtle shape, mostly determined by the shape of the wares.

Twisting method. This method includes single, double, triple, direct, and reverse twisting. The method of single twisting is that two adjacent weft strips are crossed every time they're woven into a warp strip. If you cross and then rotate it 180°, it is called double twisting. When three weft strips are twisted together and woven into warp strips at the same time, it is called triple twisting.

Picture 8.32 Wicker basket woven using the "cross"-shaped warp-setting method

A method unique to wickerwork is "the binding method", which uses hemp thread, thick string or tendon rope as the warps, wicker as the wefts, and then weaves them using the flat pattern. The binding method uses white wicker, but uses it while the wicker has not completely dried yet. When weaving, it is necessary to tighten the warp thread, rein in the wicker, or even put the warp thread into the surface of the wicker to make it concave, hence the name "binding". Tight-textured wickerwork, such as shallow baskets, dustpans, and willow pots, are all made this way. A shallow basket is a typical tightly woven piece of wickerwork. Its shape is designed for stacking and holding a variety of crops. There is also a small shallow basket with a diameter of about 30 cm, which is used to hold tobacco leaves or needles and thread (as shown in Picture 8.33).

A water bucket is also a popular item made of wicker. Also known as a "wicker pot", it is much lighter than wooden or iron barrels of the same capacity (as shown in Picture 8.34).

The wicker safety helmet was once the symbol of Chinese workers and was widely used at construction sites, workshops and in the mines.

Wicker suitcases with iron or cowhide strips woven into the waistline to make it firm yet light, were the most commonly used before.

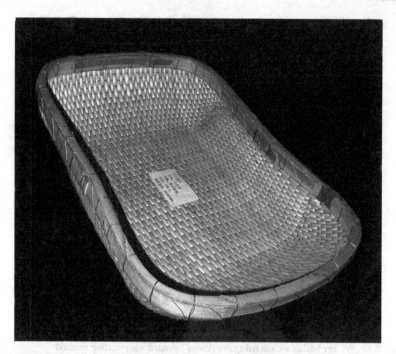

Picture 8.33 Shallow basket woven using the binding method in Beijing

Picture 8.34 Water bucket woven with wicker in Shandong province

8.5 The History of Tying

Tying generally refers to products that use bamboo and wooden frames and attach paper to it. The most typical varieties are lanterns, kites and paper models. In *List of the First Batch of Intangible Cultural Heritage* proclaimed by the State Council in 2006, there are kites, lanterns and colorful paper models, more specifically the Weifang kites of Shandong province, the Banyao kites of Nantong, Jiangsu province, the Lhasa kites of Tibet, the Xianju lanterns of Zhejiang province, the Xiashi lanterns of Haining, the Quanzhou lanterns of Fujian, the One Thousand Angle lanterns of Dongguan, Guangdong province, the Huangyuan Pai lanterns of Qinghai province, the Paper Models of Fenghuang, Hunan province, the Straw Model Carvings of Yongqing, Hebei province, the Color Cloth Ningtai of Handan, the Plastic Paper Lion Heads of Pizhou, Jiangsu province and Foshan, Guangdong province.

Since 1984, the International Kite Festival has been held in Weifang, Shandong province from the 20th to the 25th of April every year. Up to 2010, the festival had already been held for 26 consecutive years, accompanied by colorful traditional art activities, attracting a large number of Chinese and foreign enthusiasts and tourists to compete and visit. In 1989, the International Kite Museum was established in the city, which is currently the largest one in the world. Weifang kites actually drove the entire region's economy, which quickly made it spread from a local kite culture to a regional one.

When it comes to lanterns, those in the Nanjing Confucius Temple Lantern Market are the most representative ones. Since 1987, this market has been held during the Lantern Festival every year, which means that in 2010 it had already been held for 23 consecutive years. In 2007, on the 15th day of the first lunar month, 450,000 people visited this Lantern Market. Taiwanese folk lantern artists brought 7,000 lanterns to the Lantern Festival in 2010, forming the biggest lantern festival in history (as shown in Picture 8.35).

8.5.1 Lanterns

Lanterns are a unique kind of Chinese craft and festival paraphernalia. In the Han dynasty, sacrifices were made to the Taiyi deity (the emperor of Heaven and the supreme Deity of the Han dynasty) from dusk till dawn on the first lunar month's *shang xin* day, which refers to the first *xin* day (there was a *xin* day every ten days). On that day, lanterns were hung up and children would sing songs around them. This was how the Shangyuan Lantern Market began.

During the Southern and Northern Dynasties period, there were many different kinds of lanterns at the Shangyuan Festival, but the steps for constructing a standard, structurally sound one, were not created yet. In the first month of the fourth year of the Jinglong period (710) in the Tang dynasty, Emperor Zhongzong and his Empress visited the Lantern Market accompanied by thousands of palace maids, many of

Picture 8.35 Nanjing Confucius Temple Lantern Market

whom took the opportunity to run away. In the year of 713, a 20-*zhang*-high (about 67 m) lantern was built, which contained 50,000 lamps. This huge lantern was the rudimentary form of *Aoshan* lantern in later dynasties. Mao Shun, a lantern maker, made a 150-*chi*-high (50 m) colored lantern building with 30 rooms, displaying lanterns with the shapes of a soaring dragon, phoenix, tiger, and leopard.

In the fifth year of the Qiande period under the reign of Emperor Taizu (967) in the Song dynasty, the Emperor ordered the lanterns to be lit for five days during the Lantern Festival. During the Xuanhe period (1119–1125), the capital Kaifeng had a great deal of lanterns. The imperial palace even had huge lanterns built by Kaifeng government such as the Aoshan Lanterns. During that period, each lantern was a novelty. Lanes, incense shops, drugstores, tea houses, and wine shops all made their own lanterns to attract tourists, among which the "Lotus Wangjiaxiang Shop" was considered the most outstanding. In *Old Things about Wulin*, there are records of "boneless lamps", "bead lamps", "sheepskin lamps" and "horse riding figures" built during the Southern Song dynasty. In the painting *Watching Lanterns* from the Song dynasty, rabbit lanterns, melon-shaped lanterns and small lanterns can be seen, among which the square lantern was the earliest type of lantern to contain rotating shadow figures, i.e. trotting horse lantern.

In the Ming dynasty, when the capital was set in Nanjing, the imperial palace ordered the lanterns to be lit for 10 days during the Lantern Festival. During the Yongle period (1403–1424), despite the capital moving to Beijing, the traditions of the Lantern Festival were retained. Right before the Lantern Festival, a seven-floor-high archway lantern would be built in the Palace of Heavenly Purity, and a square "Aoshan Lantern" built in front of Shouhuang Hall, reaching twelve floors in height. On the day of the Lantern Festival, fireworks were lit as well. *Painting of Emperor Xianzong of Ming Enjoying Lantern Festival* depicts the scene of setting up the "Aoshan Lantern" at the imperial palace. The Lantern Market was outside the Donghua gate, stretching for one kilometer. It was at its most lively on the thirteenth day of the first lunar month. At night, the lanterns were bright and colorful.

During the Qianlong period (1736–1796) of the Qing dynasty, many lantern festivals were held at the Old Summer Palace, at which 3,000 people would hold lanterns to form the Chinese characters "太平万岁" (Tai Ping Wan Sui), meaning long life and peace. The scene of rich mansions decorated with lanterns during the Lantern Festival is depicted in Chap. 53 of *A Dream of Red Mansions*: "The window panes and portals are taken off, and all kinds of palace lanterns put up. Both the inside and outside of the building are decorated with paper lanterns and artifacts, which are embroidered, painted, piled and carved from horn, glass, yarn and silk." The Beijing Lantern Market gradually moved to the Qianmen and Liulichang areas. During the Lantern Festival, the silk shops, ginseng shops and polishing factories along these streets all put up lanterns to attract customers. National Palace Museum in Taipei contains the *Painting of Joyous Matters of Approaching Peace*, which can also be called "paintings of lanterns in the *Qing dynasty*", because they depict a variety of lanterns, such as rolling lanterns, vehicle lanterns, flower basket lanterns, Taiping drum lanterns and New Year lanterns.

In addition, ceramics, embroidery carvings and common New Year paintings all have Pictures of lanterns (as shown in Picture 8.36).

Picture 8.36 *Painting of Chetian Lantern Market* by Huang Yue (1750–1841) of the Qing dynasty

8.5.2 Kites

Some people say that the first kites were wooden kites made by Mo Zi and Gongshu Ban, while others say that they were paper kites made by Liu Bang after he got the idea from Han Xin to use it to measure the distance to the Weiyang Palace. Because both accounts are based on stories, they can be used as vague references at best.

The earliest record of Chinese kites is from the third year of the Taiqing period under the reign of Emperor Wudi (549) in the Nanliang dynasty. When Hou Jing rebelled and attacked the Taicheng city of Nanjing, Emperor Jianwen was trapped inside the city. There is an old story about Yang Che'er, who made paper crows to hide imperial orders in, and made them fly using the northwestern wind. This story indicates that kites had been used for communication about 1500 years ago. In the Tang dynasty, kites were still used for military communication. In the second year of the Jianzhong period under the reign of Emperor Dezong (781), Tian Yue attacked Linming city and the Garrison General Zhang Pi was besieged. In desperation, he used paper kites to hide urgent documents and made them fly over 100 *zhang*, or more than 300 m. Ma Sui received the documents because of this, attacked from Huguan county and thus, was able to save the besieged city.

Gao Pian in the Tang dynasty wrote a poem called *Kite*: "In the silent darkness of the night, kite strings resound. The wind, trusted by the palace and merchants to hold them up. Faintly they glide, like a song. Listen as the sound is carried by the wind, away." This is the earliest known poem about kites.

Flying kites was a very common pastime in the Song dynasty. Many youngsters often held kite competitions along the West Lake in Hangzhou. The loser of the competition had to give the winner 2–3 *liang* (200–300 g) kite line. Most of the kites used in competitions were vermilion or black, and they were reeled in and released using "medicine thread".

Painting of Peddler by Li Song, a painter in the Southern Song dynasty, contains bird-shaped kites, catfish kites and tile kites. We can see that these kites were equipped with bamboo poles with a round string at the top, which could be swung around (as shown in Picture 8.37).

Kite making and flying techniques improved in the Ming dynasty. Commonly used blue and white porcelain bowls with patterns of children playing as well as conflicting color bowls and plates contain images of children flying kites—most of them tile kites. *Fang's Ink Collection* includes a Picture named *Nine Children Ink*, in which a boy is flying a diamond-shaped kite with four tails.

In the Qing dynasty, many works of literature and poetry incorporated kites into their narrative. The kites in Chap. 70 of *A Dream of Red Mansions* are described as having the shapes of butterflies, beautiful people, phoenixes with soft wings, strings of seven geese, and an exquisite Chinese character "喜" (*xi*), which means "happy events". Two special kites, namely "food delivery kites" and "firecracker kites", were also mentioned. Yangliuqing and Mianzhu woodblock New Year paintings also depict kite-flying scenes.

Picture 8.37 A copy of the kites in Li Song's *Painting of Peddler*

8.5.3 Paper Models

Paper models are also called "color models". When they are used in funerals, they are called "burning models". To create these models, bamboo, wood strips, rattan or reed, and sorghum stalks are tied into the model's frames or skeletons. These are then covered with cloth or paper, and finally painted with colorful ink.

The earliest known paper model is the paper coffin unearthed from tomb No. 506 in Astana, Turpan city, Xinjiang in 1973. The frame was made of fine wood stalks, supported by five arc-top holders, with paper pasted onto its surface and then painted in red. The same text that was found in the tomb was written in the fourth year of the Tang dynasty's Dali period (769), which means it has a history of more than 1200 years.

The Taoist Temple of Kaifeng in the Northern Song dynasty set up the "Lone Soul Taoist Rite" during the Zhongyuan Festival on the fifteenth of the seventh lunar month on the fifteenth of the seventh lunar month, also known as the Festival of the Dead

Spirits, and burned paper model "Money Mountains", also known as joss paper, to pay homage to the soldiers who died in battle. Paper models were more abundantly present among the common people. Before the Zhongyuan Festival, markets sold paper-made "spirit money" and "colorful clothes", such as funerary objects, boots, shoes, and head scarfs. They also made triangular shelves with bamboo poles, which were then woven into "Yulan Pots". The paper-made items, or joss paper, were incinerated on these shelves in order to be given to the dead. On Tomb-Sweeping Day, paper models in the shape of pavilions were tied all along the streets of Kaifeng. The traditional "Ghost Festival" is held on the first day of the tenth lunar month. Near the end of the ninth lunar month, markets along the streets of Kaifeng would sell funerary clothing, boots and shoes to be incinerated as sacrifices. After the imperial palace of the Song dynasty moved southward, the convention of burning these paper models was kept. *A Dream of Lin'an* states that the market in Lin'an, the capital of the Southern Song dynasty and nowadays Hangzhou, had a "Shujia Paper Model Shop" and a "Xujia Paper Model Shop" on Lion Lane.

In the Ming and Qing dynasties, the joss paper models became more delicate and diverse. *A Brief Introduction to the Scenery of Emperor Capital* states: On the first day of the tenth lunar month "ghosts ask for clothing". Craftsmen in paper model shops cut five-colored paper into winter clothing and bundle it together. This bundle is inscribed with the name and seniority of the deceased and incinerated in front of the house. On the fifteenth day of the seventh lunar month, both the Taoist Zhongyuan Festival and the Buddhist Feast of All Souls are held. Important folk activities are held, such as "burning boats", in which boats are built with bamboo and wood, covered with silk, and incinerated on the embankment. This custom was observed until the founding of the Republic of China (as shown in Picture 8.38).

Nowadays, funerary customs have changed a lot. Besides the traditional joss paper models, joss paper televisions, computers, refrigerators, water heaters, houses, cars, mobile phones, and other modern items are also used.

8.6 Lanterns

8.6.1 Varieties of Lanterns

In addition to tied lanterns, there are also lanterns made of silk, straw, lotus leaves, horn, glass, metal, and wormwood. Below, only tied lanterns will be discussed.

8.6.1.1 Figure Lanterns

Figure lanterns can be made in various shapes, such as dragons, lotus flowers, flower baskets, pavilions, dramatic scenes, and figures in ancient costumes.

8 Weaving and Tying

Picture 8.38 Burning boat in *Wu You's Picturesque Treasure*

Nie Fangjun is well-known in Hunan province for creating figure lanterns. His most distinctive lanterns are marine animal lanterns (as shown in Picture 8.39).

The Nanjing Lantern Market is located around the Confucius Temple on the north bank of the Qinhuai River. The main varieties seen there are lotus lanterns, carp lanterns, rabbit lanterns, and trotting horse lanterns. Unique to the Confucius Temple is their toad lanterns. To make them, an oval-shaped bamboo circle is tied into a three-dimensional structure and a T-shaped bamboo frame is attached. Then the strings are lifted and the bamboo frame twisted, which will cause the toad to stretch its legs, making it look more lifelike. Aircraft lanterns are a new type of figure lantern. A small windmill is installed onto the front of the fuselage, representing the propeller (as shown in Picture 8.40).

8.6.1.2 Needle-Punched Lanterns

Needle-punched lanterns are mostly produced in Xiashi, Zhejiang province. In the late Qing dynasty, Yan Yuanzhuang and his son, Yan Shaozhuang were famous for it. Other craftsmen were known for making calligraphy lanterns, like Fang Tanyuan, or

Picture 8.39 Crab lantern made by Nie Fangjun

for making goldfish and carp lanterns, like Huang Shiqin. In order to make a needle-punched lantern, the frame is made with bamboo strips and covered with lantern sheets. The so-called lantern sheets are pieces of *xuan* paper pasted on the frame. This paper has holes punched in it with needles to form the main pattern, after which the same process is repeated for the background color with closely packed pinholes. Nowadays, the manufacturing process of these lanterns has improved. First, holes are punched in colored paper. Then the paper is folded and each shape is connected to each other to form a three-dimensional shape without the need of a frame. Therefore, it is called a "boneless lantern" (as shown in Picture 8.41).

8.6.1.3 Palace Lanterns

Palace lanterns were originally just palace lighting fixtures, which later became popular among the common people. Pine, willow or sorghum stalks were commonly used for the frame. The lamp surface was made of *xuan* paper, gauze, silk, or flat glass, which was in turn painted with a wide variety of content, such as figures, flowers, birds, and landscapes. Palace lanterns are divided into pentagonal, hexagonal and octagonal shapes. They are divided into two layers: top and bottom. The first layer is called the lantern canopy. The point sticking out of the canopy is usually in the shape of a dragon head or crested head. The second layer is the lamp body, which consists of various patterns made by paper cutting (as shown in Picture 8.42).

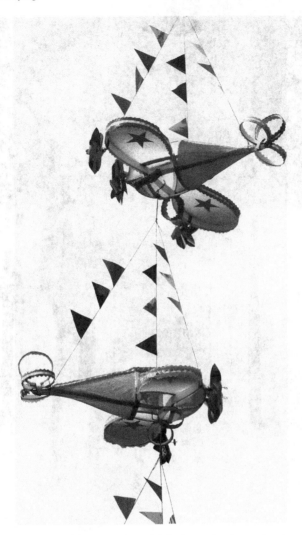

Picture 8.40 Aircraft lanterns found at the Confucius Temple Lantern Market in Nanjing

8.6.1.4 Trotting Horse Lantern

These lanterns are made with thin wood strips or sorghum stalks. It has an axis in the middle installed onto a pinwheel. The papercut figures or horses can be twisted onto the axis with iron wire or bamboo strips. Candles are burned underneath so that the heated air rises up and makes the pinwheel drive the vertical axis. This is how the papercuts trot or revolve (as shown in Picture 8.43). There also exists a trotting horse lantern that uses movable pictures. The heads, hands and feet of the paper figures are made separately. Thin iron wires attach them to the surface of the

Picture 8.41 Boneless lantern in Xiashi, Zhejiang province

lantern. A shifting block is connected to the vertical axis with iron wires. When you turn the shifting block, the figures will shake their heads, raise their hands and move their feet. According to this principle, a wide variety of trotting horse lanterns can be created.

8.6.1.5 Kongming Lanterns

Kongming lanterns, also known as flying lanterns, were first recorded in the beginning of the Qing dynasty. They are lanterns that use the same principles as hot air balloons.

Picture 8.42 Palace lanterns

When the lanterns are lit, the air becomes heated, the air density decreases and the lantern expands until it begins to lift off. During the Lantern Festival, people in the coastal fishing villages of Fujian province have the custom of flying Kongming lanterns with a diameter of nearly one-meter and a tail of four meters. Dozens of Kongming lanterns taking off into the sky at the same time, soaring upward, is a magical sight to behold.

Kongming lanterns are set off on the night of the Songkran Festival in the regions of the Dai people, Xishuangbanna, Yunnan province. The frames of these lanterns are woven with thin bamboo strips. These are then covered with 48 sheets of corrugated paper, which makes it look like a big paper ball with a large hole underneath, and four ropes are then tied around that hole. Wooden strips are then used to create a big cross, which, in turn, is twined with cloth strips soaked in oil. Many people lift the paper ball and hold it above the bonfire. When the paper ball starts to take off, the wooden "cross" is ignited, quickly hung underneath the paper ball using the four ropes, and let go at the same time. The paper ball will take off slowly using the lighted "cross".

Picture 8.43 Trotting horse lantern made by Zhang Shuangzhi in Beijing

8.6.2 The Craft of Making Lanterns

8.6.2.1 Frames

Most lantern frames are made of bamboo. A special process is needed to properly bend bamboo strips. First, soak bamboo strips in clear water for a few minutes, then heat the bamboo with the fibres underneath a lamp so that the heated bamboo strips become soft and pliable. The next step is to finalize the shape. Build a solid shape according to the desired structure, and tie up the bamboo slats with twine or cotton thread where they intersect and overlap.

There are four commonly used connection methods: One is the overlapping joint method, in which the two ends overlap and are bound with threads. The second is the oblique joint method, in which two ends of the bamboo strips are cut out with

8 Weaving and Tying

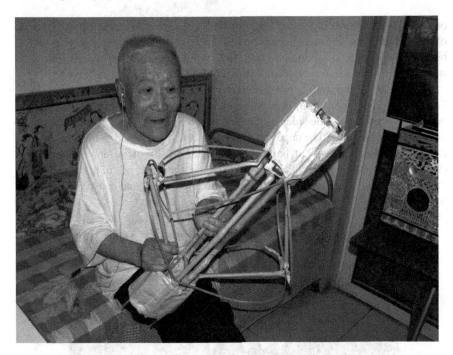

Picture 8.44 Huang Songqing, an over-90-year-old descendant of the Beijing lantern craft

an oblique surface, after which they are connected and tied together. The third is the mortise and tenon joint method, where a mortise is cut out at one end of each bamboo strip and a tenon at the other. Both ends are then plugged together after being rubbed with glue. The fourth is the casing joint method. First, a casing is made of paper and cloth or of iron and copper. Then two butted bamboo strips are inserted at both ends of the casing. After binding them together, apply a little glue to avoid any loosening and dislocation (as shown in Picture 8.44).

8.6.2.2 How to Craft a Trotting Horse Lantern

It is best to use sorghum stalks to tie up the frame of the trotting horse lantern and to paste paper on the stalks to serve as the surface of the lantern and to prevent the stalks from cracking. The length of the horizontal and vertical stalks should be equal. It is suggested that each set of these stalks should be cut at the same time. In addition to applying glue, the joints of each stalk must be fastened with bamboo sticks. Any place where three stalks intersect should be separated to avoid the bamboo sticks from tangling. After the frame is finished, the pinwheel at the upper end of the vertical axis should be made first. Take a round piece of paper and cut out a number of small triangles pointed to the center of the circle. Fold each triangle down to form a pinwheel. The rotation speed of the pinwheel can be adjusted by changing the angle of the folds (as shown in Picture 8.45). Install a steel needle at the top and bottom of the vertical axis. The steel needle is then placed onto a small piece of glass at the bottom of the lantern to reduce friction. In the middle of the vertical axis, thin iron wires or thin strips are used to fasten papercut silhouettes and paper horses.

The content of the trotting horse lantern is concentrated on the lantern's surface. To ensure a clear projection, thin paper is used to cover the outside frame of the lantern. However, only the sides of the lantern are covered. The top and bottom are not covered with paper. The papercut figures and horses attached to the vertical axis can be either painted or cut out, but the shapes on the outside of the lantern can only be cut-outs. The candles are installed on both sides of the lower beam, juxtaposed with the vertical axis. After being ignited, the warm air will rise and cause the pinwheel to rotate.

Picture 8.45 The vertical axis of a trotting horse lantern

8.7 Kites

8.7.1 Types of Kites

8.7.1.1 Classification by Subject

There are bird-shaped, insect-shaped, fish-shaped, human-shaped, Chinese character-shaped, equipment-shaped, and geometric-shaped kites.

8.7.1.2 Classification by Structure

(1) Hard Wing Kite. The two wings are composed of upper and lower horizontal bamboo strips, which are placed perpendicular to the body. It can neither be folded nor disassembled. The Beijing "swallow" is the most typical example (as shown in Picture 8.46).
(2) Soft Wing Kite. The wing only has one bamboo strip, and the lower outline is made of cloth, which floats with the wind. Common shapes include eagles, butterflies, dragonflies, and swallows (as shown in Picture 8.47).
(3) Bat Kite. It is shaped just like a flat board. It can be divided into the "soft bat kite" and the "hard bat kite". The outer frame of the hard bat kite is made of bamboo strips. The whole frame has a regular flat surface and does not need to be tied with string. It needs strong winds to fly (as shown in Picture 8.48). The lower part of the soft bat kite is not supported by bamboo strips, and it is bent into a backward bow shape when flying.
(4) Long Stringed Kite. It is commonly called "centipede" since it is basically a long string of connected round pieces with a dragon-shaped head in front. This is why it is also known as a "dragon kite". Every round piece is called a "centipede round sheet". Each sheet has a thin bamboo strip, commonly known as "centipede legs", which needs to be tied with chicken feathers or paper scraps at both ends.
(5) Detachable Kite. Components are manufactured separately and assembled before flying. The production process is more complicated, since the shape and how each part is connected needs to be carefully designed. It is commonly connected with accessories such as mortise and tenon joints, bolts, sleeves, and hinges. The most famous detachable kite is from the "Wei family of Kite" in Tianjin (as shown in Picture 8.49).
(6) Simple Kite. It refers to the kites made by common people for the main purpose of flying. They do not pursue the perfection of shape and color. Despite some being made of waste paper, they can still be set to fly. Common shapes include "tiles", "diamonds", and "delta wings".

Picture 8.46 Hard wing kite "swallow" in Beijing

8.7.1.3 Special Kite Attachments

The special attachments of kites include sounding, luminescence and "kite touch".

The sounding attachments include wind organs, whistles and wind drums. A wind organ is made by bending bamboo into the shape of a bow and tying strings through it made of silk thread or bamboo chips, commonly known as "wind organ ribbons". When the kite is flying, the wind blows on the strings and makes a sound. When it comes to whistle attachments, the "*banyao*" kite in Nantong, Jiangsu province is the most exemplary one. A big *banyao* can be installed with more than 100 whistles of different sizes. The big whistles are made of gourds, and the smallest whistles are made of ginkgo pits. These whistles are tied to the windward side of the kite (as shown in Picture 8.50).

Another kind of sound device is literally called "backside gongs and drums". It uses bamboo strips to form a frame on which small gongs, drums, paddles, and drumsticks are installed, which are then driven by wind wheels. When the wind wheel rotates, it drives the paddles to tap the drumsticks, which then hit the gongs and drums (as shown in Picture 8.51).

Picture 8.47 Soft wing kite "butterfly" in Nantong, Jiangsu province

The "kite touch" attachment, also known as the "food delivery" attachment, consists of a bracket, a "lunch box", a "wing" and a trigger mechanism. The attachment is first tied onto the lower end of the kite line. Using the wind, it then rises along the line until it touches the kite bar, triggering the mechanism, which opens the box and makes it rain confetti. At the same time, a rubber band makes the two wings close again, which, in turn, causes the "lunch box" to slide down along the kite line and back into the hands of the kite runner.

In the past, the lanterns hanging on the flying kites at night were mostly homemade little red lanterns, which were tied to the kite line and soared with the wind. Nowadays kites can be equipped with light-emitting diodes and batteries, connected in a string in the night sky, shining and magical.

8.7.2 How to Build a Kite

Traditional kites use bamboo for their frame. Contemporary kites use a wider variety of synthesized materials for their frames, such as aluminum alloy tubes and strips, carbon fibre tubes and glass fibre.

Picture 8.48 Hard bat kite "fishing boy" in Weifang, Shandong province

8.7.2.1 Structure

Kite structure follows two principles: First, it needs to be designed to be as compact as possible and the decoration should be as simple as possible. Second, it should be made as light as possible, using the least amount of bamboo necessary. The size and proportion of various parts of a well-designed kite are fully optimized and stabilized. Take a Beijing "swallow" kite as an example: The whole body is square. The length of the head is one-fourth of the total length, the abdomen two-fourth and the tail one-fourth. One-seventh of the total width of the wings equals the width of the abdomen. This way, as long as the widths of the wings have been determined, the dimensions of all the other parts can be easily obtained.

Picture 8.49 Detachable kite "eagle" in Beijing

8.7.2.2 Trimming Bamboo Strips

Bamboo strips used for making kite frames should be smooth, tough, and uniform in thickness. To achieve this, put the split bamboo strips flat on a chopping board, with the green bamboo facing down and the bamboo culm, the "meat", facing up. Hold the knife with your right hand and pull the bamboo stick with your left hand to scrape the bamboo culm off. Repeat this process to make it gradually thinner. It can also be planed with a carpenter's plane. The edges of the bamboo strips are very sharp, which causes it to not only cut your fingers, but also the twine or the fabric. File off the edges with glass or porcelain fragments, or polish them with sandpaper to make them smooth.

In order to ensure that the two wings of the kite are symmetrical and balanced, the wide bamboo strips are usually cut and trimmed and then split, so that the length, thickness and shape of the two bamboo strips, even the positions of the bamboo joints, are exactly the same.

8.7.2.3 Bending the Bamboo Strips

Many kite parts are made by bending bamboo strips and shaping them. First, the predetermined shape is drawn on paper or a wooden board, this will serve as the "pattern". Bamboo strips are then baked with an alcohol lamp or a candle by having the bamboo culm face the fire. When it becomes soft, it is gently bent and continuously adjusted to fit the pattern.

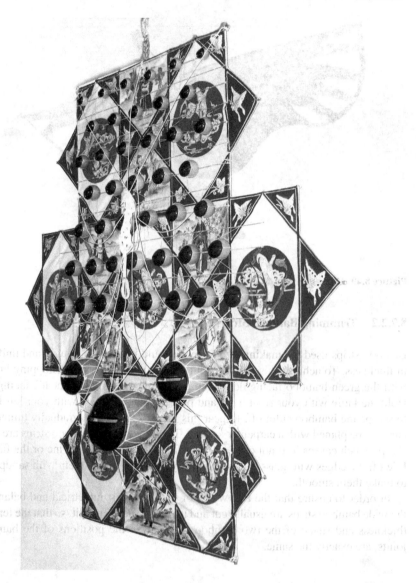

Picture 8.50 *"Banyao"* kite in Nantong, Jiangsu province

8.7.2.4 Binding

There are two types of kite frame binding: joint binding and cross binding. Common methods of cross binding are: (1) The Round Tenon Cross A round hole is drilled in the middle of a bamboo strip, the end of another bamboo strip is cut into a cylindrical shape, glued and inserted into the round hole. This method is also known as "cut and

Picture 8.51 "Backside gongs and drums" kite

insert". (2) The Mortise and Tenon Cross Each of the two bamboo strips has a gap about half its thickness, which perfectly makes one fit the other, while its surface remains flat. (3) Splint Cross A thin wooden board is attached to each side of the bamboo strip, and another bamboo strip is pressed between the two boards, glued tight, and then it's all bundled with thread.

8.7.2.5 Pasting

It is commonly known as "masking". When silk is used as fabric, it is necessary to "hang pulp" to make it flat, tough and convenient for processing. Be it paper or cloth, they both should be laid flat first and then the kite frame is put on top of it. The kite shape is drawn on the fabric by tracing the frame's outline with a pencil. Then the frame is taken off and, using scissors, the pattern cut out about 1 cm outside the pencil line.

As the cut fabric is facing down, the front side of the frame is coated with paste or latex to adhere to the fabric and fix the two together. Select the force-bearing part of the kite, make a small incision that starts 3–5 cm from the outside material, then fold and stick this outside material onto the frame. This is commonly known as "packing the edge". A medium-sized kite commonly has 5–6 packing edges. After the kite has dried, all the redundant edges are cut off, which is commonly known as "cleaning the edges".

8.7.2.6 Painting

Painting is an important step in the four procedures of kite creation, "tying, pasting, painting and putting", and its procedure is unique or at least very particular to a type of kite, which makes every kite a beautiful decorative painting.

Chinese painting pigments are the first choice for painting, because they are consistent, resistant to fading, and have a good pay-off. Amateurs can also use commercial pigments and watercolor pigments to make multicolored kites. Thin brushes are used for outlines or patterns, thick brushes are used to paint the block surfaces, and wall brushes are used if there are larger blocks. There are two methods of painting. One is painting before pasting and the other is painting after pasting. After painting, the fabric may be wrinkled and in need of being ironed flat.

Some semi-three-dimensional shapes are not suitable for painting before pasting, so they are painted after pasting. In order to do this, the fabric is first pasted on the frame, commonly known as a "blank canvas", and then painted. Painting is an art. Only through study and practice can a beautiful kite be created (as shown in Picture 8.52).

8.7.3 The Principle of Flying a Kite

In order to fly, kites must overcome gravity. Therefore, they must rely on wind or airflow to produce a "difference in pressure". The pressure difference is related to the pulling force of the lead line and the windward angle of the kite. The kite forms a windward angle against the airflow under the traction of the line. The airflow is separated from the upper edge of the kite, which generates turbulence on the back of the kite. As the flow rate increases, the pressure lowers. Once the airflow gets blocked by the underside of the front of the kite, the pressure becomes higher than that on the back, thereby generating lift.

Flying kites does not mean it just hangs and stays in the air. It is necessary to master some skills and use "reel-in" and "reel-out" techniques to keep them aloft. There are three commonly used methods to make a kite take off: The first method is called fixed-point releasing. You stand still or move only a little, with your back facing the wind. Hold your line-warping equipment in one hand and lift the line half a meter away from the kite in the other hand, lifting it above your heads. When you feel the wind, you pull it back, and the kite will take off pulled against the wind.

Picture 8.52 Making a kite

Once you feel the tension on the line, you reel out the line, and the kite will continue to rise. When the pull becomes weaker, reel it in again. It is necessary to reel it in quickly but release it slowly. After several times of repeating this process, the kite will be flying high in the sky.

The second method is called double fixed-point releasing. One person stands with their back to the wind holding the line-warping equipment while another person, standing 20 m or more away, holds the kite in their hands. When the latter feels the wind pick up, they release the kite and call the former to reel in the line. The one holding the line immediately reels the line in as quickly as possible, making the kite rise and fly towards them. When it reaches a certain height, they stop reeling the line in, pull it backwards, and reel it out when they feel a pull on the line. This is repeated until the kite has attained a high altitude.

The third one is running and releasing. If the wind on the ground is too weak, it can be strengthened by running against it with the kite held up. If the right side is windward, the right hand holds the kite, and the left hand holds the kite line two meters away from the kite. While running into the wind, look back and observe the kite. When you feel the wind pull on the kite, reel out the line while you keep on running. When the pull weakens, stop reeling out and continue running. After the kite reaches a certain height, you stop and then pull on the line. When the kite has reached the higher-altitude winds, you can pull back the line and slowly move back with the wind. Keep it aloft while walking, you can go back to your starting position (as shown in Picture 8.53).

Picture 8.53 *Ten Beauties Flying Kites*, a Yangliuqing New Year painting

8.8 Paper Models

8.8.1 Categories of Paper Models

Paper models used for funeral etiquette are also called "burning models" or "burning paper models", and its subjects can be divided into the six categories below.

8.8.1.1 Ghosts and Gods

According to Chinese legend, the ghost that can carve out a way for spirits to go is called *fangxiang*, who was posted in front of funeral processions to frighten away all kinds of demonic evils. There are many Buddhist gods, such as Guanyin, Manjusri, Maitreya, and Skanda, but Taoist gods as well, such as the Officer of the Living World, the Officer of the Netherworld, the Earth God and the Dragon God (as shown in Picture 8.54).

8.8.1.2 Figures

The most commonly used figures are Jintong, the golden boy, and Yunu, the jade girl, as well as figures from famous dramas. Jintong holds the banner and Yunu

8 Weaving and Tying

Picture 8.54 Paper model of the Officer of the Netherworld, made by Nie Fangjun in Hunan province

lights the way, both leading the way for the deceased to the western netherworld. "Figures from famous dramas" are the characters from dramas and stories. The most exciting figures in dramas are produced in Caoxian county, Shandong province, in Phoenix county, Hunan province, and in Putian, Fujian province. They are often used to display filial piety, such as the figures in the drama *Third Madam Teaches Son*.

Picture 8.55 Paper model of chariot, produced in Junxian county, Henan province

8.8.1.3 Horses and Chariots

Horses and chariots are the main items used for burning and ranges from first to third class. The first-class chariots are made true to size—they can even be pulled and walked—while the third-class chariots are the simplest, usually measuring <3 ft. The wheels and the drivers are roughly cut out of one single piece of paper, painted a little, and pasted on the chariots. The economic status of bereaved families can be seen from the scale of the paper models and the class of paper chariots and horses (as shown in Picture 8.55).

8.8.1.4 Buildings and Warehouses

"Buildings and warehouses" is the collective name for the palaces in the netherworld. The netherworld buildings are modeled after palaces with black-and green-tiled double-eaved roofs. They are divided into upper and lower layers that can be easily tied and pasted together. They are assembled once they arrive at the incineration site. The netherworld warehouses have multi-eaves *xieshan* roofs. Standing in front of the warehorses are the figures of the Officers of the Living world and the Netherworld. Burning models are still popular in the countryside, and there are even paper models that are almost two-meter high and can even be entered. The binding

Picture 8.56 Paper model of building, made of printed card paper in Hebei province

process has changed over time. Now, the doors, windows and walls are simply printed on cardboard and assembled without binding frames (as shown in Picture 8.56).

8.8.1.5 Animals

The paper models of animals are mainly models of poultry, livestock, rare birds, and exotic animals. Paper horses and paper pigs symbolize wealth, while lions and cranes symbolize good fortune. If the deceased person is female, a paper cow is added. According to folklore, a woman consumes too much water during her lifetime. When she arrives in the netherworld, the King of Hell will punish her by having her drink a lot of water. The paper cow can drink the water for its master.

8.8.1.6 Utensils

Paper utensils can include golden and silver towers, potted flowers, the four treasures of study, silk, satin, and antiques. The 92-year-old Beijing paper model artist Huang Songqing said: "When it comes to the models used for burning, you can make whatever you want. Anything in the real world can be made into a paper models."

8.8.2 The Making of Paper Models

In the past, a workshop specializing in paper models was called "Clothing Store for the Netherworld". The master of the shop could also make paper buildings and lanterns. Nowadays, there are only a few of these workshops left. These workshops make paper models with a mix of characteristics, blending home and abroad, and past and modern times.

8.8.2.1 Frames

Bamboo, wood, sorghum stalks, and straws are used to build the frames. With the large volumes and complex shapes, paper models require many kinds of materials. For example, when making paper figures, wooden sticks are used to make the torso, bamboo or stalks are used to make the arms, legs and feet, and bamboo strips are used to make the shoes, hats, and belts. In order to increase the volume, rice straw or wheat straw is tied onto the frame. The materials need to meet a certain standard. Craftsmen need to ensure that the frame is strong and not loose, easy to burn and not expensive.

The way of binding a paper model's skeleton is similar to that of a lantern, with twines or cotton thread at the joints. For large paper models, ropes are used to pull everything taut. Then small wooden sticks are inserted through the ropes and the ropes are twisted to tighten the two ends.

8.8.2.2 Paper Types

The paper used is divided into three types: plain paper, colored paper and decorative paper. Plain paper is used for the initial shape, as a way of laying the foundation. Colored paper is used for large-area decoration, such as clothes, walls and utensils. Colored paper is also used for cutting, engraving and pasting specific parts, such as building doors and windows, tiles, robes, boots, and hats.

There are two types of decorative paper. One is factory-produced deco paper, such as "eight blender" (paper with regular patterns) and "grid brocade" (paper with regular grids). The other is flower paper printed by artists themselves, which can be pasted on requiring a minimal amount of cutting. They can save more time and effort than manual painting.

8.8.2.3 Making Faces

The surface of the head is a pulp mask made from mold. First, the head's surface is shaped with yellow mud and coated with soapy water. Then, three or four layers of soaked paper are pasted on the mold. When it is semi-dry, the clay mold is taken out, and then two or three layers of paper are pasted on the inside to become a "workblank". Then white powder is brushed onto it, followed by adding the skin color with paint, the outline of the eyebrows and lips with ink, and rouge powder applied to the cheeks with cotton. Straws are used to tie the mask onto the head. A hat made of colored paper is added as well. When assembling the paper model, the head is attached firstly, and then the clothes are pasted to the body.

Chapter 9
Lacquering

Zhou Jianshi and Hua Jueming

Ten Defects in *Han Fei Zi* states: "During the reigning period of Yao, food utensils were lacquered with black paint. During the reigning period of Yu, sacrificial vessels were lacquered with black paint on the outside and red paint on the inside". China had used natural lacquer to decorate objects as early as the late Neolithic Age. After a long-term evolution, lacquerware has become widely used, lacquer art highly developed, and lacquer culture brilliant and enduring, thus becoming an important part of Chinese culture.

Lacquering decoration is the ancient name for painting. "Lacquering" meant painting, while "decoration" meant drawing, embedding, engraving, grinding, stacking, and engraving. The skill and beauty of lacquering still enriches people's lives both materialistically and spiritually.

9.1 The History of Lacquering

The earliest known lacquerware in China is the vermilion lacquer wooden bowl unearthed at the Hemudu cultural site in Yuyao, Zhejiang province, dating back to about 7,000 years ago (as shown in Picture 9.1). Painted lacquer pottery, painted lacquer ware, and jade-inlaid lacquer cups were also unearthed from other Neolithic sites such as those of the Majiabin culture, Liangzhu culture, Daxi culture, and Qujialing culture, which indicates that lacquer art was more widely spread and also incorporated a variety of techniques.

Z. Jianshi (✉)
Academy of Arts & Design, Tsinghua University, Beijing, China
e-mail: zjsdaqi@yahoo.com.cn

H. Jueming
The Institute for the History of Natural Sciences, Chinese Academy of Sciences, Beijing, China

© Elephant Press Co., Ltd 2022
H. Jueming et al. (eds.), *Chinese Handicrafts*,
https://doi.org/10.1007/978-981-19-5379-8_9

Picture 9.1 Vermilion lacquer wooden bowl unearthed at the Hemudu cultural site in Yuyao, Zhejiang province

Tribute of Yu in *Book of Documents* states: "Between the Ji River and Yellow River lies the Yanzhou district [...] Their tributes are lacquers and silks", "Between the Jing Mountain and Yellow River lies Yuzhou [...] Their tributes are lacquers [...]". A large number of different kinds of lacquerwares seem to have been used during the Xia and Shang dynasties. *Local Officials* in *Rites of Zhou* states: "If the private workshops produce lacquer, the government will levy a tax equal to a quarter of the earning". The phrase "lacquering decoration" is first seen in the *Office of Spring* in the same book. During the Spring and Autumn period and the Warring States period, lacquer art was widely used in sacrificial vessels, tableware, vehicles, palaces, coffins, and shields.

The Warring States period to the Han dynasty saw the first peak in lacquerware production and decoration skills. The main areas where lacquer was produced, as mentioned in *Classic of Mountains and Seas*, are Jing Mountain, Guo Mountain, and Yiwang Mountain.

According to *Records of the Historian*, Zhuang Zhou (Zhuang Zi) used to be a "lacquer garden official", "thousands of wooden wares were painted [...] thousands of buckets of lacquer", and "thousands of acres of lacquer trees planted in Chen and Xia... the officer in charge of lacquer is equal to [...] a Lord". It can be seen that lacquer trees were planted on a large scale at that time, and lacquer manufacturing became an independent manual industry.

Chu State was rich in lacquer trees. Compared to the Central Plains, its artistic creation was more dynamic and the lacquerware were creatively richer with stretched ornamentation and smooth lines. The frequent use of openwork engraving and embossing greatly enriched the artistic expression, while tenon joints and bonding made the shapes more complex and delicate. For example, the painted seat screen from Jiangling, Hubei province, and the lacquer plate from Suixian county can be called the representative works of lacquer art of this period (as shown in Picture 9.2). In the Qin dynasty, there was a so-called "lacquer canal" near Xi'an, which was said to have been excavated to transport Nanshan lacquer. The famous Lishan Qin figurines were all painted with lacquer.

Picture 9.2 Painted woodcarving seat screen from the Warring States period, unearthed in Jiangling, Hubei province

In the Han dynasty, the lacquer industry prospered more, so the government enabled craftsmen to specialize in lacquer production in Shu county and Guanghan county. The inscription on the double-ear handle plate, made in Guanghan county in the fourth year of the Yuanshi period (4) in the Western Han dynasty, implies that at that time, there were craftsmen specializing in painting, drying, engraving, inlaying, gilding, drawing, and polishing. The degree of specialization was very high and the production procedures and techniques fully developed. The coffin with colorful lacquer clouds and strange patterns unearthed at the Mawangdui Han tombs in Changsha are so colorful and rich that it is able to visualize momentum and endless movement (as shown in Picture 9.3). With the increase of the variety and quantity of lacquerware, the manufacturing process greatly expanded as well. The application of ramee-lacquer and thin wooden-lacquer made the shapes more diverse. The techniques of embossed lacquering and engraving golden lines also appeared. The Northern and Southern dynasties period inherited the legacy of the Han dynasties and built on it. For example, Cao Cao's *On Sundries* listed items such as silver lacquer cases, lacquer boxes, lacquer pillows, and lacquer mirrors. Lacquerware with leather bases and bamboo bases were found at Zhu Ran's tomb in Ma'anshan, Anhui province. Techniques such as lacquer tracing and cloud engraving were used with colors such as vermilion, black red, light gray, dark gray, gold, and brown. The lacquer paint is rich in content, reflecting the scenes of real life at that time, such as feasts, hunts, and acrobatics. After the Eastern Han dynasty, Buddhism prospered, and ramee-lacquers were used on Buddha statues. They were used for their light weight and high quality. The *Luoyang Jialan Record* states that the huge Buddha statue, when exhibited, drew tens of thousands of visitors.

Picture 9.3 Coffin with colorful lacquer clouds and strange patterns in the Han dynasty

Ramee-lacquers continued to prevail in the Tang dynasty. In the second year of Tiabao period (743), Jianzhen, the eminent monk, crossed the sea to Japan accompanied by masters such as Yijing and Si Tuo. They introduced this technique to Japan. The 3-m high Vairocana Buddha statue has a bamboo base and was coated in 13 layers of cloth and paint, which is why it is still in good condition now. *National History Supplement* reads: "People in Xiangzhou prefecture are so good at lacquering that their exquisite lacquering patterns are called Xiang patterns". The prefectures of Pu, Xing, Wu, Taiwan, and Jin were also important in the production of lacquerware in the Tang dynasty. The creation of burnished work with gold or silver inlay shows the prosperity of the Tang dynasty and is rich with characteristics of that era. It is inlaid with pieces of gold and silver and engraved with fine patterns, which were repeatedly painted and sanded to even them out, which make the lacquerwares gorgeous, noble, and beautiful (as shown in Picture 9.4). *Youyang Notes* and *Anecdote Novel of Tang Dynasty* record dozens of lacquerwares that were bestowed upon An Lushan by Emperor Xuanzong in the Tang dynasty. Engraved lacquering was also applied during this period. The helmet and armor from the Tang dynasty, found in Milan, Italy, is made of camel skin, coated with multiple layers of black, red, and brown lacquer, and engraved with dot patterns to expose the red on the surface of the black lacquer, resulting in an interesting style.

There were many lacquer workshops in the Song dynasty, such as the Youjia Lacquer Shop on Liren Lane and Qi's Cloud Engraving Shop near Qinghu River in Hangzhou. The imperial court set up an exclusive agency, imperial kitchens, to make lacquer bowls and dishes used by the royal family. Common, everyday lacquerware's most winning feature is its shape, not its decoration. However, the craftsmanship is very particular. The bases and lacquer are painstakingly crafted. Although it is plain,

Picture 9.4 Burnished work with gold or silver inlay in flower and bird patterns from the Tang dynasty, kept at the National Museum of China

it is refined and durable. It is known as "one-color lacquerware". The widespread use of lacquerware also led to the emergence of a business that specializes in repairing lacquerware.

The most distinctive feature of this period is their circularly overlapped framework. Cut good wood into strips; soften it with hot water; stack it from small to large, bond, and dry; and trim the edges and corners to form the shape. Its biggest advantage is that the base can be made very thin, but the joints can be staggered to disperse the stress, and its firmness is far beyond the ability of engraving or combining from a solid piece of wood. It is a great example of simple, ancient mechanic ingenuity (as shown in Picture 9.5). The engraved lacquering techniques using cinnabar, black, and cloud engraving reached a very high level in the Song dynasty. Gao Lian's *Eight Notes for Health* states: "Most of the boxes used in the palace are made of gold and silver, piled up to dozens of layers with vermilion lacquer, and engraved with figures, terraces and flowers". He also praised it with "exquisite knife work, engraved skillfully". Engraved lacquering became an important way to decorate lacquerware in the Yuan dynasty. The representative figures of that time are Zhang Cheng and Yang Mao. *Annals of Jiaxing Prefecture* reads: "Zhang Cheng and Yang Mao, born in Yanghui, Xitang town, Jiaxing, are best known for engraved cinnabar lacquering". Their works are kept in Japan and in the Palace Museum in China. For example, Zhang Cheng's gardenia pattern lacquer plate (as shown in Picture 9.6) was lacquered nearly 100 times, almost 1 cm thick and engraved in high embossing. Red and black lines are exposed on the inclined plane. The lacquer is hard and the workmanship is exquisite.

Ming and Qing lacquer art was built on the legacy of the previous dynasties, presenting a situation of mutual overlap yet great ambitions. During the Ming dynasty's Hongwu period (1368–1398), there were more than 5,000 registered painters at court, and Zhang Degang, son of Zhang Cheng, and Bao Liang were awarded the post of deputy calligrapher. The Qing dynasty Manufacturing Office set up a lacquer workshop. The most famous areas for producing lacquerware were

Picture 9.5 Purple brown lacquer lamp from the Song dynasty, kept at the Nanjing Museum

Picture 9.6 Gardenia pattern engraved cinnabar lacquer plate made by Zhangcheng in the Yuan dynasty, kept at the Palace Museum

Jiaxing, Yangzhou, Fuzhou, and Luling, where masters appeared in large numbers and branched into different schools. Huang Dacheng and Fang Xinchuan in Huizhou, Yang Ming and Zhang Degang in Jiaxing, Jiang Qianli and Zhou Zhu in Yangzhou, Yang Wei and Yang Shilian in Changshu, and Shen Shaoan in Fuzhou were, respectively, famous for engraved cinnabar lacquering, mother-of-pearl inlay, eight treasures inlay, gold tracery, bodiless lacquering, and thin lacquering with gold and silver. During the Yongle period (1403–1424), imperial lacquer workshops were set up in the district of Guoyuanchang and lacquerers were recruited from Yunnan. The influence and popularity of lacquer extended to the Qing dynasty: The Imperial Palace, Buddhist temples, garden houses, and even the covers, window lattices, partitions, columns, plaques, and couplets of residential houses were all decorated with lacquer art. *Yangzhou Original Boat Record* contains a detailed description of the lacquer art process and materials. Another type of lacquering is furniture lacquering. The Ming dynasty's *Antique Authentication* divides it into seven types: cloud engraved lacquering, engraved cinnabar lacquering, piling red lacquering, gold inlay, drilling cloud lacquering, and mother-of-pearl inlay. But, in fact, hundred treasures inlay, gold tracing, filling lacquering, marble inlay, bamboo inlay, and other techniques were also used by craftsmen, depending on the conditions of the materials as well as the customer (as shown in Picture 9.7).

Colored engraved lacquering is a new technology that appeared in Ming dynasty. This entailed first lacquering the wood, then engraving the patterns into it and finally filling the engraved parts with various colors. Now it's also called "engraved lacquering" for short. During the Qianlong's period (1736–1796), Shen Shao'an improved the bodiless lacquering techniques by replacing the thick linen with summer cloth and silk, thus creating "Fujian bodiless lacquerware", which is delicate, light, and

Picture 9.7 Square engraved lacquer table from the Qing dynasty

coveted all over the world. Hundred treasures inlay became one of the representative crafts of the Ming and Qing dynasties (as shown in Picture 9.8). Famous craftsmen included Zhou Zhu, Lu Yingzhi, and Lu Kuisheng. Lu Yingzhi was also good at making lacquer sand inkstones. His grandson Lu Kuisheng still cooperates with scholars to make lacquer inkstone boxes and arm rests that are simple, elegant, and smooth. All these study utensils display a refined artistic taste.

The publication of *Lacquering Decoration Record* in the Ming dynasty was a major event in the history of lacquer art and the highest achievement of lacquer art of this period (as shown in Picture 9.9). Its author is Huang Cheng, with the pseudonym of Dacheng, from Pingsha district, Xin'an town, Huizhou city. *Antique Authentication* and *Collections of Qing dynasty* state that "his engraved cinnabar lacquer is comparable to the techniques of Guoyuanchang", and that he is famous for his "round and clear engraving method". The book is divided into two series, *Heaven and Earth*, with 18 chapters and 186 articles. Starting with raw materials, tools, and equipment, it goes on to cover a wide range of topics, such as methods, principles, taboos, and classifications of lacquerware production, techniques used in various lacquerware, repair, and antique. *Lacquering Decoration Record* is a historical summary of the lacquering craft, fully exploring its long tradition and rich

Picture 9.8 Flower pattern treasure inlay with black lacquer pen holder from the Ming dynasty

cultural value, which makes this a work of high academic value. The whole book runs through the creation philosophies seeking harmony between man and nature, and also puts forward strict requirements for a painter's moral character. For example, the "four losses" refers to "the rules are not followed", "the methods are incorrect", "the object is damaged", "doing incomplete work due to tiredness or laziness", and the "three diseases" refer to "the unique skills not being passed down", "no consistent improvement", and "unsuitable talent". This type of book is rare in any kind of crafts, even in the whole world. In the fifth year of the Ming dynasty's Tianqi period (625), Yang Mingwei, a famous craftsman, wrote a preface and, bit by bit, added notes to make the content richer and more informative.

Its fate was the same as *Heavenly Creations*, which spread to Japan after publication, but is rare in China. In 1927, Mr. Zhu Qiling discovered this book in Japan and published it immediately, which is now known as the "Zhu Ming edition of commentaries". After Mr. Wang Shixiang visited the old lacquerers, he carefully researched and observed them, and wrote the *Interpretation of Lacquering Decoration Record*, which was published in 1983. Only then did this great work become well known and passed on. Professor Zhang Yan from Southeast University has been engaged in the practice of lacquering since she was a teenager. She even changed her profession to teach and design. She is still rigorous in her studies. She has visited almost all of the places in China that produce lacquerware and wrote the book *Illustrated Handbook of Lacquering Decoration Record*. "This book tackled and solved many problems that no-one had been able to solve before".

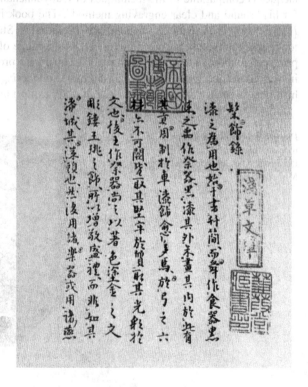

Picture 9.9 Printed page of *Lacquering Decoration Record*

9.2 Materials and Tools

9.2.1 Natural Lacquer

China is a huge lacquer-producing country, and lacquer trees can be found in many different regions, such as Liaoning province, Hainan province, coastal cities, and Tibet, especially in the north–south arc section of the Qinling Mountains and Wumeng Mountain. However, the provinces of Shaanxi, Gansu, Sichuan, Hubei, Yunnan, and Guizhou are the most important for the production of lacquer (as shown in Picture 9.10).

There are two types of lacquer trees: large and small. In order to collect raw lacquer, we need to cut into the lacquer trees. Small lacquer trees are generally cut twice every 3 years, yielding 150–250 g per tree annually. The cutting cycle of large lacquer trees is longer and their yields are lower, but they dry better. The lacquer cutting season varies from place to place depending on their respective climate, from summer solstice or summer heat to first frost or cold dew, which lasts 90–120 days. Lacquer trees can be cut when they are 4–5 years old to 10–12 years old. The tools used for this cutting are a paint knife, scraper, clam shell, lacquer bucket, and lacquer basket (as shown in Picture 9.11). Before being cut into, lacquer trees are investigated in detail. The operation areas are divided and the cutting is carried out continuously (though intermittently every few days). There are two types of lacquer cutting: curve cutting and straight cutting, and each type is further divided into various shapes. When the shape of the cut is better, the tree heals faster and the lacquer yields will be larger. After cutting, it cannot be cut again for 1–4 years. So there are sayings as "cutting three times in seven years" and "cutting at both ends in three years". Lacquer trees with a good growth and large lacquer flow can be cut every year (as shown in Pictures 9.12 and 9.13).

Picture 9.10 Lacquer trees

Picture 9.11 Cutting tools and tool baskets

The main components of natural lacquer are urushiol, nitrogenous compounds, and gums, as well as moisture and a small amount of organic matter. The higher the urushiol content, the better the quality of raw lacquer. The content of raw lacquer in China is as follows: urushiol 40–80%, nitrogenous compounds and gums about 10%, moisture 15–30%.

The properties and quality of raw lacquer are mostly identified by the senses and through experience, including concentration, color, filaments length (length of suspended filaments), smell, dryness (drying speed), rice stars (size and number of particles inside the lacquer), and color changes (when shaken). There is a saying in *Compendium of Materia Medica*: "It is as bright as a mirror; the suspended filaments turn as a hook; when shaken, it looks like the color of amber; when scooped up, it foams". These are also the criteria for testing lacquer. In the process of purchase, storage, and use, inspection methods such as sensory recognition and the "frying pan method" (heating the lacquer in a special small pan to measure its properties and residual weight) must be used to distinguish its authenticity and quality.

Natural raw lacquer has a variety of excellent properties that makes it incomparable to synthetic lacquer, earning the name of "the king of lacquer" mainly because of its hard and wear-resistant film, rich luster, tight structure, good impermeability and insulation, strong adhesion, solubility resistance, water resistance, and heat resistance.

Natural lacquer must be processed and refined to remove some of the moisture and impurities, increase its shade, transparency and glossiness, and adjust the dryness.

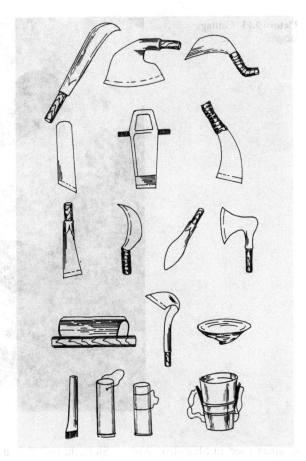

Picture 9.12 Cutting knives and collecting buckets

Sometimes, depending on what it is used for, other materials need to be added. The process of refining lacquer has always been profoundly studied. *Interpretation of Lacquering Decoration Record*, written by Wang Shixiang, a contemporary Chinese antiquities collector, quotes the methods of refining lacquer in the literature of the past dynasties, such as *Book of Guqin Music at Yuguzhai*, which states: Filter the raw lacquer, put it in the dish to dry in the sunshine, and stir it to remove the moisture. When the color is as bright as soy sauce, add borneol or pig bile to mix evenly, so that the lacquer becomes clear and not stagnant. If you need black paint, you can add rust water. Then use linen to filter the impurities several times. If you don't insolate it, you can heat it with a slow fire, then turn down the fire and stir it with a fan. After it is cold, you can heat it and stir it again. If you repeat it several times, the color of the lacquer will be bright, which is especially better than the lacquers insolated in the sunshine. This traditional method of refining lacquer is still used by lacquerers in Suzhou. The types and production methods of traditional refined

Picture 9.13 Cutting lacquer

lacquers used in Shanghai, Wuhan, Sichuan, and Fujian are different. The quick-drying lacquer from Sichuan, for example, just needs three steps: filter, stir, and insolate. The varnish lacquer is made of quick-drying lacquer and base oil (tung oil); the translucent lacquer is sun-dried and stirred until brown, which is then used for blending with black paint and multi-color paint. Lacquer, after adding tung oil, is used as a clear lacquer; transparent lacquer can be produced by adding yellow gardenia juice or garcinia to translucent lacquer; black polished lacquer is made by adding ferrous hydroxide to translucent lacquer.

Raw lacquer, also known as Chinese lacquer, is a strong sensitizing substance. In order to assassinate the enemy, Yurang of the Jin State during the Spring and Autumn period and the Warring States period covered his body in raw lacquer, which gave him paint sores, scabs, and disfigured him. This is a famous story in Chinese history. People who are allergic to lacquer should beware of allergic dermatitis. Most people will increase their resistance to allergies after being exposed to lacquer. "Being bitten by lacquer" is very uncomfortable, but it is mostly limited to affecting the skin. Most lacquer farmers and painters who have been dealing with a lot of lacquer for a long time are healthy and have no occupational diseases later in life.

9.2.2 Auxiliary Materials

9.2.2.1 Tung Oil

The price of raw lacquer is high and the lacquer's color is not as bright as the oil's color, so it is often necessary to add drying oil to increase the color and reduce the cost. Oil and lacquer are part of the same family. There are many kinds of drying oil, in which the most commonly used is tung oil. According to *Interpretation of Lacquering Decoration Record*, perilla oil (refined from white perilla) may have been used for mixing paints in the Shang and Zhou dynasties as well as the Warring States period. Sesame oil and walnut oil were used in the Wei, Jin, and Southern and Northern Dynasties period, and tung oil was mainly used ever since the Song dynasty.

Tung oil is a special product in China, and it is mainly produced in Sichuan, Hunan, Hubei, Zhejiang, Yunnan, and Guizhou Provinces. It dries quickly, it is firm, non-sticky, resistant to water and alkali, and is widely used in the lacquer and paint industry.

9.2.2.2 Pigment

It refers to the pigments and dyes that are compatible with raw lacquer and used for toning. Traditional pigments include realgar (As_2S_2), orpiment (As_2S_3), silver vermilion (HgS), indigo (extracted from indigo), turquoise [$CuCO_3 \cdot Cu(OH)_2$], oil fume, and pine smoke, which are used to, respectively, obtain the colors yellow, red, blue, green, and black (as shown in Picture 9.14).

9.2.2.3 Metals

The most commonly used metal materials are gold and silver foils, sheets, wires and powders, as well as copper sheets and tin sheets. The tin powder used for the "tin ornaments" engraved lacquering in Yichun, Jiangxi province, is obtained by repeatedly frying, cooling, and sieving refined tin in an iron pan.

9.2.2.4 Inlay Materials

There are many kinds of lacquered inlaid objects, including mother-of-pearl, bones, stones, and treasures. According to *Lacquering Decoration Record*, inlay materials include coral, amber, agate, precious stones, hawksbill, mother-of-pearl, ivory, and rhino horn. *Words in Lu Garden* also lists real pearls, jasper, jadeite, crystal, *chequ*, green gold, green pine, and yellow amber. Eggshell inlaying is popular in modern times. Before inlaying, the eggshell must be soaked in rice water or weak acid to remove the film in the shell. It can also be crushed into a powder for decoration.

Picture 9.14 Grinding, cited from *Heavenly Creations*

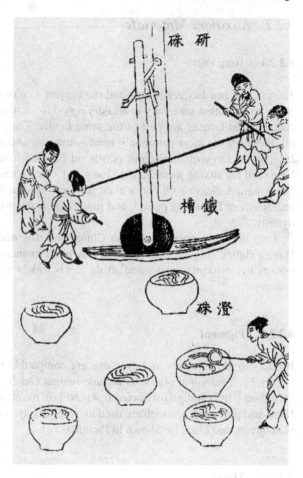

9.2.2.5 Grinding and Polishing Materials

The traditional lacquer industry uses cyan river stones and red sandstones to grind rough paint and lacquer into dust, and use carbon powder to sand down the floating light (the floating light on the lacquered surface is the original light of the paint film). After pine or beech is burned into charcoal, it can be used to sand the lacquered surface, which is called grinding charcoal.

In ancient times, lacquerware was first coated with coarse dust and then sanded with shark skin. Now coarse sandpaper is used instead of shark skin. The lacquered surface is then polished with ground and sifted brick powder or porcelain powder. As stated in *Book of Guqin Music at Yuguzhai*: "For the so-called polishing [...] dip the old sheepskin in sesame oil, and then dip in brick powder or porcelain powder and rub the surface of the lacquerware [...] The more times you rub, the brighter it becomes, so that you can use it as a mirror". In Fuzhou and Beijing, painters mix

Picture 9.15 Lacquer scrapers

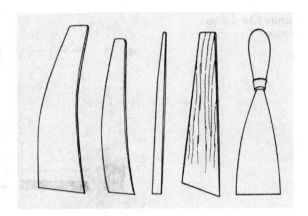

brick powder with pig blood, lime or tung oil, lead and soil to make gray strips, also known as artificial grindstones. In the northern regions, lacquerware is polished with horsetail grass.

9.2.3 Tools

9.2.3.1 Scrapers

The wooden body, bamboo body, strip body, and leather body all need to be scrapped. The tools used for this are called scrapers, or scrape boards, made of bamboo, bone, iron, copper, horns, or rubber. Its shape is narrow and thick at the top, for easy holding, and wide and thin at the bottom in the shape of a blade, which can be long or short depending on a worker's needs. A rigid scraper is used for square lacquerwares and a rubber scraper with good elasticity for round lacquerwares (as shown in Picture 9.15).

9.2.3.2 Lacquer Brushes

The upper part of the Chinese character for painting with a lacquer brush "髹" (*xiu*) contains the radical "髟" which looks like long hair drooping down. It shows that lacquer brushes were originally made of human hair. The brush should be elastic with an appropriate length, and the bristles should be strong and not break easily. In addition, oxtail, horsetail, and pig bristles are often used to make brushes. Because the hair is harder, brushes made with such hair are used to apply lacquer dust or thick lacquer (as shown in Picture 9.16).

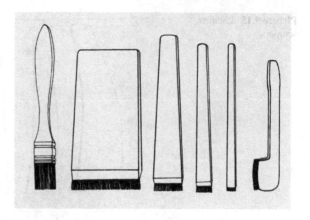

Picture 9.16 Lacquer brushes

9.3 Techniques

9.3.1 Body Techniques

Lacquerware bodies are mostly made of wood and cloth, as well as bamboo, bamboo strips, leather, metals, ceramics, and porcelain.

9.3.1.1 Wood Body

As wood is light, resistant to wear, stable, and not easily deformed or cracked, it is commonly selected to make lacquerware bodies. Most common are nanmu, pine, beech, and camphor (as shown in Picture 9.17). When the workblanks are joined together, they should be firmly glued together with "fat paint" (raw lacquer mixed with cattlehide rubber or flour), tied with rope, and then wedged tightly. After trimming, it is brushed with primer, pasted with cloth, and scraped with lacquer ash. It is first scraped with coarse lacquer ash (made of raw lacquer, brick powder, or glutinous rice paste); secondly, it is scraped with medium lacquer ash (made of raw lacquer, brick powder, soil powder, and glutinous rice paste); thirdly, the gaps at the edges and corners are filled in; fourthly, it is scraped with fine lacquer ash; and finally, the edges and boundaries are trimmed. Generally speaking, scraping lacquer ash is the same as polishing or sanding in order to make the lacquerwares smooth. The ash scraping process of large screens made in Fuzhou is as follows: Firstly, it is scraped with coarse ash and the burlap is pasted with raw lacquer; secondly, it is scraped with the medium ash and the coarse ash; thirdly, it is scraped with the medium ash again and then scraped with fine ash. After scraped with ash, it is lacquered, dried in the shade (as shown in Picture 9.18), and then ground with a sharpening stone. If there are still uneven spots, fine ash can be used to smoothen those out.

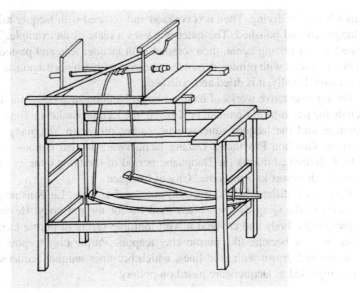

Picture 9.17 Wood-working lathe

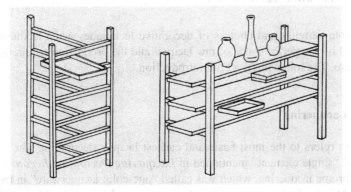

Picture 9.18 Trestles in shade house

9.3.1.2 Demolding

The techniques of demolding make use of the viscosity of lacquer and the tension of cloth to paste cloth onto a lacquer body layer by layer, which technically makes it a cloth body. In ancient times, it was called ramee-lacquer. Its advantages are that it is light, does not deform nor crack, convenient to model, and low in cost with a variety of sizes. Naturally, this makes it popular. Commonly used techniques for demolding are *yin* demolding and *yang* demolding. The former one uses a bottle as an example. It takes a gypsum mold as its body, then pastes on cloth, coats it with primer, and demolds it after drying. Then it is covered and scraped with lacquer ash. Finally, it is lacquered and polished. The latter one uses a plate as an example. The mold is coated with a parting agent, then scraped with lacquer ash, and pasted with cloth. Then it is coated with primer, demolded, scraped with ash once again, and lacquered once again. Finally, it is dried and polished.

The representative works of bamboo body, bamboo strip body, and leather body include the pen tube of Shanlian in Huzhou, the lacquer basket in Yongchun, Fujian province, and the lacquer suitcase and leather pillow in Yangjiang, Guangdong province. Guizhou Province's Dafang lacquerware is also famous for its leather body. It flourished during the Daoguang period of the Qing dynasty (1821–1850), when Guizhou was known as the "City of Lacquer".

They made dishes, bowls, mirrors, pots, and teapots. Lu Kuisheng, a famous lacquerer in the Qing dynasty, often used tin for the pot body. He painted black lacquer on the body and covered it with multiple layers of purple lacquer, so that the color of it became like purple clay teapots. Purple clay teapots can also be lacquered and drawn with gold lines, which becomes unique lacquered pottery or can be regarded as lacquerware based on pottery.

9.3.2 *Decorative Techniques*

The multiple varieties and choices of decorative techniques when lacquering are determined by the performance of raw lacquer and the production requirements of lacquerware. The following is a brief introduction.

9.3.2.1 Lacquering

Lacquering refers to the most basic and earliest lacquer decoration. The "natural color" and "single element" mentioned in *Lacquering Decoration Record* belongs to monochrome lacquering, which was called "one-color lacquerware" in the Song Dynasty. This kind of lacquerware's winning feature is its shape. Since it doesn't have any decoration, it focuses on the beauty of the texture and has extremely high requirements for the lacquering techniques used.

Thick lacquering. The lacquer used in this method is usually oily. It will not wrinkle nor leave brushmarks after lacquering. It is convenient and finished in one go, which makes thick lacquering still popular today. To make vermilion lacquerware, for example, the body must first be washed and dried, then the vermilion and tung oil are mixed to create the vermilion lacquer and then the lacquer is evenly brushed on the ware. Usually, it must be lacquered three times before being put into the shade room.

Polishing varnish lacquering. Take black lacquer for example. You can choose black polished lacquer to lacquer the first layer and send to the shade room. After it has dried, you can polish it. Then apply the second coat of lacquer, shade and polish.

Picture 9.19 Long eyebrows arhat covered with golden lacquer, made by Zhang Ying and Zhang Qiu

After polishing away the wear marks with fine finishing paper, dip the palm of your hand in vegetable oil and tile powder and buff the lacquer surface repeatedly to let the luster come out. This technique is called "making lacquer shine". Finally, dip cotton into good, raw lacquer, and oil to thinly lacquer the wares. When it is half-dry, apply the polishing lacquer technique again and again to make the lacquer shine like a mirror. This is commonly called the "lustering method".

Thin Lacquering. This method was initiated by Shen Shao'an, a lacquerer in the Qing dynasty. Grind gold and silver foil into a paste, then add lacquer and pigment, and repeatedly pat lacquer onto the surface of the objects with your palms and fingers. This way you apply just a thin layer. Next, cover it with a layer of tung oil to protect the lacquer surface. Because of the addition of gold and silver, it makes it look more luxurious and valuable.

Clear lacquering. *Lacquering Decoration Record* refers to clear lacquering as covering gold, silver, colored, or wood lacquerware with transparent lacquer (as shown in Picture 9.19). Because of its simplicity and convenience, it is widely used in construction, furniture making, and common lacquerwares, which are plain and elegant. Golden lacquering can be applied as golden foil or powder, which is prepared with raw lacquer, bright oil (thicker tung oil), and pigment.

9.3.2.2 Drawing

Drawing is the oldest method of decoration. As far as lacquer art is concerned, it refers to drawing patterns directly on the bases, without covering, sanding, or polishing.

Color drawing. It is also called "drawing with lacquer". If the lacquer is replaced by oil, it will be called "drawing with oil". Drawing patterns with lacquer or tung oil and applying pigment on the wares is called "dry-colored lacquering" (as shown in Picture 9.20). *Lacquering Decoration Record* states: "There are all kinds of dry pigments, and they don't luster". Because they don't lacquer it again after applying the pigment, there is no luster in the varnish.

Drawing golden lines. This refers to pasting golden foil or drawing golden lines on lacquerwares (as shown in Picture 9.21).

Rendering golden lacquering. In order to do this, you first lacquer with pigments, and apply gold and silver powder when the lacquer is about to dry. This will create a rich or light rendering effect due to the density of the powder. If combined with drawing golden lines, it is called "halo gold painting" (as shown in Picture 9.22).

Picture 9.20 Dry-colored lacquering, designed by Qiao Shiguang and made by the Lacquers Factory in Pingyao

Picture 9.21 Vermilion handwarmer lacquerware with golden dragon and phoenix patterns, made in the Qing dynasty

Picture 9.22 Halo gold painting, made by Ning Baoping

9.3.2.3 Inlay Techniques

Mother-of-pearl Inlay. Mother-of-pearl is another name for shells. The mother-of-pearl inlaid lacquer *lei* of the Western Zhou dynasty, which are a kind of ancient drinking cup, were excavated in Liulihe town in the Fangshan district of Beijing, which pioneered this technique. The mother-of-pearl must be ground flat, cut with a bow saw and filed down. Then the mother-of-pearl must be pasted on the surface of the lacquerware, scraped, leveled, and milled with water. To harden the mother-of-pearl, black paint must be applied over it and then sent to the shade room. The black paint on the pearl must be polished off to reveal its texture, and then the surface is processed further (as shown in Picture 9.23).

Metal inlay. According to *Lacquering Decoration Record*, metal inlays can be divided into "inlaid gold", "inlaid silver", and "inlaid gold and silver"; however, it is also said that "there are other kinds of metal inlay, in which metal chips and wires are used. The inlays can either include various metals, or just one kind of metal". In Weifang, Shandong province, they embedded silver wires on lacquered furniture, which causes it to be categorized as "metal inlay".

Picture 9.23 Thick mother-of-pearl inlay

Picture 9.24 Eggshell inlay: Scenery in Jiangnan, made by Qiao Shiguang

To reduce costs, silver inlays were replaced by tin. There are two methods for creating patterns: knife engravings and etching. The latter uses strong acid to corrode the tin. Because the raw lacquer is corrosion-resistant, the painted pattern remains.

Eggshell inlay. It is unknown when eggshell inlay started. The popular wall paintings and pot paintings in Ansai district, Shaanxi province, are mostly made with eggshell inlays, which may be one of its origins. After this technique was applied to lacquer art in the 1920s and 1930s, it spread quickly and became an important staple in modern Chinese lacquer art. It makes the color tone of lacquer art turn from dark to pure white, and its cracked texture gives it a gentler and simple kind of beauty than can be accomplished with other techniques (as shown in Picture 9.24).

Bone inlay. All kinds of bones can be inlaid onto the surface of the lacquerwares to form patterns (as shown in Picture 9.25).

9.3.2.4 Engraving and Filling

Engraving the lacquer and then filling it with gold, silver, or colored lacquer is called "engraving and filling".

The technique of engraved golden lines, also known as sunken golden lines, means engraving into the lacquer's surface, brushing it with golden lacquer, and filling it with golden powder when it is nearly dry (as shown in Picture 9.26). Engraved silver lines and engraved colored lines use the same method.

Engraving and filling. Engraving decoration with colored lacquer on its surface is called *"diaotian"*, while, in Japan, it is called *"jujiang"*, and *"luqi"* in Chengdu,

Picture 9.25 Bone flat inlay pattern, made by Luo Qingping

Picture 9.26 Engraved golden lines, made by Ning Baoping

China. Golden lines must be engraved after the lacquerware has been applied and polished; however, when engraving and filling, it is the other way around (as shown in Picture 9.27).

Engraving lacquering is also known as "engraving dust". It is also called "colored engraving lacquering" in *Lacquering Decoration Record*. "Engraving" in the term refers to it being engraved directly into the lacquer (Yin Wen) as opposed to being piled on the lacquer (Yang Wen). In order to facilitate the engraving, blood ash is used instead of general lacquer ash.

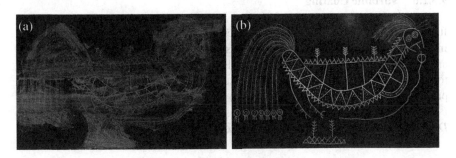

Picture 9.27 Engraved and filled by lines, made by Ning Baoping **a.** Filling it with color **b.** Finished

Picture 9.28 Golden fish, varnish lacquer in polishing and painting, made by Zheng Yikun

9.3.2.5 Polishing and Painting

Another technique is to (1) scatter lacquer powder, or gold and silver powder, (2) brush on color lacquer, and (3) paste gold and silver foil on the surface to make it uneven, and then covering or filling it with color lacquer. This causes various lacquer layers to overlap with each other and blur together. These layers are then ground down and polished until the surface is smooth and the patterns appear. This method uses sanding and polishing as the main technique, which can also be used as a special way of applying color, so it is called "polishing and painting".

Colored polishing and painting. It can be divided into filling and covering lacquer (as shown in Picture 9.28).

Maki-e is a technique in which you scatter metal powder, mother-of-pearl, and dry lacquer on the lacquerware's surface, then varnish, sand, and polish. "Maki-e" is a term used in Japan, and this technique represents the typical characteristics of Japanese lacquer art (as shown in Picture 9.29).

9.3.2.6 Variable Coating

Its ancient name in China is "zhang lacquering", while in Japan it is "variable coating", which means ever-changing, and has been widely recognized by the lacquer art world. It is characterized by using tools or media to apply particles or thinners to make the lacquer surface lines. The resulting texture has an abstract beauty and produces unique artistic effects.

Tools for patterning include hairbrushes and scrapers. According to *Lacquering Decoration Record*, hairbrushes can be used to produce "brush silk" and "swaying

Picture 9.29 Cherry blossoms, Maki-e, made by Qiao Jia

brush silk". The former one are straighter lines and the latter are curved lines. When brushing, you can use a bristle lacquer brush to dip it in thick lacquer and drag it into a pattern on the lacquerware (as shown in Picture 9.30).

After entering the shade room, the lacquer will be sanded, polished, and lustered.

The media used for dragged patterns include leaves, seeds, linen, silk screens, mother-of-pearl, gold and silver, dry paint, and eggshell powders. When they are scattered on the lacquer surface, their impressions or inlays are painted into wrinkles, which look completely natural (as shown in Picture 9.31).

Diluents such as gasoline and camphor oil are used to make drifting lacquer and are sprinkled on the lacquer's surface randomly or the lacquer body is moved about to make the wet lacquer permeate and flow. This is a modern technique, which creates varied and unexpected effects.

Picture 9.30 Brush patterns, made by Ning Baoping

Picture 9.31 Pressed patterns with flax, made by Xing Kai

Picture 9.32 Surface stack, Han dynasty stone relief pattern, made by Luo Zhiqiang

9.3.2.7 Piling and Decoration

Thick paint, charcoal powder, paint ash, and dry paint powder are piled on the paint surface, and then the piled patterns are decorated, which is called "piling and decoration". In the *Lacquering Decoration Record,* they are called "yang shi" and "dui qi" (piling up) as opposed to being engraved into the lacquer (yin shi). According to its technique, it can be divided into line stack, surface stack, thin stack, and high stack (as shown in Picture 9.32).

9.3.2.8 Engraved Lacquering

The color of the lacquer used for engraving is divided into black, red, yellow, rhino (black and red), and colorful (black, yellow, green, red, etc.). Beijing's engraved lacquering is the most representative one, in which copper bases are commonly used. Besides that, there are also wooden bodies and cloth bodies. The Copper body must be burned with enamel, lacquered with ash, shaded, and repeated many times, with as few as 120 layers and as many as hundreds of layers. Then, after trimming and file shaping, the draft is set and then engraved. The works with higher requirements are divided into "*shangshou* work" and "*xiashou* work". The former refers to the engraving of landscapes, flowers and birds, figures and pavilions, while the latter is also called "brocade work", which refers to all kinds of brocade patterns being engraved using the blank space of patterns. After the engraving process is completed, it should be dried, polished, and waxed.

The art of lacquer decoration is extensive and profound, and it is ever-changing. It varies according to local conditions, equipment, being single-use or comprehensive-use, and masters' teachings as well as specific conditions. This paper only summarizes its basic working procedures and techniques to serve as a reference. Please refer to Picture 9.33 for the composition of various techniques.

9.4 Types of Lacquerware and Its Production

China's lacquerware has been developing for a long time and is produced all over the country. Depending on where in China lacquerware is produced, it will often vary in regional or ethnic characteristics, as well as in types, shapes, and techniques. Therefore, people sometimes name types of lacquerware after their places of origin, such as Beijing engraved lacquer, Fuzhou bodiless lacquerware, and Tiantai dry ramee-lacquers.

9.4.1 Beijing Engraved Lacquer and Gold Lacquer Inlay

Beijing was once ancient capital of six dynasties. After Kublai Khan built *Dadu* (modern-day Beijing), he moved Mobei craftsmen to Beijing, which was the beginning of government-run lacquering in Beijing. *History of Yuan* states that Liu Yuan, a famous lacquerer, "has a wonderful combination of thoughts, and unique skills"; "When he molds Buddha statues and applies ramee-lacquer, no-one in the world can compare". It can be seen that Beijing's lacquering decoration skills at that time had already reached a very high level. In the Ming dynasty, craftsmen worked in shifts and a large number of lacquerers came together in Beijing. After the mid-Ming dynasty, common lacquer works prospered, and innovations in lacquer engraving, engraving, and filling, and gold engraving took place. In the Qing dynasty, there were oil and wood engravings and lacquer works made in the Imperial House. Emperor Qianlong's fondness for lacquer engraving caused this skill to become even more refined. *Textual Research on Lacquerware Production in the Qing dynasty* by Zhu Jiajin (1914–2003) lists the decoration techniques and various lacquerwares used in this period. At the end of the Qing dynasty, the Beijing Arts and Crafts Bureau was established, under which there were engraved lacquer divisions and painting divisions that imitated Fuzhou bodiless lacquerware, but, overall, the trend was declining. During the Republic of China, the level of lacquer engraving picked up again, and

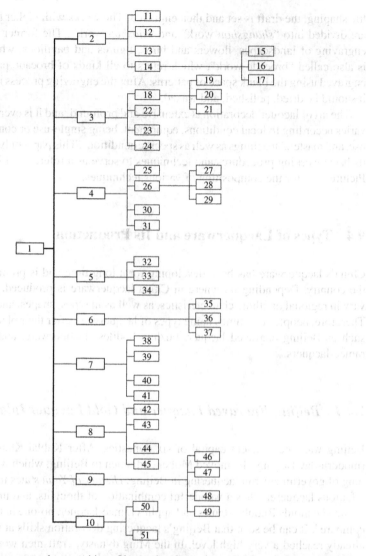

Picture 9.33 Category and composition of lacquering decorative techniques: 1. lacquering decorative techniques 2. lacquering 3. drawing 4. inlay techniques 5. engraving and filling 6. polishing and painting 7. variable coating (*zhang* lacquering) 8. piling and decoration 9. engraved lacquering 10. antiquing and modelling colored lacquers 11. thick lacquering 12. polishing varnish lacquering 13. thin lacquering 14. clear lacquering 15. on colored lacquerware 16. on gold lacquerware 17. on wood lacquerware 18. color drawing 19. drawing with lacquer 20. drawing with oil 21. dry-colored lacquering 22. drawing golden lines 23. rendering golden lacquering 24. colored drawing with gold and silver lines 25. metal inlay 26. mother-of-pearl inlay 27. thick mother-of-pearl 28. thin mother-of-pearl 29. surface inlay 30. eggshell inlay 31. bone inlay 32. engraved golden lines (sunken golden lines) 33. engraving and filling (*diaotian*) 34. engraving golden lines (engraving dust; colored engraving lacquering) 35. colored polishing and painting 36. maki-e 37. aluminum powder polishing and painting 38. tools for patterning 39. media for patterning 40. powders for patterning 41. diluents for patterning 42. line stack 43. surface stack 44. thin stack 45. high stack 46. engraving red lacquer 47. engraving black lacquer, 48. engraving rhino (black and red) lacquer 49. engraving colorful lacquer 50. antiquing 51. modelling colored lacquers

the engraved cinnabar lacquer-folding screen *Celebration of Birthday by Immortals*, designed by Wu Yingxuan, won the gold medal at the Panama Pacific International Exposition. President Li Yuanhong specially created a plaque for it that reads "Unsurpassed Elegant Art". During the Second Sino-Japanese War (1937–1945), the lacquer industry struggled to survive through the troubled times, and even settled for using lead oil and paint ash instead of big lacquer and vermilion. Traditional artists marked their work with the names of their workshops to show the difference.

After the founding of the People's Republic of China, Beijing's lacquer engraving quickly picked up again. In 1951, the Beijing Engraving Lacquer Cooperative was established, with famous artisans such as Wu Yingxuan, Sun Caiwen, Du Bingchen, and Liu Jinbo. The participation of designers and the recruitment of students as artisans have changed the personnel composition of the lacquer engraving industry. There was originally only a small workshop for gold lacquer inlay. Since the establishment of the Beijing Gold Lacquer Inlay Factory, the industry has grown to more than 1,000 people.

9.4.1.1 Beijing Engraved Lacquer

The output of engraved lacquerware in Beijing accounts for more than 70% of the country. The products include that use engraved cinnabar lacquer, engraved black lacquer, engraved colored lacquer, and cloud engraving (as shown in Pictures 9.34 and 9.35). They mainly use copper bodies, while the mouths and feet are mostly made of high-quality brass threads. The traditional four colors of red, black, yellow, and green are used, with red as the main color. There are some innovations in brocade patterns, such as new water brocade patterns and ground brocade patterns. The high-relief hollowing technology pioneered by Du Bingchen and Liu Jinbo raised the lacquer technique to a new level and expanded the field of artistic expression. The themes and patterns are also more diverse. The works have distinct local characteristics and court art styles (as shown in Pictures 9.36 and 9.37).

There are more than ten processes in creating Beijing engraved lacquer, among which lacquering and engraving are particularly important. After dipping a silk ball in paint and using it to coat the lacquer body, a fine brush is used from top to bottom and from left to right to make it level. The first three or four layers of lacquer are called matting lacquer. Then, it is put into the shade, which is called "soft drying". After drying in the shade, it is sanded down with fine sandpaper, and then repeatedly lacquered until it reaches the specified thickness. For engraving colored lacquers, the lacquer's color must be changed as required, which is called "*kaodi*". The engraving process includes a series of steps: stabbing, lifting, slicing, shoveling, hooking, clustering, and patterning. "Stabbing" means sticking a knife into the paint layer to a specified depth; "lifting" means using tools to eliminate unnecessary parts. After "stabbing" and "lifting", the lacquer surface will have a sags and crests pattern. "Slicing" means engraving the paint layer above the brocade pattern with a slicing knife, and the figures, flowers and birds, pavilions and rocks are all presented in clusters. "Shoveling" means using a shovel to make images that are difficult to show

Picture 9.34 Engraved cinnabar lacquer, Beijing

Picture 9.35 Engraved lacquer brush pot made by Wu Yingxuan

Picture 9.36 Downhill Leopard, made by Wen Gan

Picture 9.37 Peace and Bless Lacquer box with mother-of-pearl inlay, made by Liang Fusheng in Yangzhou

with a slicing knife, such as the textures and veins of rocks. "Hooking" means using a hook knife to engrave lines and silk veins, such as leaf stem veins and animal hair. Engraved lacquer works are time-consuming and laborious, often taking 6 to over 12 months (as shown in Picture 9.36).

9.4.1.2 Gold Lacquer Inlay

Gold lacquer inlays are magnificent and elegant, and another characteristic product from Beijing. It is divided into three categories: inlaid gold and silver, colored lacquer, and engravings. Each category is subdivided into several varieties. For example, inlays can be divided into pattern inlays (inlaid with ivory, jade, and other materials according to different patterns such as landscapes, flowers, and birds in the artwork), flat inlays, and whole inlays (the whole is inlaid without leaving any base lacquer). Gold and silver lacquer includes engraving gold (rendering golden lacquer), gold sweeping (brushing the body with gold lacquer, requiring no brush marks, and then brushing gold powder after drying it in the shade), silver sweeping, gold foil varnish lacquer, silver foil varnish lacquer, flat gold-and-black lacquer (after pasting gold foil, lacquer it with black lacquer, and then varnish), flat gold and colored painting, flat silver-and-black lacquer, and flat silver and colored painting. Engraving and filling included engraving ash, engraving and filling, engraved golden lines, engraved silver lines, and engraved colored lines.

9.4.2 Yangzhou's Mother-of-Pearl Inlay Milling and Engraved Lacquer with Jade

In the Tang dynasty, Yangzhou was the center of commerce and handicrafts throughout the country and engraved lacquerware had already emerged. In the Yuan dynasty, the center of lacquerware production shifted to Yangzhou and Jiaxing, and in the late Ming dynasty, famous varieties such as mother-of-pearl inlay, treasure inlay, and engraved cinnabar lacquer were created as literati participated in lacquerware design. However, the Qing dynasty saw the heyday of Yangzhou lacquerware and its influence spread all over the country. After the opening of the Shanghai Port, Yangzhou's lacquerware was transformed into "foreign goods", making "works for export" a new production focus. The sales volume of famous lacquerwares made by Liang Fusheng achieved half of the sales in the whole industry, and his works won the silver prize at the Panama Expo (as shown in Picture 9.37). In the warring periods of modern times, lacquer art was on the verge of extinction in many areas, but Yangzhou's lacquer workers flocked to Shanghai and passed it on through these exported works. After the founding of the People's Republic of China, the Yangzhou lacquer industry continued to update and develop towards exporting to the rest of the world, causing the Yangzhou Lacquerware Factory to became a well-known lacquerware company in China. Even now, it is still held in high regard.

The greatest feature of Yangzhou lacquerware is the cultural characteristics formed by the combination of lacquerers and literati, which gives it its fresh and elegant aesthetics as well as advanced techniques of high-quality engraving and inlaying. For example, the lacquer and calligraphy fold screen made by Liang Fusheng's lacquer is based on the works of Yangzhou's famous calligraphy and lacquer masters such as Zheng Xie, Hua Yan, and He Shaoji. The lacquered floor of the engraved folding screen is mostly a black and chestnut-shell color on which the words are engraved in convex, while the lines are engraved in concave. The lacquer color is blended, which is simple and elegant. The embossed lacquer screens do not contain golden lines or varnish lacquer, which gives it a classic style. They are quite different from the luxurious styles of the jade inlaid or golden screens. Nowadays, Yangzhou lacquerwares are most known for milled mother-of-pearl inlay and jade-inlaid engraved lacquer. The former one is ethereal, with fresh and meticulous patterns; the latter was introduced to the court by the literati, becoming the most decorative and "palace" style of Yangzhou lacquerware. Lacquer art furniture is based on the Suzhou style, with concise shapes, symmetrical knots, long and straight lines, and decorated with various lacquering decorations. Famous artisans include Liang Guohai, Sun Guoyun, and Gao Yongmao. One of its most representative works, *Ode to Peace*, is painted by Chen Zhifo, produced by Liang Guohai and Wang Guozhi, and hung in Jiangsu Hall of the Great Hall of the People.

9.4.3 Weifang's Silver-Embedded Lacquerware

The silver-embedded lacquerware from Weifang, Shandong, was first produced during the Daoguang period of the Qing dynasty (1821–1850). It embeds silver threads into brush holders or inkstone boxes, which are made out of red sandalwood or mahogany, to form characters and patterns. Once formed, they are lacquered and polished, which makes the lacquerware elegant, dexterous, and full of bejeweled charm. This technique was first created by Yao Gongfu and Tian Shefan. Tian Zizheng and Tian Ziyou, sons of Tian Shefan, opened Treasure House in Jinan to manage lacquerwares inlaid with silver threads. *Book of Luochuang* states: "Tian Zizheng and Tian Ziyou are famous. They use ebony as their lacquered body and the silver threads embedded in it are so firm that they will last for a long time. Among their lacquerwares, the stationery is the most refined". In 1915, Tian Zijing's lacquer inlaid gold and silver threads won the gold medal at the Panama Expo. In 1928, the Tongyinshan Workshop enrolled more than 40 apprentices and established branches in Beijing, Nanjing, and Jinan, and its products were exported to Southeast Asia, the United Kingdom, and the United States. At present, there are more than 300 kinds of lacquerware produced by the Weifang Silver Inlaid Lacquerware Factory. In addition to stationery, they produce tea tables, flower tables, screens, and antique shelves. The well-known artisan Wei Yugong created the method of combining thick and thin threads, and also used embossing, round engraving, and other techniques to enrich the expressiveness of inlaid silver threads. In 1979, he was awarded the title of "Expert of Arts and Crafts".

9.4.4 Pingyao Polished Lacquer and Xinjiang Engraved Three-Color Lacquer

Folk songs sing about Pingyao that: "Since ancient times, Pingyao has three treasures: lacquerware, beef, and yams". The lacquerwares of Hongjinxin Lacquer Workshop in Pingyao, Shanxi, were exported to Singapore, Russia, and the United States as early as the Qing dynasty's Qianlong period (1736–1796). During the Guangxu period (1875–1908), there were as many as 16 lacquer workshops in Pingyao. In 1958, the *Pingyao Polished Lacquer Factory* was established, which became famous for its varnish-lacquered furniture with stacked golden lines.

One of Pingyao's characteristic lacquerwares are black lacquered. Every time you lacquer it, you have to dry it in the shade, sand it until the surface is smooth, and then use the palm of your hand to spread the quick-drying lacquer over it. After it dries, you can use oil sludge to polish it, and add decoration in the forms of colored lacquer or golden lines.

Stacking is the local appellation of embossed lacquer. After the stacked pattern is dry, the gold glue lacquer is applied thinly and gold and silver foil or powder added, and finally gold lines or lacquer brushed on. After this has dried, it is covered with transparent lacquer, which gives it the name "varnish-lacquered furniture with stacked golden lines". The screens and cabinets are resplendent, gorgeous, and festive, and are loved by the people. Xue Shengjin, a famous artisan, followed by Qiao Quanyu and Ren Maolin, was responsible for some of the meticulous figures and clay sculptures at the Yongle Palace and Shuanglin Temple, enriching their lacquered headdresses and clothes patterns with strong local characteristics. His technique for creating varnish-lacquered furniture with stacking patterns of golden lines on the surface is full of novelty. After his years of hard work, Pingyao lacquer was seen for its own bright and splendid aesthetic style, and his works have won many awards. Xue Shengjin himself was awarded the title of "Master of Chinese Arts and Crafts" in 1997 (as shown in Pictures 9.38 and 9.39).

Xinjiang engraved cloud lacquer was first created by the lacquerer Zhang Fanwa during the Jiajing period (1522–1566) in the Ming dynasty. The utensils created this way are exquisitely engraved, rich in layers, and changing colors, which are widely appreciated.

Picture 9.38 Xue Shengjin painting

Picture 9.39 Hanging screen of mansions (partial), varnish-lacquered with stacked golden lines, made by Xue Shengjin

9.4.5 Fuzhou Bodiless Lacquerware

Many pieces of ramee-lacquer and wooden-based lacquer were found in the Huangsheng tomb in Fuzhou, dating back to the Northern Song dynasty. *Collection of Suichu Hall* reports that most of the best lacquers came from Tingzhou, Fujian. The set of pots and bottles the author ordered from Tingzhou contains 72 bowls, saucers, cups, and basins, in which 61 pieces are lined with silver. Based on this, Wang Shixiang, a contemporary Chinese scholar, believes that there is a long tradition of lacquer in Western Fujian and that their professional lacquerers have high standards. During the Kangxi, Yongzheng, and Qianlong period (1662–1796), the government of Ryukyu sent craftsmen to Fujian to learn lacquering techniques such as "lacquer vermilion", "silver and vermilion", and "pink and vermilion". Li Yu's *Leisure* records Wei Lanru and Wang Mengming, the best lacquerers in Fujian, stating: "Hundreds of years ago, the lacquers in Fujian were famous. Customers came from all directions to buy them and were surprised at the ingenuity of the lacquerers. Then they spread and developed their techniques far and wide" (as shown in Picture 9.40). Shen Shao'an, a famous artisan, restored the long-lost technique of ramee-lacquer during the Qing dynasty's Qianlong period (1736–1796) and skillfully decorated the lacquerware with a golden and silver coating, which established Fuzhou bodiless lacquerware as gorgeous and unique. During the Guangxu period (1875–1908), Empress Dowager Cixi bestowed his sons Shen Zhenghao and Shen Zhengxun with four grades of honors. Bodiless lacquer was known nationally as "Fujian lacquer" and became the representative of Fujian lacquer art. At that time, there were more than 40 lacquer shops in Fuzhou, which made it a major local industry. Its products sold well, both at home and abroad, and won many awards. The Shen family's contribution to lacquer art is tremendous. However, the sixth generation's Sun Shenyuan changed his field to aerodynamics research, ending a family inheritance that had lasted for more than a hundred years. Earlier, the Shen family's original craftsmanship was secretive, with various taboos throughout the production process. After entering the twentieth century, due to the changes of the social, political, economic, and cultural environment, more artisans have unveiled and mastered these secrets.

Fuzhou bodiless lacquerware has a long production cycle, requiring at least 20 days. It is characterized by its thin body, light weight, slim precision, and being smooth and bright, like a shining jade (as shown in Picture 9.41). To create this type of lacquerware, silk and fine grass cloth are used instead of burlap as the body. The fine wood body is made of aged nanmu, and the round turning body is made of beech and camphor wood. The wood body should be lacquered with sealing lacquer before scraping it with lacquer ash, so as not to let water seep in, and then scraping it with coarse, medium, and fine ash in turn. Lacquer decoration is also very important. Highly skilled artists can make the lacquer shine like a mirror without needing any polishing, pressing, or wiping, which the artists call "hard paint". The application of a variety of techniques can make the effect of the decoration plain or gorgeous. There are ten categories such as color lacquering, coating gold and silver,

Picture 9.40 Bodiless lacquer statue of Kwan Yin (Avalokiteshvara) made in the Qing dynasty

flowers, brocade lacquering, engraved lacquering, varnishing, stacking, inlaying, shining cover, and imitating colored drawing. Among them, brocade lacquering uses incense ash powder, lacquer, and bright oil to make brocade and then brushing it with gold lacquer. Silver inlays are now replaced by tin sheets. Shining cover means to increase the luster and transparency of lacquerware, so that it becomes more gorgeous yet subtle, and more durable so it may be enjoyed for longer. The "imitating colored drawing" technique did not aim at seeking the sameness, but seeks the elegance of drawings. There are many famous lacquer artisans in Fuzhou, among which Li Zhiqing was one of the greatest. Contemporary descendants include Wang Weiyun, Huang Shizhong, Zheng Yikun, and Wu Chuan.

9.4.6 Tiantai Dry Ramee-Lacquer

Zhejiang's Tiantai has a long history of doing dry ramee-lacquer art. The ramee-lacquer statue of Jianzhen made by the artist monk Situo has been well preserved. This skill is mostly used on wooden statues, and is divided into bodiless and wooden body dry ramee-lacquer. It uses natural raw materials and uses comprehensive techniques, such as sculpting, varnishing, ramee-lacquering, gilding, and colored lacquering to achieve its artistry. The products are strong and durable, and can be

9 Lacquering

Picture 9.41 Bottle with drawings of mountains and lakes lacquered with colored lacquer, made by Lin Liyan and Zheng Zengxun

waterproof, fireproof, and weatherproof. Since the 1970s, Tang Jingchao and Jin Renchu, descendants from Tiantai's Buddhist city, have been engaged in the creation of gold ramee-lacquers. Their Buddha statues and figures have been sold to many countries and regions. Their representative works *Avalokiteshvara with Thousand Hands* and *Eighteen Arhats* are kept at the Palace Museum and the National Treasure Museum of the Tiantai Sect in Japan, respectively (as shown in Picture 9.42). At present, this technique has been successfully passed on and has good prospects for further development.

Picture 9.42 Manufacturing Buddha statues with dry ramee-lacquer

9.4.7 Chengdu Lacquerware (Luqi)

The Ba-Shu area, which were two ancient states in Sichuan, has been an important area for producing lacquer since ancient times. The unearthed Sichuan lacquerwares from the Warring States period and the Qin and Han dynasties have inscriptions of "Chengting" and "Chengshi", both of which refer to having been made in Chengdu. Later lacquerwares contained inscriptions of the "Shu County Workshop" and "Guanghan County Workshop", which were mostly ramee-lacquers, decorated with gold and silver buckles, colored paintings, and engravings. *Ode to the Capital of Shu* by Yang Xiong of the Western Han dynasty said: "When it comes to engraved lacquerwares, there are hundreds of talents". It can be seen that Chengdu was already an important town for lacquerware production at that time. In the Tang dynasty, the Lei family in Sichuan was famous for being good at making *guqin*. Ge Luofeng invaded Shu (today's Sichuan province), captured Sichuan's lacquer workers, and sent them to Yunnan. This is how the techniques of filling and engraving of lacquer were introduced to Yunnan, which brought the province a prosperous period for hundreds of years. In Sichuan, lacquerwares made with these techniques were called "*luqi*" (halo lacquerware). Furniture, plaques, and statues were all lacquered these ways. In the early years of the Republic of China, there were more than 50 lacquer workshops in Chengdu, such as Kejiaxiang, Duanchang on Taiping Street, and Tongfahao.

In 1956, the *Luqi* Society was established in Chengdu, which was later renamed Lacquerware Factory. People like Shen Fuwen and Hu Kaixin have been studying and improving lacquer art for many years, contributing many innovative achievements (as shown in Picture 9.43).

The term *luqi* (halo lacquer) has different interpretations. *Annals of Dongchuan Prefecture* reads: "The milk lacquer table is as bright as a mirror". Shen Fuwen believes that "*luqi*" can be referred to as "milk lacquer" because "*lu*" has similar pronunciation to "*ru*" (milk) in Chinese, and Qiao Shiguang also believes that "*lu*" refers to lacquer being as thick and mellow as milk. Chengdu *luqi* is characterized by engraving, filling, and tin strip pasting. Engraving and filling include two methods. One is engraving patterns and filling them with colors, and the other is engraving and filling hidden patterns. The former method is different from that of Beijing and Yangzhou. The black paint is painted twice on the gray body and then lightly engraved, by which only the lacquer film is cut into, not the lacquer body. Then the colored paint is used to fill the sunken areas, which can then be smoothened and polished along with the lacquer. The engraving is shallow, so the knife must be steady, which depends on the craftsman's exquisite skill. The latter is gilded and covered after the lacquer engraving and filling. The pattern is hidden underneath the lacquer layer, which gives it a sense of being embossed. Tin strip pasting or tin-embedded mercerizing refers to sticking tin strips in a pattern profile on the lacquer surface, which is then lacquered and dried, and finally sanded to show the tin. The fine engraving pattern on it is meticulously done, making the inside and outside of the engraving pattern appear a bright and beautiful silver. This method requires the craftsmen to pay attention to the angle and direction of the knife, so

Picture 9.43 Lotus, hair engraving with embedded tin, made by Su Ji

as to form a refractive effect. In addition to the above, there are other techniques such as engraving golden lines, engraving colored lines, Maki-e, multiple painting, mother-of-pearl inlay, and eggshell inlay.

9.4.8 Wenzhou's Ou Mold

The lacquer in Wenzhou, Zhejiang Province, creatively changed their lacquerware models into clay models. With its stronger plasticity, freer and more diversified colors, it became an independent category, known as the Ou mold, which was seen as the best during the Song dynasty. *Dreams of Splendor of the Eastern Capital* and *A Dream of Lin'an* record that there were many lacquer shops selling Wenzhou Ou molds in Kaifeng and Hangzhou. After the founding of PRC, the Wenzhou Ou Molding Lacquer Factory was established, through which many excellent works by famous artisans such as Xie Xiangru and Zhang Guoqiu have been handed down from ancient times. Ou molding artisan Yang Wenguang apprenticed with his father since he was a child and became an expert in lacquer art at the age of 15. After retiring, he remained devoted to his former career. With a sense of historical responsibility to protect the inheritance of the Ou mold, he taught himself the physical and chemical knowledge of lacquering and wrote a book about it, for which, sometimes, an article had to be edited more than ten times. Finally, the book *Decoration Technology of Lacquer* was completed and published in 2004. This book comes directly from an artisan of lacquer and is a real monograph on crafts; the content is substantial, while the text is easy to read. It contains some professional questions, such as "how to give the lacquer better transparency" and "how to preserve the lacquer's color for a longer period". Thus, he was praised by the famous lacquer artisan He Haoliang. Mr. Wang Shixiang wrote about the book: "This book is an old lacquerer's own experiences, and craftsmen in the lacquer industry should have his professional ethics".

9.4.9 Yi Lacquer

The most distinctive and popular lacquerware among ethnic minorities are those of the Yi people. These are mainly produced in Xide, Meigu, Bhutto, and Ganluo counties in Liangshan Yi autonomous prefecture, Sichuan province. There are about 1.55 million Yi people in the prefecture. Almost every family uses lacquerware. Their tableware includes cups, plates, alms bowls, bowls, and pots (as shown in Picture 9.44 and 9.45) and their utensils include saddles, quivers, armor, and scabbards. The Jiwu lacquer art family in Xide county has passed down their technique for 15 generations, up to their current descendants Jiwu Wuhe and his son Wuhe Wudong. The Jiwu lacquer art family in Meigu county has passed down their technique for 13 generations, up to Baishi Fuji.

The lacquerware of the Yi nationality is mostly made of wooden-based lacquer. The round bases are spun with a turning-lathe (as shown in Picture 9.46). Their non-circular wooden spoons and jugs are cut with an axe and a machete. Their leather bases are made from buffalo skin that is soaked for 20 to 30 days, and used to cover wood to finalize the design. Horn cups are lacquered directly on the horn. The lacquerware with the most national characteristics is the artificial flask, which has a simple shape, but an ingenious design. The wine injection port is located at the bottom and there is a switch so that the wine will not flow out from the bottom. You

Picture 9.44 Bowl and spoons

Picture 9.45 Painting the pattern, selected from Song Zhaolin's *Lacquering Decoration in Liangshan Yi Nationality*

Picture 9.46 Turning-lathe, selected from Song Zhaolin's *Lacquering Decoration in Liangshan Yi Nationality*

can suck the wine out from the mouth when drinking (as shown in Picture 9.47). Raw lacquer is not dried or boiled, but directly painted on the surface of the vessel, and silver vermilion and yellow stone are added to make the color red or yellow. The lacquerware is mostly black, red, and yellow, and most of them are color-painted. The lacquer brushes are made of goat's beard, so that the lacquer could be applied smoothly into abstract geometric patterns, which are quite elegant.

9.4.10 Tianshui's Carved and Filled Lacquer

Lacquerware is mainly carved and filled, which originated in Sichuan. In the early years of the Republic of China, Zhang Jihong, Governor of Tianshui city, started Longnan First Craft Factory, hired Sichuan artisans Wang Jicheng and Wang Junjie to make carved and filled furniture and food wares. He also hired calligraphers and painters to draft new things. By doing this, Tianshui's carved and filled lacquer

Picture 9.47 Liquor flask

art formed a unique style similar to its painting and calligraphy. In 1992, Longnan First Craft Factory was renamed Minsheng Craft Factory and increased its varieties of lacquerware. In 2007, Tianshui's carved and filled lacquer was listed in the *Second Batch of National Intangible Cultural Heritage List*, with good developmental prospects. Tianshui's carved and filled lacquers include wood, bodiless, leather, and bamboo bases. It is subtle in its carving technique, elegant in its color filling, and interesting in its lacquering (as shown in Picture 9.48). One of its representative works, a round table, imitates Qi Baishi's painting of maple leaves and grass insects, and boasts about the subtle veins of the leaves and insects and the transparent cicada wings. It even won the praise of Shen Fuwen, a famous lacquer artisan. The descendants of Shen Fuwen are Ma Junmu and the brothers Guo Lixue, Guo Bingxue, and Guo Duxue.

Picture 9.48 Flower table, inlaid with mother-of-pearl, designed by Wang Xiaoya

9.4.11 Dafang's Leather Base Lacquerwares

Guizhou's Bijie lacquer is one of the four famous lacquers in China. Dafang lacquerware began in the early Ming dynasty, making daily-use lacquerware such as hat tubes and tea pots for ethnic minorities. In the Republic of China, Zhang Boqing's Treasure Store's leather lacquerwares were famous for their elegant shape, tight seams, and meticulous workmanship, while the leather lacquer plates, boxes, boxes and sets produced by Bai Zhongyi's Minsheng Lacquer Store were famous for their beautiful patterns.

During the anti-Japanese period, Dafang lacquerwares were sold throughout the southwest of China. After the founding of PRC, lacquer processing groups, cooperatives, and the Dafang Lacquerware Factory were established one after the other, and lacquerware production continued to develop.

Dafang lacquerwares can be identified by their leather base. Buffalo or horse skin is soaked in water, baked in fire, covered with wood and sand, buried under the earth, and sanded with stone. There are two shaping methods: ring cutting and ring binding. After dusting, shaping, and fixing, the utensils are molded, the seams are tightly closed, and then you can lacquer or paint. Due to the limitations of the molding method, the shapes of the wares are mostly square or round. Leather lacquerware is heat-resistant, drop-resistant, and moisture-proof, and can be used for storing tobacco, tea, and wine to preserve their color and flavor, causing it to be loved by the public (as shown in Picture 9.49). During the Republic of China, Dafang lacquerware was characterized by (1) embedding lacquer and (2) creating hidden patterns. The former used yellow stone lacquer to paint patterns on the middle of the lacquer surface and then, after it dried, applying transparent lacquer. This technique was named after the pattern: embedding lacquer. The latter is to draw golden lines on the middle lacquer surface and then polishing it. The golden patterns are hidden underneath the lacquer, and their patterns are divided into concave and convex types to present different shaded, which is why they are called "hidden seals" or "hidden patterns". In the 1950s, Dafang lacquerers introduced techniques such as bodiless lacquerware, brocade lacquering, Maki-e and needle carving from Fuzhou, which enriched the expressive force of their lacquer art. Tan Rongzhong was the first master of Dafang leather lacquerwares in the Republic of China.

Picture 9.49 Cigarette case, leather-based lacquerware with hidden patterns

Yang Chenglong specialized in decorative techniques. His nephew, Yang Shaoxian, was the Director of the Dafang Lacquerware Factory and known for developing lacquerware for Maotai Liquor's containers.

9.4.12 Taiwan Lacquerwares

In the Qing dynasty, some Fuzhou and Quanzhou lacquer masters moved to Taiwan to make wood and bodiless lacquerwares, which kickstarted the lacquer industry in Taiwan.

During the Japanese occupation of Taiwan, Vietnamese lacquer trees were introduced and a lacquerware factory was set up in Taichung. The products were used by the Japanese in Taiwan and sold to Japan. After Taiwan's recovery, the Meicheng Crafts Company, founded by Liu Shufang, made bodiless lacquerware and hanging screens, which were extremely popular. Since the 1970s, the cost of local lacquerware in Japan became too high, so people moved to Taiwan to continue to develop the production of wooden lacquerware there. The finished products were mainly sold in Japan. Later, with Taiwan's own economic growth, the method of OEM's production was eliminated. In the 1980s, Taiwan's labor-intensive industries moved to mainland of China, but there were few successors in the traditional lacquer art. Only the technique of bodiless statue lacquer remains intact.

Taiwan's lacquerware created its own characteristic products by integrating Chinese culture and Japanese lacquer art. Famous lacquerers include Chen Huoqing, Wang Qingshuang, Lai Gaoshan, and Huang Lishu (as shown in Picture 9.50).

Picture 9.50 Painted can, made by Huang Lishu

9.5 Lacquer Painting

Lacquer painting refers to the innovation and development of traditional lacquer art into the present era, which evolved the techniques of lacquer decoration from being practical to pure art. The early lacquered Chinese *guqin* and lacquered coffins were painted with vivid scenes such as hunting, music and dance, gods and monsters, which can be said to be the precedent of lacquer painting, but it was still an accessory rather than an independent art variety. In 1963, Vietnamese lacquer paintings were exhibited in Beijing and Shanghai, which gave the audience a brand-new feeling and shocked the Chinese lacquer art circles. Using this as a starting point, lacquerers such as Qiao Shiguang, Wang Heju, Huang Weiyi, and Cai Kezhen began to create lacquer paintings. Later, lacquer paintings were often exhibited in national and local art exhibitions. In 1979, the Fifth National Art Exhibition awarded the winning works, lacquer paintings, for the first time. In 1984, lacquer painting was exhibited as an independent type of painting in the Sixth National Art Exhibition for the first time, with as many as 120 submissions. In 1986, the Chinese Lacquer Painting Exhibition was held at the National Art Museum of China, with more than 700 lacquer paintings on display. That same year, the Chinese Modern Lacquer Painting Exhibition was exhibited in Petrograd and Moscow, with a total of six exhibits. In 1991, the Modern Chinese Lacquer Painting Exhibition toured Tokyo, Kanazawa and Fukuoka, Japan. This indicates that lacquer painting had established and made considerable progress in China, even to the point of having an international influence (as shown in Pictures 9.51, 9.52).

After nearly half a century's exploration and development, Chinese lacquer art formed several schools revolving around different art colleges or lacquer production areas:

In Beijing, Qiao Shiguang from the Central Academy of Arts and Crafts (now Tsinghua University Academy of Arts and Design) has made great efforts to strengthen the paintability of lacquer painting for many years, and he still pays attention to the inheritance of the traditional materials and techniques. Scholars from the Minzu University of China and the Central Academy of Fine Arts are also engaged in creating lacquer paintings.

Fujian's lacquer paintings are represented by the first generation of lacquer painters Wang Heju, Zheng Liwei, Zheng Yikun, and Wu Chuan, and the second generation of lacquer painters Tang Mingxiu, Wu Jiashuan, Chen Lide, and Zheng Xiuqian. Relying on the strong lacquer art tradition, they have formed their own characteristics, excellent workmanship, and given full play to the advantages of natural lacquer, causing them to win many awards in national art exhibitions.

Centered on the Sichuan Fine Arts Institute in Chongqin, Huang Weiyi, Yang Fuming, and Xiao Lianheng participated in the creation of lacquer painting very early and achieved outstanding achievements under the advocacy of Mr. Shen Fuwen.

Representative figures of Jiangxi lacquer painting include Chen Shengmou, Gong Sheng, and Yin Chengzhong. They have good artistic accomplishments, pay attention to the artistic expression of their works, and have created many new techniques based on applying traditional techniques, such as embedding cover, engraving and grinding, onto synthetic lacquer.

Picture 9.51 Contemporary lacquer paintings: 1. Fog, by Wu Chuan, selected from *On Lacquering and Painting*. 2. Yue Pond in Hong Village, made by Ning Baoping. 3. Fishing Boat in the Evening, made by Zheng Xiuqian. 4. White House in front of a Snow Mountain, made by Cheng Xiangjun. 5. Still Life, made by Zhu Xihua

Picture 9.51 (continued)

Guangzhou is centered on the Guangzhou Academy of Fine Arts, most represented by Cai Kezhen. He studied in Vietnam in the early 1960s, specializing in lacquer painting. After returning home, he devoted himself to promoting lacquer painting and once painted large-scale lacquer murals for the Guangdong Hall of the Great Hall of the People.

Nanjing University of the Arts established the lacquer art major under the advocacy of deans Bao Bin and Feng Jian. Their lacquer painting developed very well, causing the emergence of a group of young and middle-aged authors, such as Li Yongqing, Wang Hu, Li Shu, and Zhang Chengzhi. Nanjing Normal University's Wu Keren is also quite an accomplished lacquer painter.

Huang Weizhong of Tianjin Academy of Painting pioneered the aluminum lacquer painting that uses aluminum plates as the base and nitrolacquer as the paint. The

Picture 9.52 Qiao Shiguang's lacquer paintings: 1. Village near the Stream. 2. Fujian Extra-thin Noodles. 3. Believers

process is simple, the color is bright, and it is especially suitable for decorating planes.

Different from oil painting and traditional Chinese painting, lacquer painting's uniqueness lies in the characteristics and application of lacquer. Qiao Shiguang once pointed out: "Lacquer painting is not easy", and lacquer painters should "clearly realize the arduousness of further improving the artistic level of lacquer painting and strengthening the language of lacquer painting". "Starting with natural lacquer and traditional techniques, lacquer painters should first master the language of lacquer painting, and then create lacquer painting articles". While developing artistic lacquer painting, we should not neglect decorative lacquer painting, which can be combined with murals and reliefs to enter the creative field of modern environmental art in order to achieve better survival and development.

9.6 The Beauty of Lacquering and the Way of Lacquering

The introduction and conclusion of the *Complete Works of Traditional Chinese Handicrafts: Lacquer Art* edited by Qiao Shiguang discuss the connotation and extension of lacquer art, the beauty of lacquering, the way of lacquering, the relationship between traditional lacquer art and modern life, and the prospects of Chinese lacquer art. Later, in the name of *Lacquer Art Theory*, those parts were included in the book *Talking about Lacquering and Painting*. The following is an excerpt to conclude this chapter.

9.6.1 The Beauty of Lacquering

The aesthetic value of lacquer is based on its practical utility. At beginning, lacquer was found and applied to the daily utensils. As time passes, people pay more attentions to the aesthetics value of lacquerwares.

Lacquer has an excellent appearance and beauty. As the national lacquer, Chinese lacquer was developed by skillful lacquers in all ages. It boasts both the natural beauty of the lacquer and the superior skills of the lacquers. The two are mutually causal.

The phrase "as black as lacquer" points out the most essential aesthetic appeal of lacquer, as Huang Yongyu, a contemporary Chinese artist, said, "Lacque-black is the darkest and most beautiful black in the world". Wu Guanzhong, a contemporary Chinese artist, also said, "Traditional Chinese culture has two beautiful colors, one is the whiteness of *xuan* paper, and the other is the blackness of lacquer". Black lacquer is refined to a purer black. After grinding and polishing, it becomes even more exquisite. Black is the soul of lacquer art.

Vermilion is the second. The lacquer is mixed with silver vermilion to become vermilion lacquer, which is bright, heavy, steady, and long lasting. Since ancient times, vermilion lacquer has become the second largest color next to black lacquer in lacquer art. The expression "vermilion lacquer does not need to be modified, just as white jade does not need to be carved" underlines its natural beauty. Black and vermilion are the most representative and symbolic colors in lacquer art. The viscosity of liquid lacquer enables gold, silver, mother-of-pearl, jade, bone, and horn to be inlaid in lacquerware. The nobility of gold and silver, the multi-color of mother-of-pearl inlay, the beauty of jade, the simplicity of bone segments and the beauty of black and vermilion lacquer complement each other.

The translucent lacquer can be applied to polishing. After being polished, it has been sanded down to the point that it brings out the changes of light and shade, and gives birth to the hazy beauty of the lacquerwares.

The textures on the lacquer can be drawn with tools or printed with mediums to produce various patterns. Besides that, thinners can be added to flow and permeate into textures. After repeated lacquering and grinding, an ever-changing abstract texture is formed, which is even more beautiful and fascinating.

After the liquid lacquer has fully dried, it can be carved with a knife or needle to further draw dots, lines, and plane. It can also be filled with colored paint, gold, or silver, thus producing a sharp, simple, and bright aesthetic feeling.

Lacquer is viscous so that we can use it to brush or draw lines, just as an architectural color painting. Charcoal powder and tile ash can also be added to lacquer, so that it has a strong plasticity. On the hundreds of layers of lacquer, engravers can be employed, resulting in techniques such as engraved cinnabar lacquering and three-color lacquering, which constitute and amplify the beauty of *yin* and *yang* engravings.

The above extensive and in-depth exploration of the properties of natural lacquer has created various techniques, such as painting, depicting, inlaying, engraving, grinding, painting, varnishing, piling, and lacquer engraving, forming many corresponding lacquer art varieties and the unique beauty of various lacquerwares. Lacquer grade is closely related to lacquer properties and its techniques. Lacquering and decorating, under various techniques, enable the beauty of materials to be put on full display. It also makes lacquer art an independent discipline in itself. Lacquer art is based on beautiful materials and good craftsmen.

9.6.2 The Way of Lacquering

Lacquer art is not only built on the quality of lacquer and its skills, but also infused with human spirit and emotion. It is the unity of technology and art, material and spirit, tools and The Way (Tao).

Lacquer craftsmanship contains the beauty of materials and the spiritual beauty of decoration. The deep richness of the lacquer, the grace and luxury of gold and silver, the beauty of mother-of-pearl inlay, and the gentle simplicity of eggshells provide a rich material prerequisite for the spiritual beauty of lacquer art.

Lacquer art also contains the beauty of materials and the spiritual beauty of decoration. The processes of polishing, grinding, engraving, and stacking make the spiritual beauty of lacquer really shine.

From the initial use of natural lacquer to the development of various techniques, its aesthetic value and practical function are always connected as a whole. Colored lacquer *zhi* (ancient cups) from the Warring States period had 20 snake patterns engraved on the body and the cover. These patterns are so intricately interspersed and intertwined, and their shapes are so peculiar and varied that it makes us feel like the patterns are full of movement and strength. The lacquer cases unearthed from the Han tombs at Mawangdui consist of seven small ones with different sizes in square, rectangular, round, oval, and horseshoe shapes, assembled in a large set, designed to perfectly fit, making them beautiful and coordinated. The rationality of their function reflects a high level of ingenuity, which is by no means dependent on a technique.

On the long journey of lacquer art, we also, of course, encounter mediocre works. The reason is the imbalance of technology and materials. Along with the traditional thought of following The Way (*Tao*) and ignoring tools, there is also the tendency of solely focusing on technology and materials. These two seemingly contradictory tendencies violate the principle of the unity of technology and art, and the unity of tools and The Way (*Tao*). It should be noted that materials and technology are important, but there is also an artistic standard to uphold. Good techniques and luxurious materials do not automatically result in excellent lacquerwares. While the ordinary eggshells can sometimes make masterpieces that are on par with gold, silver, and jade.

The tendency of only focusing on materials and technology is related to court art. This kind of aesthetic view takes preciousness as the beauty, infiltrating a strong concept of hierarchy, mixing beauty and glory into one, and even becoming a symbol of distinguishing the nobility from peasants, thus turning to the opposite of beauty. Because of the precious materials and complicated technology, the aristocratic tendency of lacquer art has always been very obvious. We can find many vulgar works which use the precious materials extravagantly and indiscriminately. Heng Kuan of the Han dynasty opposes to use the luxurious lacquer cups in *Talks on Salt and Iron*. Burnished works with gold or silver inlay in the Tang dynasty cost both money and labor. After the An-Shi Rebellion (755–763), Emperor Suzong in the Tang dynasty ordered a ban to that kind of luxurious lacquerwares. However, the extravagance of the past dynasties remained the same, since it was determined by the nature of autocratic rule. Huang Cheng takes "being obscene and clever" and "being indiscriminate and dazzling" as the precepts of lacquers in *Lacquering Decoration Record*, which is a famous saying and even now, it still rings true with the same practical significance.

On the other hand, folk lacquer art is rooted in the daily lives of the people, with practicality as the leading factor. It takes local materials and does not apply elaborate decorations, imbuing it with pure and simple beauty. In order to achieve the unity of practicality, good quality, and low prices, civil lacquerware must have an ingenious design, which fully shows the wisdom and skills of lacquers. This is the correct way of creating lacquer art.

The beauty of lacquer art, in the final analysis, lies in the beauty of art and the beauty of spirit, no matter if it is practical lacquerware, furnishing lacquerware, lacquer paintings, or lacquer sculptures. Artistic beauty is on the strength of material and crafted beauty, while beautiful materials and crafting aim to achieve artistic beauty. Art is sublimated by materials and craftsmanship, and craftsmanship is immortalized through art. We should not only pay attention to the beauty of lacquer decoration, but also hold true to the way of creating lacquer art.

Chinese lacquer art has come a long way on its 7,000-year journey. Since its birth, it has been deeply imprinted by the national culture and has a distinctive oriental charm. The combination of traditional lacquer art with modern art and industrial designs will surely gain broad development prospects towards pluralism, blending, integration, and contemporary life.

Chapter 10
Furniture Making

Hua Jueming

Human beings made furniture to meet the needs of a sedentary lifestyle. The mats for sitting and a long desk used for writing and conducting official business are all furniture used by Chinese people in their early years. Since the Tang and Song dynasties, sitting on a chair with your feet naturally perpendicular have become the mainstream of Chinese people's daily lives. So chairs and stools, console tables, beds, and cabinets all form the same pattern throughout all kinds of traditional Chinese furniture.

Furniture is necessary for home life. For example, volume 15 of "*All Kinds of Daily Groceries*" in *A Dream of Lin'an* by Wu Zimu, a scholar of the Southern Song dynasty, said that "The furniture in an average home includes a table, stools, a bamboo bed used to cool off in summer, and an ancient folding chair". Up to now, people living in the area south of the Yangtze River still regard furniture as "*jiasheng*". Furniture is closely related to people's lives. Large pieces of furniture such as a bed and a storage cabinet, and small pieces of furniture such as suitcases and benches, are a necessity. Their shape, structure, production methods, decoration, function, and application reflect the living customs and social features of different times and regions. These became an important part of national culture. When you take a look at the history of furniture, you can see the history of human civilization laid before you in terms of utensils, crafts, folk customs, aesthetics, and society (as shown in Pictures 10.1 and 10.2).

The production of furniture is also related to the national economy and people's livelihoods. Throughout history, it is not just a relatively independent part of handicrafts performed on a small scale, but it is a major category within traditional handicrafts. Although most Chinese traditional furniture is made of wood, there's also furniture made of bamboo, rattan and other materials.

H. Jueming (✉)
The Institute for the History of Natural Sciences, Chinese Academy of Sciences, Beijing, China
e-mail: huajueming@163.com

© Elephant Press Co., Ltd 2022
H. Jueming et al. (eds.), *Chinese Handicrafts*,
https://doi.org/10.1007/978-981-19-5379-8_10

Picture 10.1 The furniture store recorded in *Qingfeng Painting* by Jinkun, a famous painter in the Qing dynasty, in the fifth year of the Qianlong period (1740), which is currently stored in the Palace Museum

Picture 10.2 A woman's dowry in her bridal chamber in Ninghai, Zhejiang province, cited from *Girl Dreams of a Rich Dowry* by He Xiaodao

This book will cover the evolution of wooden furniture, its different categories, its production techniques, and different schools.

10 Furniture Making

10.1 Evolution

In ancient times, Chinese people sat on the ground. Since the crops varied from place to place, people used a variety of products such as reeds, bamboos, and straw to weave mats. People also laid felt and blankets on the ground to sit on. Everyday expressions, such as "banquet", "taking one's seat at a banquet", "rolling up like a mat" and "chairman", all include the Chinese character "席" (*xi*), which means "mat", and thus can all be traced back to the distant past of sitting on the floor.

It is known that the earliest bed in China was unearthed from the Warring States tombs in Xinyang, Henan province (as shown in Picture 10.3). The bed surface is covered with wooden boards, surrounded by a frame, and there are two openings on the sides so people can get in and out of the bed. The height of the bed's legs is only 19 cm. An *an* is a piece of furniture that people used when they ate, read or wrote. The earliest one was unearthed from the Warring States period tombs in Liuchengqiao, Changsha and Hunan province (as shown in Picture 10.4). Beds were widely used in the Han dynasty. They were used both for sleeping and sitting. People could receive guests that would sit on the bed during a feast and play games. Some beds were also equipped with draperies or screens. The *ta*, similar to a couch, was small. There is a saying that "if it's three point five *chi* (about 1.1 m) long, it's called a *ta*; if it's eight (2.7 m), it's called a bed". Lacquer art prevailed in the Han dynasty and furniture painting became the fashion of that era. Furniture was painted in red ink, with gorgeous patterns, bright colors, and meticulous craftsmanship. The most representative pieces of that era are the lacquered *an* and *ji* (a kind of small table) that were unearthed from the Mawangdui Tombs of the Han dynasty in Changsha, Hunan province.

Picture 10.3 Bed of the Warring States period

Picture 10.4 *An* of the Warring States period

The evolution from sitting on the floor to sitting on a seat with your feet perpendicular to the ground encompasses a long historical process. The *Huchuang*, a cross-legged stool that originated in *Xiyu* or the Western Regions, and the ancestor of the folding stool and the folding armchair, started being used in China's Central Plains in the Eastern Han dynasty due to the development of the Silk Road and the Western Regions.

The great integration of nationalities in the Southern and Northern dynasties promoted a change of lifestyle. Square seats such as chairs and stools were used in upper-class families, as well as *dun*, which is like a stool and is made of thick wood. The height of beds increased noticeably, and the body of the bed also became larger. On top of all that, a new type of three-legged stool also appeared.

The Tang dynasty was the transitional period of the changes in lifestyle. High-type and low-type furniture coexisted at that time, and armchairs entered the homes of common people, but sitting on the floor was still customary for most people. During this period, the chairs and seats were wide, heavy and squat, while the beds, with their vertical screens, were big. Large *an* and chairs were often decorated with elaborate carvings, showing the magnificent atmosphere of the prosperous Tang dynasty.

The furniture of the Five Dynasties period was different from the Tang style. From the paintings of Gu Hongzhong (910–980) and Zhou Wenju, it can be seen that furniture at that time was light, simple and subjectively beautiful (as shown in Picture 10.5). This new furniture boasted round-backed chairs and railing beds. Their size was consistent with how people sat, with their feet down, instead of sitting on the floor.

The Song dynasty was a key period for the transformation of Chinese people's lifestyle. During this period, tall furniture developed greatly and grew in popularity. There were all kinds of desks and tables, such as painting desks, *xiang an*, which was a table that could hold a candlestick, desks, and rectangular tables. There are also many kinds of chairs, such as (1) the *jiao* chair, which was an ancient folding chair, (2) Zen chair, which was a relatively large chair and was named after Zen masters who could sit on it cross-legged, (3) backrest chair, and (4) round-backed armchair. The structure of furniture was influenced by wooden architecture. For example, the

Picture 10.5 *The Night Revels of Han Xizai* painted by Gu Hongzhong

structure of boxes and cabinets mostly adopted the beam-column frame structures seen in wooden buildings. During the transitional period of the Northern and Southern Song dynasties, the way people sat changed from sitting on the floor to sitting with their feet down, which means that people's lifestyle had undergone a historic change (as shown in Picture 10.6). Furniture shapes became more uniform during the Song, Liao, and Jin dynasties. New pieces of furniture came into existence such as kitchen drawers, folding tables, high tables, and folding chairs. Decorative elements were added, such as the horse-hoof-shaped legs, bent legs and high girdle waist. The Yuan dynasty was relatively short and therefore did not have a great impact on the development of furniture. Its only contribution was the creation of drawer tables.

The Ming dynasty ushered in the golden age of Chinese furniture making. The design, production and materials of furniture reached its peak, forming a unique style praised and famous all over the world, known as the "Ming-style furniture" (as shown in Pictures 10.7 and 10.8).

Especially during the Jiajing, Longqing and Wanli period (1522–1620) and about two hundred years after that, the use of hardwood furniture became fashionable all over the north and south of China (as shown in Picture 10.9).

Because the best furniture is made of hardwood, many items have been passed down to this day and become the treasures of major museums and collectors.

The style characteristics of furniture in the Ming dynasty can be summarized by concise modeling, rigorous structure, moderate decoration and subjectively beautiful texture. The shape of furniture matches its function, with an elegant appearance and scientific design. The parts with a large span are inlaid with firm and beautiful parts that serve as both support and adornment of the furniture's frame, which was called apron-head spandrel, apron, four-sided inner frame, arch-shaped inner frame, and

Picture 10.6 Furniture in a Song dynasty painting

humpbacked stretcher during the Ming and Qing dynasties. Decorative techniques such as carving, inlaying and tracing, were often combined with a variety of materials, such as enamel, mother-of-pearl inlay, bamboo, ivory, jade, and stone, all of which were made and processed to an exacting first-class standard. For example, the back of a chair had carved or inlaid patterns, while some parts of the table were also carved in flower patterns, which resulted in a simple and comely style. Exquisite and high-quality wood with natural color and beautiful texture, such as fragrant rosewood and red sandalwood, did not require any lacquer decoration and thus displayed its natural beauty. All these characteristics are related to the direct participation of literati in

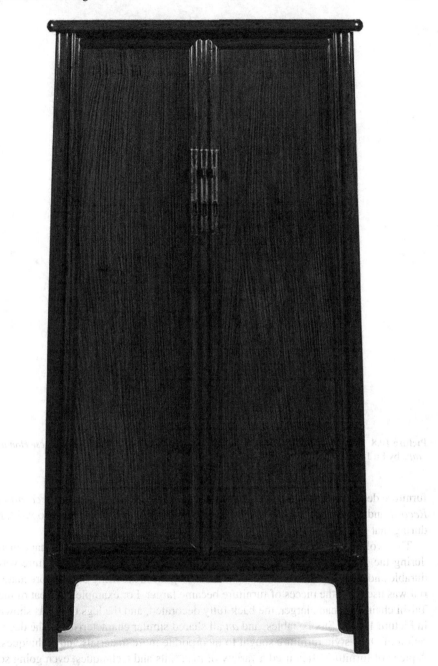

Picture 10.7 Ming-style corner cabinet, cited from *Furniture Making* in *Chinese Handicrafts* by Lu Jun

Picture 10.8 Ming-style round-backed armchair, cited from *Furniture Making* in *Chinese Handicrafts* by Lu Jun

furniture design and production of that time. Works such as *Lacquering Decoration Record* and *Mirror of Craftsmanship and Guidelines by Lu Ban* were also written during that period.

The evolvement of furniture style from the Ming to the Qing mainly happened during the Kangxi and Jiaqing period (1662–1820). The Qing-style furniture was durable and dignified instead of being tall and elegant as the Ming-style. More material was used and the pieces of furniture became larger. For example, the seat of the Taishi chair was made larger, the back fully decorated, and the legs thick (as shown in Picture 10.10). Stools, tables, and *an* all shared similar characteristics. The decoration of Qing-style furniture sought to incorporate more materials and techniques. A piece of furniture often used a variety of materials and techniques, even going so far that the whole body was carved, without leaving blank, so that it became burdensome and impractical. From Emperor Daoguang's reign (1782–1850) to the end of the Qing dynasty (1912), furniture production went from bad to worse because of political and economical turmoil. Due to foreign influence, the style of furniture was

Picture 10.9 Furniture in the illustrations of *The Golden Lotus*

Picture 10.10 Ming-style Taishi chair, cited from *Furniture Making* in *Chinese Handicrafts* by Lu Jun

affected and widely produced furniture started showing signs of the French Rococo style. Since the Republic of China, furniture has undergone great changes under the impact of modernization. Since this book only covers traditional handicrafts, that change will not be explained here.

10.2 Categories and Styles

Wang Shixiang (1914–2009), in his book *Research on Furniture in the Ming Dynasty*, divides furniture into five categories: chairs and stools, console tables, beds, cabinets, and others. This book follows that classification but with some slight changes, and introduces different styles of each category according to their functions:

10.2.1 *Chairs and Stools*

Chairs and stools are tools that people sit on. Those with a backrest are called chairs, while those without a backrest are stools. Stools have pedals. In the beginning, pedals were simple foot boards. After the Han dynasty, their height gradually increased and became bigger before they became tools for people to sit on.

The stool depicted in the mural in the Eastern Han tombs in Liaoning province is wide at both ends and narrow in the middle, which makes it look like a waist drum. The porcelain stool model from the Sui tombs, which has a rectangular surface with the same width as the legs, is the earliest known physical evidence of a stool. Stools were very common in the Song dynasty. There were square stools, round stools, benches, low stools, and round piers stools. The stools of the Ming and Qing dynasties were more exquisite in workmanship and materials, but there was no big change in their shape.

10.2.1.1 *Wu* Stool

It is commonly known as a "small bench", which is widely used and can be placed in the courtyard and on the warmer end of a *kang* (brick bed) (as shown in Picture 10.11). The material of the stool's surface can be, but is not limited to, wood, rattan or leather.

Picture 10.11 *Wu* stool, cited from *Furniture Making* in *Chinese Handicrafts* by Lu Jun

Picture 10.12 Square stool, cited from *Furniture Making* in *Chinese Handicrafts* by Lu Jun

10.2.1.2 Square Stool

The square stool comes in various styles. The surface of its seat ranges from one *chi* to two *chi* (33 cm to 67 cm) (as shown in Picture 10.12) and it can be used in conjunction with a square table. The importance of the square stool in people's daily lives is second only to that of chairs. In the Ming and Qing dynasties, the surface of square stools was mostly made of marble or softly woven silk rope and rattan, which was both practical and gorgeous.

10.2.1.3 Round Stool

The round stool has a round surface. It became popular in the late Qing dynasty with its elegant shape. It is often decorated with painted and engraved flowers. The legs of the stool are reinforced with a stretcher and *yaban*.

10.2.1.4 Drum Stool

The drum stool's legs are bent and its feet are connected to the supporting base, which makes the overall shape look like a waist drum, hence its name.

If the legs and the supporting base are on the ground and the shape of the stool's bottom and surface is the same, it is called a drum stool. This delicate and ornate seat is often used indoors or in the courtyard.

10.2.1.5 Bench

Its surface is narrow and long, like a long strip. It is a popular seat among common people due to its convenience and practicality. It can also be used to support boards, boxes, or simply for work. For example, a knife sharpener often walks through the streets with a small bench. A spring bench (as shown in Picture 10.13), lighter and more delicate, is mostly used indoors, and usually has drawers woven with rattan underneath.

The earliest known seat was a folding stool. However, its original name was "*hu chuang*" (*hu* bed), as recorded in the *Book of Later Han:* "Emperor Ling of Han (156–189) likes *hu* clothes, *hu* hangings and *hu* beds […]" This shows that the folding stool was introduced into the Central Plains from the Western regions. In *Ma Weidu Talking about Collection: Furniture*, the author quotes Li Bai's poem *Silent Night Thinking*: "Before my bed a pool of light, I wonder if it's frost aground", and Bai Juyi's poem *Yongxing*: "There is a *hu* bed in the boat". The "bed" and "*hu* bed" in the poems both actually refer to the folding stools. Today, a folding stool is still called a "small bed" in Shanxi and Hebei provinces. This kind of seat is light and convenient, but it is not that comfortable. In the Song dynasty, a back and armrest were added, thus resulting in the ancient folding chair.

Picture 10.13 Spring bench, cited from *Talking about Mahogany Furniture in the Ming and Qing Dynasties* by Pu Anguo

Picture 10.14 Ancient folding chair of dalbergia odorifera in the Yuan dynasty, collected by Mrs. Chen Mengjia, and cited from *Classic Chinese Furniture: Ming Dynasty* by Wang Shixiang (1914–2009)

10.2.1.6 Ancient Folding Chair (as Shown in Picture 10.14)

Also known as *Jiaowu*, meaning "foldable", the ancient folding chair was first used in northern China. It fit the needs of nomadic and military life, as well as trekking, which is why it is also called a walking chair. In ancient times, the folding chair was only used by dignitaries, so it gradually became a symbol of power. It is recorded in the *Outlaws of the Marsh* that people use the available seats to divide people into their different ranks.

10.2.1.7 Round-Backed Armchair

Commonly known as an easy chair, it is named after the circular and curvy shape formed by the back and armrest. Its design is ergonomic and comfortable, and it is a good example of traditional Chinese furniture.

10.2.1.8 Armchair

There are two closely-related kinds of armchairs: hat chairs and *nanguan* hat chairs. The hat chairs, also known as *sichutou*, get their name from their shape, which resembles the hat of an ancient official. The *nanguan* hat chairs are one of the main pieces in Ming-style furniture. The former has a top rail and armrest, while the latter

Picture 10.15 Armchair

does not (as shown in Picture 10.15). The oversized armchair, also known as *taishi* chair, is a kind of armchair with flower patters, which was widely used between the Song and the Qing dynasties, by the upper-class and common people alike. It can be said that it suited both refined and popular tastes. They are often arranged in pairs, for example, two chairs for each small table. Eight chairs with four small tables form a hall, which subsequently became the fixed pattern for halls.

10.2.1.9 The High-Backed Chair

The high-backed chair has a backrest but no armrest. It is widely used and can be divided into screen back chairs and lamp hanging chairs.

10.2.2 Beds

Beds used to be objects both for sitting and lying down. The couch, was originally used as a seat and afterward as a bed. Generally speaking, a couch was smaller and shorter, and there was no fence around it (as shown in Picture 10.16). When beds became larger and taller, fences were installed to prevent people from falling off as they slept. In future generations, people even installed frames on their beds.

10.2.2.1 Couch

Both single-seated and double-seated couch were widely used in the Han dynasty. In the Tang dynasty, the height of the couch increased, while the couch in the Ming and Qing dynasties inherited the style of the Song and Yuan dynasties, which were narrow and mostly without fences.

10.2.2.2 Arhat Bed

The Arhat bed (as shown in Picture 10.17) were surrounded on three sides and often displayed in the hall, where it could be used to sit or lie on. In the era when Buddhism was widely upheld, ordinary people and monks often sat on the bed to read scriptures and talk about *Tao* (The Way), hence the bed's name.

10.2.2.3 Shelf Bed

The shelf bed (as shown in Picture 10.18) has four sides with four corner columns. The top of the bed is covered with lintel boards around the cover, which creates a private space and gives people a sense of safety and comfort. For the common people, it was mostly used as a bed.

Picture 10.16 Elm couch in the Qing dynasty, cited from *Ma Weidu Talking about Collection: Furniture*

Picture 10.17 Arhat bed, cited from *Furniture Making* in *Chinese Handicrafts* by Lu Jun

Picture 10.18 Shelf Bed made of dalbergia odorifera, cited from *Research on Furniture in the Ming Dynasty* written by Wang Shixiang

10.2.2.4 Alcove Bedstead

It evolved from the ancient shelf bed. The front side of the bed had a space where washing utensils and a chamber pot could be placed and thus formed a multifunctional private space. It was very popular because of its large size as well as its concealment, which makes it almost like a small room inside a large room (as shown in Picture 10.19).

10.2.3 Table and An

Tables are widely used for dining, reading, painting, and office work. An "*an*" can be an altar *an*, when it is used as a table to place fruit as an offer to dead people (usually ancestors). It can be a painting *an*, when it is used for writing and painting. It can also be an eating *an*, when it is used as a dining table. The *an* comes in two

Picture 10.19 Wedding bed used by the Ninghai people in Zhejiang province, cited from *Girl Dreams of a Rich Dowry* by He Xiaodao

Picture 10.20 Square table, cited from *Talking about Mahogany Furniture in Ming and Qing Dynasties* by Pu Anguo

common styles: with a flat top and with an upturned top. A *ji*, on the other hand, is both smaller and lower than a regular table, and is commonly known as a tea table or flower table. There are many kinds of tables, *an* and *ji*, so we should pay special attention to how to arrange them properly when we use them.

10.2.3.1 *Kang* Table

It is a unique piece of furniture placed on the *kang* (brick bed) in northern China. In warm weather, it can be moved outside sometimes.

10.2.3.2 Square Table

A square table is the most commonly used table in China. Both its production and placement are very simple. The big table (as shown in Picture 10.20), known as "*baxian* table" (square table), can hold eight people, and the small one is called "*liuxian* table" (medium-sized square table).

10.2.3.3 Round Table

This kind of table started gaining popularity in the Qing dynasty. The small one serves as a tea table, while the big one as a dining table. Some round tables consist of tabletops and table racks.

10.2.3.4 Desk

It is similar to *ji* and *an* in shape, but it has drawers for easy reading and writing. A drawing table is similar to a desk, but it usually has no drawers. In addition, there are half round tables, also known as crescent tables, half tables, also known as half square tables, and long tables as well.

10.2.3.5 Altar

Most of them are placed in halls. They have a long body and an upturned top, which gives them a solemn and dignified look. Candle holders, incense burners and other offerings are often placed on alters.

10.2.3.6 Flat-Top *An*

There is no upturned top on either end of this kind of long table. It is often placed in the study or hall, where antiques, stationery and books are placed.

10.2.3.7 Tea Table

Most of them are square or rectangular, and are used for placing cups, plates and tea sets on, hence the name. They are often combined with armchairs to make a set, so their shapes, decorations and colors must be coordinated. The surface of the tea table is sometimes decorated with marble and mother-of-pearl inlay.

10.2.3.8 Flower Table

The flower table is usually high, since it is used to display flowers or bonsai trees. This kind of table usually has an elegant shape. The incense table is an ancient piece of furniture for holding incense burners (as shown in Picture 10.21).

10.2.4 Box and Cabinet

The box and cabinet are pieces of furniture used for storage. The former is smaller, while the latter is larger. The box is not as tall as it is wide, and can be easily moved. It is, however, usually placed in a fixed place, with a base underneath. The cabinet is larger. The height of a corner cabinet is greater than its width (as shown in Picture 10.22), while a sideboard cabinet is the opposite. They are not easy to move; thus they remain in a fixed location. Kitchen cabinets were introduced after

Picture 10.21 Incense table made of dalbergia odorifera in the Ming dynasty, cited from *Classic Chinese Furniture: Ming Dynasty* by Wang Shixiang

the Jin dynasty. Their functions were similar to those of regular cabinets but they were generally wider.

10.2.4.1 Round Corner Cabinet and Square Corner Cabinet

The former has a cap on the top and the doors extend into the axis by the frame, which closes by themselves due to gravity. The latter has no cabinet cap on it, with the same size from top to bottom, and its four corners form a square. The doors are connected with the cabinet walls by hinges (as shown in Picture 10.23).

Picture 10.22 Sideboard cabinet, cited from *Furniture Making* in *Chinese Handicrafts* by Lu Jun

10.2.4.2 Top Cabinet

It is composed of a bottom and a top cabinet, also known as top box corner cabinet, which is an important combined storage tool and is furnished in pairs, so it is also called a four-piece cabinet.

10.2.4.3 Showcase Lattice Cabinet

It is a combination of lattice and a cabinet. The lattice part does not have a door, so it is called "showcase lattice". Usually, books are placed in the lower cabinet and antiques are stored in the upper lattice. They are mostly placed in halls and studies.

10.2.4.4 Clothing Box

It is a box for storing clothes and is mostly made of plates. The lower and upper parts of the box are often covered with copper ornaments to make the corners of the box firm. The front face plate and *paizi* of the box are used for locking (as shown in Picture 10.24). The clothing box is usually made of camphor wood, which acts as a repellant and is insect-proof.

10.2.4.5 Stationery and Medicine Box

This kind of box is small and convenient for people to carry around. It contains multiple smaller boxes or drawers for storing various stationery or medicinal materials. A large medicine box is one cubic meter and needs to be carried by two people. There are also treasure chests, which are equipped with a number of small drawers to hold jewelry, and some contain ciphers and secret boxes. The treasure chest mentioned in the story *Lady Du Angrily Sinks Her Treasure Chest* is an example of it.

Picture 10.23 Square corner cabinet, cited from *Furniture Making* in *Chinese Handicrafts* by Lu Jun

10.2.4.6 Bookshelf

Also known as a book grid, it is a necessary item for any study, with a concise shape and little decoration (as shown in Picture 10.25). There are also book boxes that are used for storing books. Some books that are both large and thick, such as the *Twenty-four Histories* and *Lei Shu* (reference books), are stored in special book boxes.

Picture 10.24 Embossed clothing box, cited from *Girl Dreams of a Rich Dowry* by He Xiaodao

10.2.4.7 Antique Shelf

Also known as *boguge* or *baibaoge*, antique shelves are shaped similarly to book shelves, but are divided into several smaller grids with different styles. This is because they are used for displaying a collection and should be pleasing to the eye. It became popular immediately after the rise of the Qing dynasty and has remained popular ever since (as shown in Picture 10.26).

10.2.5 Other Categories

10.2.5.1 Screen

A screen is a special piece of furniture. It can be used to divide space, block the wind, and shield objects or people. There are two categories: a seat screen and a foldable screen. A seat screen is mostly displayed indoors and is divided into single-leaf and multiple-leaf. A foldable screen is movable. The application of a screen, which is filled with decoration and visual effects, is closely related to the traditional values and customs of Chinese people.

Picture 10.25 Bookshelf, cited from *Furniture Making* in *Chinese Handicrafts* by Lu Jun

10.2.5.2 Dressing Table

The dressing tables or dressers evolved from mirror racks. They are necessary pieces of furniture in a traditional Chinese home and mainly come in two styles: (1) the folding style and (2) the throne style. The materials, production technology, and decoration are exquisite.

Picture 10.26 Gold-painted antique shelf made of rosewood, cited from *Fine Workmanship: A Brief History* of *Chinese Furniture* by Zhang Yu

10.2.5.3 Brazier Rack, Hanger and Washbasin Racks

These pieces of shelf furniture are made for storing clothes and utensils. Their overall appearance, shape, and decoration can be simple or intricate, unadorned or opulent. It all depends on the family that uses them. In many areas of Zhejiang province, there is a wedding custom of preparing a rich dowry. This dowry may include a wedding bed, a wardrobe, a crate, or other items. Even a washbasin rack that is used in daily life can be beautiful and elegant (as shown in Pictures 10.27 and 10.28).

10.3 Design and Production

There are many steps in furniture production, such as design, material selection, material preparation, components preparation, splicing and assembling, painting decoration, and making accessories. We will briefly take a look at each step.

Picture 10.27 Hanger, cited from *Girl Dreams of a Rich Dowry* by He Xiaodao

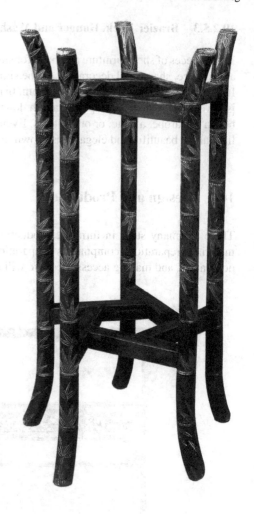

Picture 10.28 Washbasin Rack, cited from *Girl Dreams of a Rich Dowry* by He Xiaodao

10.3.1 Design

The first and most important process of furniture making is designing. The shape, size, structure, materials, and decoration of furniture are all determined in advance, and will not change anymore during construction.

Mirror of Craftsmanship and Guidelines by Lu Ban of the Wangli period's version records the making of a canopy bed as follows. "The bed is 2 *chi* and 2.2 *cun* high from the floor to the *chuangfang*, which is 7.7 *cun* or 5.5 *cun* in size. The *shangping* is 4 *chi* 5.2 *cun* high. There is a *houping* at each end of the bed, one wider than the other. The slightly wider *houping* is 4 *chi* wide, the narrower one is 3 *chi* 2.3 *cun* wide, and both are 6 *chi* 2 *cun* long. The *zhengling* of the bed is 1.4 *cun* thick, made up of both large and small boards. It has to be ensured that the middle parts should be fixed to each other. The front step is 5.6 *cun* high and 1 *chi* 8 *cun* wide. The qianmei, together with the canopy, is 1 *chi* 0.1 *cun* tall. There are four lower doors, each mearuring 1 *chi* 0.4 *cun* tall. The *shangnaoban* (headboard) is about 8 *cun* long, the *xiachuanteng* (lower rattan footboard) is 1 *chi* 8.4 *cun* long, and the rest is the *xiabanpian* (lower footboard). The door frame is 1.4 *cun* wide and 1.2 *cun* thick. The doorsill is 1.4 *cun* tall, which consists of three sections. The inside *zhuangzhimen* at both ends the bed is 9.2 *cun* or 9.9 *cun* wide, but never wider than 1 *chi*. Later craftsmen made canopy beds according to Lu Ban's size." Wang Shixiang further explains some of the terms mentioned above. The *chuangfang* refers to the bed board, *shangping*, *houping* and *liangtou liangpian* all refer to the surrounding screens. The *zhengling* refers to the front of the canopy, and *qianmei* refers to the lintel of the canopy's front eave. There are altogether four doors at both ends in front of the bed, which look like partition boards. These doors are composed of a *shangnaoban* (headboard), rattan and lower footboard. The inner *zhuanzhimen* may refer to the two doors at both ends of the bed's front. Wang also restored the diagram of the canopy bed (as shown in Picture 10.29) in his book. The above description shows that a complete and careful design process was needed to make traditional Chinese furniture. Although they did not have complete sets of design books or drawings, craftsmen were able to accurately express the design, shape, structure, and size of furniture with their own jargons and drawings based on their rich experience and professional skills passed down from generation to generation. When at work, they had a clear and vivid image of what they are creating in their minds.

10.3.2 Tools and Equipment

As the old saying goes: "A handy tool makes a handy man." "Good workmanship requires good tools." "You will get double result with half the effort if you use the right tools the right way." These proverbs emphasize the importance of tools and equipment. Tool and equipment used in furniture production are closely related to the woodworking industry, with some quite specialized. Generally speaking, they can be categorized as follows.

10.3.2.1 Workbench

It is commonly known as "*leng*" in China. When setting up a workbench, it is necessary to pay attention to the appropriate height. The professionals usually follow these rules: The distance from the tip of a carpenter's middle finger to the ground should be the exact height of the workbench when he stands up straight with his hands drooping naturally. The legs should be firmly placed on the ground, with one end supporting the middle of the board. The workbench should be very stable and the board should be straight and flat. The sawtooth of the board (the so-called "*banqi*", i.e. planing stop used for affixing materials) should be able to hold on to the wood. These seemingly simple things are of great importance to a carpenter. As the saying

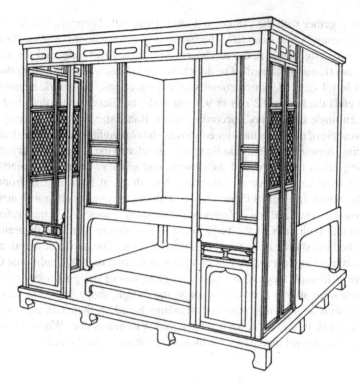

Picture 10.29 Restoration diagram of a canopy bed, cited from *Research on Furniture in the Ming Dynasty* written by Wang Shixiang

goes, a clumsy carpenter will make a scene by kicking the workbench, blaming the saw and throwing his carpenter plane. If the workbench is unstable or the parts being processed are unstable, it will result in uneven planing and sawing, and the carpenter will surely take out his anger on the workbench.

10.3.2.2 Line Drawing Tools

They include a carpenter's ink marker and any kind of cutting gauge (as shown in Pictures 10.30 and 10.31).

10.3.2.3 Saw

Sawing takes the longest time and energy. When sawing along the direction of the wood grain, a longitudinal saw with three rows of teeth must be used at a small vertical angle. When sawing perpendicular to the wood grain, a cross saw with two rows of teeth must be used at a large vertical angle. Dual saws are used for cutting

Picture 10.30 Carpenter's ink marker, cited from *Chinese Handicrafts—Furniture Making* by Lu Jun

Picture 10.31 Cutting gauges, cited from *Chinese Handicrafts—Furniture Making* by Lu Jun

materials, tenon saws are used for tenoning, and digging and bending saws have a narrow blade that they can be used to cut arcs. There are also bow saws, jointing saws, and arm saws (as shown in Picture 10.32).

10.3.2.4 Adze

Before the plane was invented, the adze was used to make wood even. Now it is mostly used for peeling round logs and making them even.

Picture 10.32 Saws, cited from *Chinese Handicrafts—Furniture Making* by Lu Jun

10.3.2.5 Axe

The blade of the adze is horizontal, and the blade of the axe is straight. The axe is used to split materials, shape wood and make the wood even.

10.3.2.6 Plane

The function of a plane is to straighten and smoothen lumber, which is extremely important for the production of furniture, especially hardwood furniture. In China, hand planes appeared quite late and were not widely used until the Ming dynasty. There are many kinds of hand planes, such as the jointing plane (for splicing wood board), the *erhutou* (for smoothing), the *tianbao* plane (for curved surface), the smoothing plane (for fine processing), the rabbet plane, the centipede plane, and the router plane (as shown in Picture 10.33).

Picture 10.33 Carpenter's planes of various colors

10.3.2.7 Chisel

Different type of chisel has a cutting edge that is used to carve various patterns, but just in different shapes. A variety of shovel-like tools were also used in the various cutting and carving processes.

Besides, there are tools and equipment for the purpose of boiling fish glue and melting wax.

Tools are vital to craftsmen. They can only fully display their craftsmanship if they have the right ones. Out of necessity, many carpenters often design and make their own tools to fit their own hands.

10.3.3 Material Selection

Furniture materials are divided into non-hardwood and hardwood.

All types of pine, Chinese fir, manchurian catalpa and basswood belong to the non-hardwood category, collectively referred to as miscellaneous wood or inferior wood, which are widely used among the common people for furniture.

Beech is quite suitable for furniture making since it is firm and beautiful in texture. It was widely used in Ming-style furniture made in Suzhou, and was referred to as "*nanyu*" (southern elm) in the north.

There are two types of nanmu (phoebe zhennan), namely phoebe nanmu and gold-rimmed nanmu. The former is mainly planted in Yunnan and Sichuan provinces, and the latter in Jiangsu, Zhejiang, Anhui, and Jiangxi provinces. They are cherished by people for their elegant look and low shrinkage percentage.

Birch is produced in the north, which is firm and shiny in texture. It is on par with red sandalwood in the south.

There are two types of cedar wood, namely, northern cedar and southern cedar. The latter, also known as oriental arborvitae, is better. Golden cypress is a valuable type of material for furniture making besides hardwood.

Camphorwood, which is common in southeast China, namely Fujian, Taiwan, Jiangxi, and Hunan provinces, has a natural fragrance and is moth-proof. Therefore, it is commonly used for boxes and cabinets, which are cherished by people.

Red sandalwood, produced in Indochina and Yunnan province, is considered the best hardwood. Being firm, dense and beautiful in texture, it has been cherished since ancient times and cost one to several times more than other hardwood. It is used to make valuable furniture and exquisite devices.

There are two kinds of Chinese rosewood: dalbergia odorifera and henryi prain. The former, produced in Hainan province, is solid and beautiful in texture and fragrant. It is locally known as "*jiangxiang* wood", which is second only to rosewood. Compared to dalbergia odorifera, henryi prain is coarser in texture, less varied in texture and has no fragrance. It is produced in Zhejiang, Fujian, Guangdong, and Yunnan provinces. Dalbergia odorifera was so scarce that furniture was mostly made of henryi prain after the mid-Qing dynasty.

Ceylon ironwood, on the other hand, was easily available. Therefore, it was often used to make large furniture in the Ming and Qing dynasties and also used in buildings in Guangdong province.

There are mature and immature wenge, with the former pale-colored, light, and shiny, and the latter dark-colored, heavy, and untextured.

After the mid-Qing dynasty, mahogany was imported in large quantities due to the lack of dalbergia odorifera and mature wenge. This kind of wood shares some characteristics with rosewood and dalbergia odorifera, and it has been widely used and been the most important hardwood for the furniture making industry (as shown in Picture 10.34).

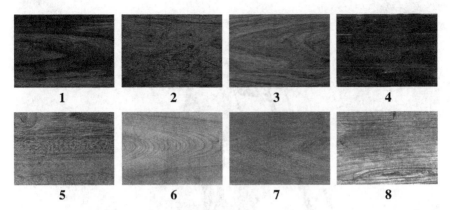

Picture 10.34 All kinds of wood, cited from *Chinese Handicrafts—Furniture Making* by Lu Jun
1. rosewood 2. dalbergia odorifera 3. mature mahogany 4. ebony 5. ceylon ironwood 6. nanmu 7. beech 8. elm

10.3.4 Material Preparation

After wood is selected, a specialized carpenter will saw it with a big saw (as shown in Picture 10.35), and then process it into board-shaped, rod-shaped and square-shaped wood planks for later use.

10.3.5 Component Preparation

Wood blanks are then processed into various components according to the requirement of different kinds of furniture. Attention should be paid to the suitability between texture and color during this step. For example, pairs of cabinets or chairs should be as symmetrical as possible. This means that the grain on door panels must go upwards. All components are made by drawing lines (as shown in Picture 10.36), planing, chiseling (as shown in Picture 10.37), tenoning and engraving.

10.3.6 Splicing and Assembly

Before components of a piece of furniture is assembled, they should be tested. If the joints are not flush, mortise and tenon are not compatible, proper adjustments need to be made. Large-sized panels and door panels are spliced by gluing, pinning, punching, or penetrating with a transverse brace. Because of the high humidity in the south, the splicing method commonly used is the combination of planted pins, iron pins, and mortises and tenons, while in the north, boiled fish glue is commonly used. The purpose of penetrating transverse braces is to prevent cracking and warping.

Picture 10.35 Sawing wood

For example, when attaching a mortised-and-tenoned frame to a floating panel, four wooden edges and cross-grid shoulders are combined into a frame, into which the wooden board is tied and embedded with transverse braces. All the coarse side of the wood is embedded in the mortise and tenon joints, which makes the panel firm and good-looking (as shown in Picture 10.38).

Picture 10.36 Drawing lines, cited from *Dayalou Studio*

10.3.7 Mortise and Tenon Structure

Chinese carpenters connect different components with each other firmly by skillfully using various fine mortise and tenon structures instead of nails or glue. This is a major feature and advantage of traditional Chinese furniture. It comes in a wide variety and goes all the way back to the Warring States period (as shown in Picture 10.39). After long-term development, it reached its pinnacle in the Ming dynasty.

Picture 10.37 Chiseling, cited from *Chinese Local Handicrafts* by Gao Xing

Tenons can be divided into the through tenon (dovetail) and half tenon (dowel). The former penetrates the mortise hole and exposes the tenon. The latter does not penetrate all the way through, and therefore the tenon is not exposed. Below, five commonly used mortise and tenon structures will be introduced.

10.3.7.1 Long and Short Tenon

Long and short tenon has a long leg and a short leg, which are perpendicular to each other, and respectively match the mortises of the *dabian* (tenon-bearing bars of a frame) and *motou* (bars of a frame with mortise).

Picture 10.38 Attaching a mortised-and-tenoned frame to a floating panel, cited from *Chinese Handicrafts—Furniture Making* by Lu Jun

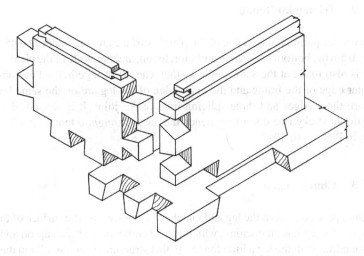

Picture 10.39 Open dovetail tenon in the Warring States period, cited from *Research on Tenon Jointing Technology of Cabinetwork in the Warring States Period* by Lin Shoujin

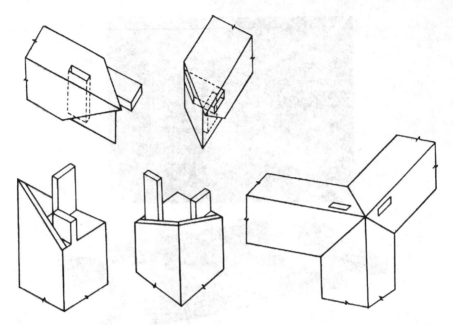

Picture 10.40 Triangular tenon, cited from *Research on Furniture in the Ming Dynasty* written by Wang Shixiang

10.3.7.2 Triangular Tenon

The two sides parallel to the edge of the plate's surface are both made at a 45 degrees incline from the bottom of the long and short tenon, and the angle of the plate's surface frame is also made at the same angle so they can all fit together. After combining, the outer edge of the frame and the outer edge of the leg are on the same level, and there are three edges and three splicing lines at the joint. It is shaped like *zongzi*, a traditional sticky rice dumpling, hence the name "*zongjiao* tenon" in Chinese (as shown in Picture 10.40).

10.3.7.3 Chuck Tenon

The tenon protrudes from the leg and matches the mortise on the surface of the table. The top of the leg has an opening, which can be embedded with an apron and apron-head spandrel, with the leg a little longer. In this structure, four legs clamp the aprons and connect them into a frame, so that the table board and legs are difficult to move, and the weight can be evenly distributed to the four legs.

10.3.7.4 Overlord Stretcher

It is mainly used to increase the sturdiness of tables and stools. The upper end of an S-shaped overlord stretcher is connected with the penetrating transverse brace of the table top, and fixed with a wood or bamboo nail, while the lower end is connected to the leg.

10.3.7.5 Four-Side Assembly

The previously-mentioned example of a penetrating transverse brace is one of these assembly methods, which is widely used, such as for the top of chairs, tables, stools, *an,* and cabinet doors. This is a method to prevent damage caused by the wood's expansion and contraction from changing temperatures throughout the year. It is not only a thoroughly tested, logical and economical assembly technique, but an aesthetically appealing one as well.

Besides the above mentioned five types of tenon, there are also other types such as half shoulder tenons, inserted shoulder joints, and hook tenons. Because there are various styles of mortise and tenon structures, even an old craftsman who has worked for decades will encounter some types that he has never seen before. Sometimes wood or bamboo nails and wedges are needed to connect or reinforce components, such as splicing and reinforcing arc components (as shown in Picture 10.41). After the furniture is assembled, its surface should be treated. Previously, furniture was polished with horsetail (also known as *jiejie* grass) dipped in water, but now it is deburred with sandpaper and then treated with wax.

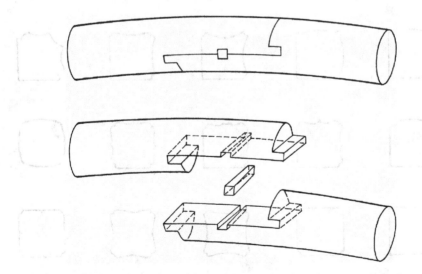

Picture 10.41 Jointing arc components, cited from *Research on Furniture in the Ming Dynasty* by Wang Shixiang

10.3.8 Decoration

Furniture decoration can be manifested in many different ways. The beauty of hardwood furniture is mainly reflected in the gorgeous texture and natural color of the wood. The doors of a cabinet have a symmetrical aesthetic feeling. Burl wood, which was called *"wenmu"* (usable wood) has a fine texture and displays special beauty, thus being regarded as a rare material for furniture by craftsmen.

The shape and structure of a piece of furniture are important for its overall aesthetic. The surface and lines of the *dabian*, *motou*, stretchers, legs, and feet form different shapes, which is called *"xianjiao"* (molding), creating a unique aesthetic feeling of the furniture (as shown in Picture 10.42). The *zanjie* (geometric patterns formed by short battens) and *doucu* (flower patterns formed by hollowed wood pieces) also add to the decorative beauty of the furniture (as shown in Picture 10.43).

Engraving plays an important role in furniture decoration. Commonly used engraving techniques include incised carving (line carving), relief engraving (bas-relief and high relief), openwork engraving, round engraving, and a combination of different methods (such as relief and openwork engraving). Engraving patterns come in a rich variety of themes and styles. Wang Shixiang, a contemporary Chinese scholar, divides them into twelve categories: (1) tendril pattern, (2) lotus pattern, (3) cloud pattern, (4) glossy ganoderma pattern, (5) dragon pattern, (6) *chi* dragon

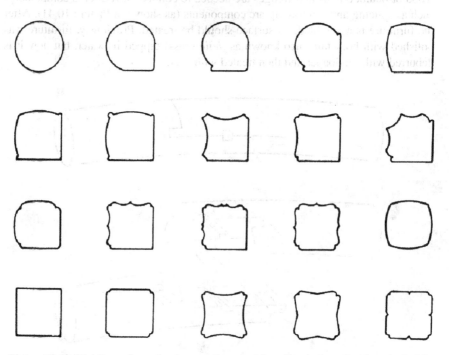

Picture 10.42 Moldings of round and square feet, cited from *Research on Furniture in the Ming Dynasty* by Wang Shixiang

Picture 10.43 *Zanjie* and *doucu* on the cabinet door, cited from *Research on Furniture in the Ming Dynasty* by Wang Shixiang

pattern, (7) flower and bird pattern, (8) landscape pattern, (9) figure pattern, (10) auspicious character pattern, (11) religious pattern, and (12) natural object patterns (such as those of bamboo node and bark). All of these patterns reflect the signs of the times they were created in. But most function as good wishes for various blessings, such as *Songhe Yannian* (longevity of pine trees and cranes), *Duofu Duoshou* (much good luck and longevity), *Yutang Fugui* (wealth and fortune, composed of magnolia, Chinese flowering crabapple and peony, as shown in Picture 10.44), and *Ping'an Jiexiang* (peace and good luck, composed of vases, saddles, halberds, and elephants, as shown in Picture 10.45).

The ways the same topics and subjects are expressed or featured change throughout history, thus providing us with enough evidence to connect them with specific locations, cultures, people, and times.

Inlays, more specifically package inlays and fillings inlays, were also commonly used in furniture as decoration. The former, as seen in the Ming-style *an* (table)

Picture 10.44 *Guose Tianxiang* (peony), cited from *Chinese Handicrafts—Furniture Making* by Lu Jun

Picture 10.45 *Pingan Jixiang* (peace and good luck), cited from *Chinese Handicrafts—Furniture Making* by Lu Jun

Picture 10.46 Ming-style *an* in Xiyuan Jiezhuanglu Temple in Suzhou, showing the water patterns formed by the spliced dalbergia odorifera chips, cited from *A Rare Ming Dynasty Big Table* by Pu Anguo

collected by Xiyuan Jiezhuanglu Temple in Suzhou, was made of 2,930 pieces of dalbergia odorifera, which makes it minimalist, unapologetic, yet stately, but more importantly still intact after hundreds of years as well (as shown in Picture 10.46). Furniture featuring the latter was sometimes inlaid with wood and porcelain, some with mother-of-pearl and bone teeth (as shown in Picture 10.47), and some were even inlaid with treasure made of various precious materials such as burl wood, jade, pearls, rhino horn, and ivory.

Valuable hardwood furniture did not need to be lacquered, but other furniture that was not made with hardwood had to be decorated with lacquer, not only for protection, but also to make it look better. Please refer to Chap. 9 of this book for the required materials and techniques, which will be omitted here.

10.3.9 Accessories

Chairs and drawers are often woven with rattan or cotton thread, both are used as furniture accessories. The surface of tables and stools, the center of the screen in the screen-type Arhat bed, as well as the center of doors of cupboards are usually made of various colors of stone, especially marble. Copper parts are used for the hinges, and face plates have different practical functions (as shown in Pictures 10.24 and 10.25). Besides general decoration techniques, other techniques, such as gilding and chiseling are also used.

Valuable furniture is decorated with intricate metalwork in gold or silver, known as filigree, and its beauty gives it the effect of being full of gold and silver (as shown in Picture 10.48). This kind of decoration is the result of an elaborate process, from spreading, covering, burning, and crocheting to painting. First, the mesh is chiseled onto the surface of an iron sheet, then the gold and silver are hammered into wires,

Picture 10.47 Mother-of-pearl inlaid lotus patterns on a square dalbergia odorifera cabinet door, cited from *Research on Furniture in the Ming Dynasty* by Wang Shixiang

heated and calendered, to then be finely chiseled into the pattern, and finally, the background is lacquered to make the pattern more eye-catching. There are even more ornaments that are made of cupronickel or cupronickel embedded with red copper.

As mentioned above, especially when it comes to the production of valuable furniture, excellent craftsmanship is required. This means that a careful design process,

Picture 10.48 Golden dragon-patterned corner of a rosewood table, cited from *Research on Furniture in the Ming Dynasty* by Wang Shixiang

meticulous coherence between all parts, exquisite craftsmanship, and the perfect realization of the predetermined intention need to come together as much as possible, while including necessary changes and adjustments every step of the way. These techniques are comparable to the concepts and practices of modern system engineering and are the essence of furniture design and production.

10.4 Ming-Style Furniture

The Ming dynasty was the golden age of traditional Chinese furniture. The development of furniture production techniques and artistic levels reached their peak in about 200 years from the Jiajing and Wanli period (1522–1620) in the mid-Ming dynasty to the Kangxi and Yongzheng period (1662–1796) in the early Qing dynasty. The high-quality furniture made during this period is generally called Ming-style furniture, but in this part, we only discuss the Ming-style furniture made in the Ming dynasty.

Lacquered wood furniture in the early Ming dynasty already made great achievements. After the mid-Ming, hardwood furniture became popular and quickly spread all over China. For example, in *What I See and Hear in the Imperial City* written by Liu Ruoyu of the Ming dynasty, it is recorded that the Imperial Household Department made hardwood tables, cabinets and shelves made for the imperial family; volume 9 of *Instructions to the Factory and Library of the Ministry of Industry* by He Shijin records that 40 beds were made in the palace in the 12th year of the Wanli period (1584), and they cost more than 30,000 gold. Most of the furniture for the folk is recorded in the following books: Tu Long's *Utensils in the Study Room*, Gao Lian's *Eight Treatises on Following the Principles of Life*, and Wen Zhenting's *Superfluous Things*. The last one records in its preface: "Tables and couches have their own sizes, the utensils and devices have their own styles. They are very delicate and convenient, simple and natural." The furniture used by literati is usually focused on its taste, which is different from that of dignitaries who prefer to show off, but both of the groups require meticulous craftsmanship in their furniture. As a result, people's pursuit of exquisite furniture soon became a social fashion, just as Fan Lian of the Ming dynasty says in *Record of Folk Life Customs*: "Since the Longqing and Wangli period (1567–1620), both ordinary people and rich families have used fine furniture. Small furniture workshops in Huizhou (nowadays Shexian county in Anhui province) vied to be opened in the larger cities, making furniture of dowries and household devices. Luxurious families have their furniture including cabinets, tables and stools made of dalbergia odorifera, gall wood, ebony, acacia wood, and boxwood. These pieces of furniture are extremely expensive and required mature technology, and this custom has been quite popular for a while." At the same time, *Guidelines by Lu Ban*, which only recorded the way of building wooden structure earlier, was added with new contents after the Wanli period, and was renamed as *Mirror of Craftsmanship and Guidelines by Lu Ban*, in which 52 furniture items with attached diagrams were attached.

Gao Feng, a contemporary Chinese scholar, says: "If traditional Chinese furniture is compared to a poem created by wood, the Ming-style furniture is undoubtedly the most wonderful and perfect part of it."

The main characteristics of Ming-style furniture are its concise shape, scientific structure, exquisite materials, meticulous craftsmanship, bright color, and elegant style (as shown in Pictures 10.15, 10.21, 10.23 and 10.46).

It also pays attention to the beauty of the whole, and realizes the perfect combination of function and aesthetics through the various combination of its camber, width, lines, and surface. The engraving is moderate, mostly with lines, and delicate ornamentation and patterns, showing an elegant and delicate style. When you take a look at round-backed armchair, you can see the upper end of the chair back is decorated with flowers, the upper components are all made of curved round materials, and the top rail and armrests are connected with beautiful and smooth arcs. The components of the chair and its tilt angle are very ergonomic, which makes people feel comfortable when sitting on it and will not feel tired even after sitting for a long time. The lower part of the chair has horizontal stretchers, and the four-sided inner frame is narrow at the top and wide at the bottom, which increases the firmness and makes it visually stable. The chair with a round top and a square bottom implies the idea of hemispherical dome cosmology, which is one of the embodiments of Chinese traditional culture.

Ming-style furniture is often made of wood with dense grain, beautiful texture and soft color, such as rosewood, dalbergia odorifera, wenge, nanmu, beech, and ebony; the furniture is fine in workmanship; the thickness of lines, the depth of grooves, the size of mortise and tenon, and the coordination of various components are all accurate. Even the chamfering of table legs are exquisite and proper in size.

Most of these hardwood furnitures are not lacquered; instead, its smoothness and natural wood grain will be displayed through meticulous polishing and scalding wax.

Literati in the Ming dynasty participated in the design and production of furniture, which played a role in the formation of its style and taste. The style, structure, production, and engraving of furniture all embody the life concept of advocating nature and pursuing the simplicity of the literati during that period. The fine character and demeanor of Ming-style furniture are categorized into five groups with sixteen characteristics by Wang Shixiang, a contemporary Chinese scholar.

The first group contains furniture with seven different characteristics: plain, simple (as shown in Picture 10.49), vigorous, solemn (as shown in Picture 10.15), majestic, plump, and dignified. Furniture of this group, mostly plain and simple, is the mainstream of Ming-style furniture. In the second group, there are three characteristics: magnificent (as shown in Picture 10.18), ornate and pretty. Furniture of this group is either luxurious, gorgeous or elegant and light. The third group has two characteristics: upright and tactful. Furniture of this group is delicate, with different fun. The characteristics of the fourth group are elusive and exquisite, and the artistic effect is displayed by proper frame and openwork engraving. The two characteristics of the fifth group are elegant and novel. Furniture of this group doesn't fall into the stereotype and shows much novelty. The categorization of the above sixteen characteristics is helpful to study and understand Ming furniture carefully.

However, there are also "eight defects" of Ming-style furniture, which are cumbersome, redundant, bloated, stagnant, slender, absurd, misplaced, and vulgar. For example, the pedestal with *chi* dragon patterns used large pieces of wood and took much work, only to look clumsy and bloated because wood is used instead of stone, which is suitable for such occasion (as shown in Picture 10.50). Pointing out the defects of the furniture in the Ming and Qing dynasties will help us improve our taste; it also shows that even in the golden age of furniture production, there are still some inferior products due to improper concepts and making processes.

In more than 100 years after the mid-Ming dynasty, many factors at that time contributed to the peak of the quality and quantity of furniture production. At the beginning of the Ming dynasty, policies such as *tuntian* system (having garrison troops or peasants open up wasteland and grow food), immigration and water conservancy construction were implemented. The system of lifelong service of craftsmen in

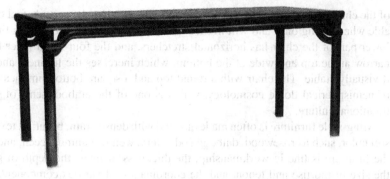

Picture 10.49 Rosewood painting table, manifesting simplicity, cited from *Research on Furniture in the Ming Dynasty* by Wang Shixiang

Picture 10.50 Dalbergia odorifera pedestal, manifesting the defect of being bloated, cited from *Research on Furniture in the Ming Dynasty* by Wang Shixiang

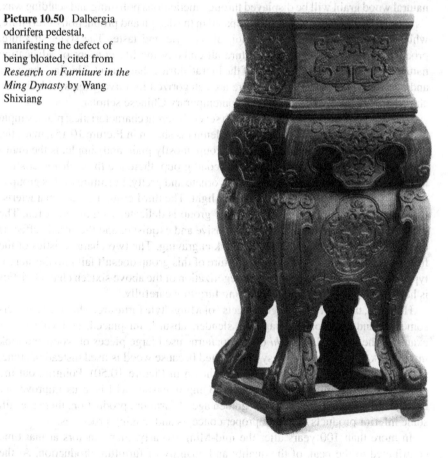

the Yuan dynasty was changed into a rotating schedule and residence system, which improved the enthusiasm of craftsmen and promoted social and economic development. After the mid-Ming, the commodity economy became more developed, capitalism sprouted, and the number of craftsmen and products increased greatly, resulting in a flourishing market. At the same time, a large number of hardwoods from Nanyang (an old name of Southeast Asia) were imported after the abolition of banning on maritime trade. All these factors helped accelerate great progress in the scale and technical level of furniture production industry in a relatively short period of time, as Wang Shixing of the Ming dynasty describes in *Record of National Geography*: "The curios placed on the table and couch in the study room is mostly made of rosewood and dalbergia odorifera…This fashion is being imitated at home and abroad, and reached its heyday during the Jiaqing, Longqing and Wanli period." The upper part of a leg of the painting table, which was made in the 23rd year of the Wanli period (1595) and is stored in Nanjing Museum, is inscribed with the following words in seal character: "The wood is excellent and hard, and the production is simple and beautiful. With this table, it seems that I can enjoy it leisurely and freely for a hundred years", which fully expresses the practical concept of furniture production during this period. Furniture in the Ming dynasty has long been valued studies by researchers. The earliest academic monograph about it is *Chinese Domestic Furniture* written by the German scholar Gustov Ecke and published in 1944; In the 1940s, Yang Yao, a Chinese scholar, wrote *Interior Decoration and Furniture of the Ming Dynasty in China*.

But Wang Shixiang is the most notable researcher in the study of Ming-style furniture. Wang showed a strong interest in traditional Chinese furniture in his early years. Since the mid-1940s, he began to collect all kinds of Ming-style furniture and related materials, visiting ordinary people, collectors, antique shops, and Luban Museum. Even when he was in his 80 s, he still worked tirelessly, travelling to Suzhou, Guangzhou and other places in search of gem furniture. He not only made great efforts to study ancient literature, but also learned from senior craftsmen. As a result, he succeeded in making an in-depth and detailed study of the furniture handed down from generation to generation based on the unearthed objects and their background, and explaining the many technical woodwork jargons as well as *Mirror of Craftsmanship and Guidelines by Lu Ban*. It was on such a solid basis that Wang completed two great books, *Research on Furniture in the Ming Dynasty* and *Classic Chinese Furniture: Ming Dynasty* in the middle and late 1980s, which were widely praised at home and abroad, and were translated into English, French and German. In the postscript of the book *Research on Furniture in the Ming Dynasty*, Wang writes about the process of his field study, research and writing. He says modestly that there are still some defects in the book, and mentions that during his 40 years of work, "several respectable craftsmen of Beijing Luban Museum have offered me a lot of help, especially Shi Hui, Li Yanyuan and Zu Lianpeng" (as shown in Picture 10.51). The completion of the two books is also indispensable for Ms. Yuan Quanyou, Wang's wife, who draws hundreds of charts, as the postscript records: "We often stayed up working, wrapped in quilts in winter and cooled by electric fans in summer." Their dedication and hard work set an excellent example for us.

Picture 10.51 Photo of Wang Shixiang (left) and Zu Lianpeng

10.5 Furniture Making in Suzhou, Beijing and Guangzhou

Suzhou, Beijing and Guangzhou house the three major schools of traditional furniture making after the Ming and Qing dynasties. Their manufacturing techniques have been listed in *Lists of National Intangible Cultural Heritage*. We will take a closer look at each one of them in the following passage.

10.5.1 Furniture Making in Suzhou

It refers to the furniture making centered around Suzhou and in areas such as Mudu, Dongshan and Changshu. They mainly manufacture Ming-style furniture, which is also known as Suzhou-style furniture.

Furniture making has a long history in Suzhou, such as the nanmu Sutra box and the small pagoda during the Five dynasties, which were discovered in the tower in Yunyan Temple on Tiger Hill in 1956.

In the Southern Song dynasty, many official families and literati lived in Suzhou. They paid attention to decorating and furnishing their houses and building private gardens, and therefore furniture making flourished during this period. After the mid-Ming dynasty, the abolition of banning on maritime trade enabled a large number of high-quality hardwood to be imported from Southeast Asia, which improved the quality of materials used for furniture.

For example, *Superfluous Things* written by Wen Zhenheng of the Ming dynasty makes a precise comment on rosewood, incense machilus and ceylon ironwood, and names them "*wenmu*" (cultured wood). Owners and literati often participated in the planning and design of furniture themselves, its style, structure and decoration permeated with the ideas, accomplishments, and aesthetic tastes of cultural elites. Every aspect of the furniture making strove for exquisiteness; thus the cultural taste was greatly improved (as shown in Picture 10.52).

Many furniture workshops as well as many skilled craftsmen gathered in Suzhuou as early as the Wanli period (1573–1620) in the Ming dynasty. For example, according to the records of *Annals of Wuxian County*: "The furniture was made with square wooden boards by Yuan Youzhu, while the swing-style furniture was made by Wu Si, both of whom were skilled craftsmen of that time." Up to the Qing dynasty,

Picture 10.52 Suzhou-style furniture

Beijing-style furniture was very popular all over China, but Suzhou still mainly produced Ming-style furniture, with its delicate and concise patterns. At the same time, small rosewood pieces made there also experienced further development. Most of the workshops in Suzhou were situated around Fanzhuangqian Lane, Wangtianjing Lane, Jingde Road, and Zhushi Temple. It is also worth mentioning that Suzhou-style furniture is still hand-made today, with a unique set of skills in each process. There are even some proverbs related to the production process, such as "the sound of handmade furniture can be heard clearly across three rooms; the furniture made is uniform from top to bottom", and "to ensure neatness, the decoration should be symmetrical and neat". Let's take a look at engraving. It is required to engrave according to drawings, and ensure that the shovel frame is consistent in depth and neat on all sides. It is three-dimensional and has exquisite lines and uniform thickness. The size of the rolled beads is uniform, elongated vertically up and down, and inlaid horizontally. The painter has to go through 16 different processes, such as making a rough shape, scraping the surface, polishing, painting, lacquering and lacquering again. Therefore, it can be said that meticulous craftsmanship is a major feature of Suzhou-style furniture.

At present, furniture made in Suzhou includes altar, *taishi* chair, tea table, flower rack, *guqin* table, bookshelf, desk, armchair, antique shelf, bed, closet, cabinet, dressing table, and square stool. In order to meet the needs of foreign users, some furniture styles, such as chairs and dining tables, became more suitable for modern life styles, but the modeling is still dominated by Ming-style furniture and has incorporated the Qing dynasty's treatment methods of engraving furniture. There are several small rosewood items, such as *ji*, screen frame, disk box, and lamp.

After years of changes, the leading enterprise of Suzhou-style furniture is Suzhou Mahogany Carving Factory Co., Ltd. The enterprise consists of four generations of craftspeople: designers include Lu Hansheng (the first generation), Xu Wenda (the second generation) and Xu Jiaqian (the third generation), carpenters include Peng Along (first generation), Wu Huankun (second generation) and Chen Guoqing (third generation), carvers include Zhao Zikang (first generation), Zhao Fengyun (second generation) and Jiang Hezhen (female, third generation), and painters Qian Yonglin (first generation), Tang Jinqiu (second generation), Zhang Caihong (female, third generation) and Cao Jianping (fourth generation).

Due to modernization as well as the difficulty, high requirements, long working hours and low income that come with working in furniture making, many young people are reluctant to do it and others have even been engaged in other industries. This all results in a crisis in the inheritance of Suzhou-style furniture making techniques. In recent years, the Suzhou Cultural Bureau and Suzhou Mahogany Carving Factory Co., Ltd. have invested more than one million yuan in the protection and inheritance of these skills. After it was listed in *List of the National Intangible Cultural Heritage*, a five-year action plan was launched and the Suzhou Ming-style Furniture Research Association was established. It will systematically collect and sort out historical materials, establish archives, invite aspiring young people to be taught the necessary skills by old craftsmen, produce 30 to 50 innovative pieces, and hold exhibitions within five years to further promote the development of the industry.

10.5.2 Furniture Making in Beijing

It came up in the early Qing dynasty. Beijing's Chongwen district is the birthplace of the crafting industry. As early as the beginning of the Tongzhi period (about 1862), a carpenter surnamed Wang opened a small workshop called "Longshun" in this area to make and repair hardwood furniture for the court.

As the capital, Beijing's common crafts are greatly influenced by court workshops, and craftsmen have many connections with the Royal Workshop, which teach and pass on certain skills. Therefore, the furniture made in Beijing already had the characteristics of court art from the start, and the precious materials such as rosewood, dalbergia odorifera, scented rosewood, wenge, and santos rosewood have always been widely used. Because it is located in the north, it is often glued with fish glue. The shape of Beijing-style furniture is elegant and dignified, the structure is symmetrical, the ornamentation is elaborate, and the production is done exquisitely. The unique polish with melted wax shows off the natural beauty of the hardwood materials, giving it a rich and luxurious look overall (as shown in Picture 10.53).

Picture 10.53 Beijing-style furniture

The above-mentioned Longshun workshop lasted until the 25th year of the Guangxu period (1899). Because of its beautiful and generous products, it was praised by customers and won the reputation of "High-quality Lasting One Hundred Years". In the 28th year of the Guangxu period (1902), two masters surnamed Wu and Fu joined the workshop and held shares. They changed the name of the shop to "Longshuncheng" and hired Wei Junfu as the shopkeeper to manage it. As a result, the production management of the shop became stricter, and the reputation, as well as the variety of business of the shop, increased exponentially. In 1956, 35 traditional furniture shops, including Longshuncheng, Tongxinglong, Tongxinghe, Liufengcheng, and Songfulu, merged into one and were rebranded as the Longcheng Table and Chair shop, but their production and sales were still separate. In 1959, the name of the shop was changed to Longshun Wood Factory. In 1966, the name of the shop was changed again, this time to the Beijing Hardwood Furniture Factory, which employed about 50 craftsmen. Since the 1960s and 1970s, the business model of the furniture factory became processing raw materials into Beijing-style furniture and then exporting it overseas. These pieces are exported to many different places, such as Europe, America and Southeast Asia. In 1993, the aforementioned time-honored brand shop reopened. However, its name was changed to Beijing Longshuncheng Chinese Furniture Factory (as shown in Picture 10.54), and became the leading enterprise of Beijing-style furniture. At present, the factory still stores the official hat chair, round-backed chair and tea tables made of dalbergia odorifera during the Ming dynasty, and stool made of rosewood with five openings on the side and cabinets made of golden camphor wood that was used for storing royal robes and were made in the Qing dynasty. In the 1960s, the factory made rosewood screens for Ziguang Hall at Beijing's Zhongnanhai and Beijing-style guest room furniture for the Beijing Hotel and Guibinlou Hotel. In 2003, it made furniture for displays in Fragrant Hills Park, also known as Xiangshan Park, and the Summer Palace.

Picture 10.54 Plaque hand written by Wang Shixiang

The production techniques of Beijing-style furniture were first passed on from the first generation's Wang Mujiang, Zhang Xiuqin and Gao Fusheng to the second generation's Wei Junfu, Fu Peiqing, Li Qiyuan, and Zhang Huaigan, then to the third generation's Chen Shukao (his teacher is Wei Junfu), Li Yongfang (his teacher is Zhang Huaigan) and Li Xiyao, and finally to the fourth generation's Gui You, Tian Yanbo and Meng Fugui. Li Yongfang, a native of Nangong, Hebei province, was born in 1931. He developed superb skills during his 60-year career as a carpenter and presided over the design and production of the screen at Ziguang Hall in Zhongnanhai and the restaurant furniture at the National Hotel. He devotes his whole life to the inheritance and development of Beijing-style hardwood furniture, which is why he is strict yet careful with imparting skills to his disciples. Tian Yanbo, born in Beijing in 1955, studied handicraft art from Li Yongfang in 1979 and is currently a prominent figure in the Beijing-style furniture industry. In 2005, he presided over the design and production of the mahogany brocade box with cloisonné and nine-dragon wall decoration, which was given to Lien Chan, then chairman of the Kuomintang of China, as a gift. The guest room furniture of the Guibinlou Hotel and the screen in Yanshangzhai in the Summer Palace were also made under his supervision.

However, the costs of making Beijing-style furniture are high but the profits thin. Many township enterprises in Beijing's suburbs and Hebei province imitate Beijing-style furniture. However, their products were inferior in quality and low in price, leading to chaos on the hardwood furniture market. People in the hardwood furniture have problems surviving in this hostile environment and many people have become unwilling to carry on. Moreover, highly skilled craftsmen are ageing, which means the required skill sets are endangered. At present, the Factory has taken measures to raise the wages of those inheriting the skills. They have also made great efforts to collect and sort out technical data, and analyze and study the still existing pieces of furniture from the Ming and Qing dynasties in order to recreate them. The Factory works closely with relevant schools and universities to vigorously train a new generation of craftsmen and technicians, and ensure the orderly and sustainable development and preservation of Beijing-style furniture making skills.

10.5.3 Furniture Making in Guangzhou

Guangzhou-style furniture is primarily made in Guangzhou and its surrounding areas. Because Guangzhou is located near the sea and timber is easily available, its furniture production has a wide variety of materials to work with, creating a wide variety of shapes and elaborate carvings since the Ming dynasty, resulting in a unique luxurious style. After the mid-Qing dynasty, because it suited the royal family's taste, Guangzhou-style furniture replaced the Qing court's Suzhou-style furniture. Its themes were richer and more varied, its form was novel, and it was also popular among the common people. According to *Supplemental Annals of Panyu County: Industry*, the furniture manufacturing industry in Guangzhou during the Tongzhi and Xuantong period (1862–1912) was "specialized and refined, with its own furniture making technology. They even had more than 100 shops in Henan province that were quite renowned at the time". There was a folk saying that went: "Haopan Street sold goods produced in Zezhuang, goods sold in Xilaichu were produced domestically, and goods sold in Ximenkou were produced in Pingyi." In the early years of the Republic of China, there were more than 1,000 people employed by the Suanzhi and Huali Furniture Industry Association. Many manual workshops were expanded into furniture factories that exported goods in batches, with a store in front and a factory in the back.

A significant technical feature of Guangzhou-style furniture is paying special attention to engraving (as shown in Picture 10.55). From sculpting to semi-three-dimensional, multi-level bas-reliefs, relief engraving, high engraving, through engraving, round engraving and three-dimensional engraving, the degree at which furniture is engraved varies greatly. Some pieces of furniture are so elaborately engraved that the engravings take up 80% of its surface. It takes a lot of time and effort to create these pieces, so it can be said that every single item is a high-grade handicraft.

The Guangzhou Woodcarving Furniture Factory is a time-honored brand, which is a representative enterprise in the Guangzhou-style furniture industry. At the end of the 1960s, there were thousands of craftsmen working at the factory, including Yang Xia, Lian Liu, Hu Zhi, Huang Zhonghuai, and Zhao Zanhui, and later they were joined by Liu Shicai, Shu Ling and Guo Zhanhua.

Guangzhou-style furniture uses exquisite materials, which means that it cannot be bleached and its color is symmetrical. If there was a slight defect, it would be discarded. In addition, the level of engraving needs to meet such high requirements that it makes each item's cost soar. Because of this, it cannot compete with modern furniture. Its market share, the development of its enterprises and the inheritance of its skills have been negatively affected. In recent years, Guangzhou's municipal government and the Woodcarving Furniture Factory have invested funds to improve the treatment and support of ageing craftsmen and strengthened the training of new generations of craftsmen. On the premise of maintaining the traditional Guangzhou-style, they are focusing on innovating the shape and style of the furniture, so as to open up the market again.

10.6 Rules of the Furniture Manufacturing Industry and a Brief Introduction of Furniture Culture

Woodworking, or carpentry, is an ancient trade (as shown in Picture 10.56). There are many people who love carpentry, from rulers, such as Emperor Zhezong of the Song dynasty, Jimmy Carter (the 39th president of the United States), to literati, such as

Picture 10.55 Guangzhou-style furniture

Tolstoy (1828–1910), a Russian writer, to an abundance of common people. Furniture making is carpentry. For example, the Royal Workshop at the Hall of Mental Cultivation in the Qing dynasty had furniture making and restoration as its main business. According to the archives of the first year of the Yongzheng period (1723), 41 pieces of furniture such as rosewood dining table, low railing bed, nanmu bookcase, spring stool, and "*baxian table*" (square table) were all made in that year, while in comparison only 8 sedan chairs and seats were made. The carpenters named in these archives include Wang Guoxian, Yu Jiegong, Yu Junwan, Luo Yuan, Lin Cai, Lu Yu, and Huo Wu. Carpenters have professional classifications. For example, sawers are responsible for sawing wood, and woodcarvers are responsible for carving and engraving decorations. Therefore, people who are called carpenters can be involved in a wide range of industries but may only be specialized in one specific carpentry-related skill.

Picture 10.56 *Carpenter St. Joseph*, an oil painting by Georges de La Tour, created in France in 1660. In the picture, St. Joseph is working, while his son, Christ, is watching. It is cited from *Chinese Local Handicrafts* by Gao Xing

The furniture industry as well as the wood, stone, clay and tile industry regard Lu Ban as their ancestor (as shown in Picture 10.57), and some regard Lu ban, Pu'an and Zhang Ban as their ancestors. The ancients said: "Every profession has its master." Lu Ban, also known as Gongshu Ban, was a carpenter in the Warring States period. According to historical records, he was the inventor of tools and instruments, such as the carpenter's square and ink marker, and the founder of the woodworking industry. According to folk traditions, people regard the seventh day of the fifth month of the lunar calendar as Lu Ban's birthday. Therefore, every year, on this day, people from the woodcraft industry go to Lu Ban Hall to hold a grand ceremony to commemorate Lu Ban. A spirit tablet is set on the altar as well as various offerings. The master of each guild leads their craftsmen to pray for prosperity and peace. After the ceremony, the participants will get together to discuss the wages of craftsmen for the coming year, the admission of apprentices, and to revise regulations. Small businesses will set up outside Lu Ban Hall and form the annual grand market. Learning skills from a master is commonly known as "entering the profession" and comes with a set of established procedures. Even craftsmen who are self-taught have to do this kind of apprenticeship; otherwise they will not be recognized by their peers and will not be able to use the title of craftsman. To become an apprentice, you first have to find a prestigious person who will serve as your guarantor. A simple ceremony of apprenticeship means that this guarantor will serve as a middleman and offer the master alcohol and refreshments following meeting etiquette. Immediately after the apprentice has kowtowed to the master, the relationship between master and apprentice takes effect. During the formal ceremony of worshipping teachers, the spirit tablet of "Heaven and Earth" and of Lu Ban should be set up on the spot, with candles placed on the table, and seats for witnesses, guarantors and the master, with the master's seat in the middle.

The ceremony is presided over by a master of ceremonies and goes as follows: (1) Witnesses, guarantors and masters are invited to sit down. (2) The master of ceremonies holds a *ruyi* (an S-shaped ornamental object, made of jade, formerly a symbol of good luck) in one hand and brings the apprentice into the hall. (3) The master of ceremonies will ask the apprentice: "Have you been persuaded by others?" The apprentice must answer: "No, it is my own decision." (4) The master of ceremonies will signal the apprentice to take four steps towards the master. Each step has a specific meaning: The first step shows a willingness to be led by the master, while always keeping the words of the predecessors in mind. The second step shows that the journey to be undertaken is clear. The third step confirms one's dedication to study hard. The fourth step is a promise to be filial and respect teachers. (5) The master will also be asked by the master of ceremonies: "Are you persuaded by others to accept (name) as an apprentice?" Master will reply: "No, it is my own decision." (6) The master lights three incense sticks and inserts them in the incense table, and then leads the apprentice to kowtow and worship in front of the spirit table of "Heaven and Earth" and Lu Ban. The apprentice then kowtows and shows respect to the master, offering tea to the master as well as the witnesses and guarantors. (7) The apprentice reads the application, written by himself, the master reads the letter of consent, and the witness and guarantor sign their names on them. (8) The master gives the apprentice a set of clothes and pats him twice on his shoulder, which is meant to encourage the apprentice to study hard and complete his study with the master as soon as possible. The master will also give the apprentice a hat, which means that the apprentice should respect the teacher. A teacher for a day is a father for life. There is also a bowl filled with whole grains, which is called "*shangfan*", symbolizing "From now on, I will take you as an apprentice, leading you into the profession, teaching you workmanship, and then you will have the professional skills and means to support yourself." After the ceremony, guests are entertained, and the relationship between teachers and apprentices establishes.

Picture 10.57 The Master Lu Ban

The carpentry industry also has recognized regulations, as does the furniture industry, such as "a set of three years and more" and "eighteen punishable errors". The former refers to an apprenticeship having to last for more than three years. During this study period, the master has the right to expel the apprentice, but the apprentice does not have the right to leave the master. If the apprentice makes a big mistake, the master can invite the witnesses and guarantors, list all the mistakes of the apprentice and then dissolve the mentoring relationship. In such a situation, the guarantors will be humiliated, and will need to apologize to the master. Also, if an apprentice makes a small mistake that is among the "eighteen punishable errors", the apprentice will be beaten. After the three years of study is over, apprentices can continue to work in the master's workshop and get paid, or start their own business. According to the rules, the income of the first three months after these three years should be paid as homage to the master. With the development of society, some of these old rules have been abolished, but many customs pertaining to masters and apprentices are still retained. This way of inheriting skills and respecting teachers is valuable and worthy of being advocated by our contemporary people. Chinese furniture culture is full of symbolism and plays an important part in traditional Chinese culture.

Taking the decoration of furniture as an example, many patterns have both good wishes and aesthetic value, forming a harmonious whole with their shapes and structures. For example, plum blossoms symbolize a noble personality, peonies symbolize splendor and wealth, and bamboo symbolizes noble moral integrity. To express people's yearnings and pursuits, characters are used as well, such as *"fu"* (happiness), *"lu"* (fortune), *"shou"* (longevity) and *"xi"* (delight). On top of that, depictions of stories could be used to highlight ethical concepts and advocate loyalty, filial piety and righteousness, with *The Twenty-four Filial Exemplars*, Meng Mu's Three Migrations and the stories of the Three Kingdoms period. Other blessing patterns, such as *wufupengshou* (five bats surrounding the Chinese character "寿" for longevity) and *mashangfenghou* (a bee and a monkey on a horse), are also common.

Take decorative furniture as another example. The decorative requirements of traditional furniture used as decoration are very specific. The types, styles and arrangement of this furniture should not only match the functions of the halls or study rooms, but also match the traditional culture, theories and even *fengshui* (geomantic omen).

The main room of a residential building usually holds an *an* against the north wall, with a candle holder and incense burners on it, and a central scroll and couplets hanging on the wall. A *"baxian* table" (square table) is placed in front of the *an*, with a *taishi* chair on either side. There are two chairs and a *ji* against each wall on either side in front of the table (as shown in Picture 10.58). The visitors are seated on the right, and the host on the left. In a well-behaved family, even in daily life, the family members are required to sit according to the traditional hierarchical differences, for example, men on the left and women on the right, older people are seated higher, and younger people lower. The bed is the main piece of furniture in the bedroom and is generally placed against the wall, with one side unobstructed in order to easily get in and out of bed. Boxes, tables and chairs should be placed in the most suitable position to facilitate people's daily life. Scholars pay special attention to the layout of their study. The desk should be placed in a place with sufficient light and stuff, such as inkstone, water injector, writing brush container, brush holder, and paperweight. Bookcase and bookshelf are placed along the wall to facilitate easy access. Visitors are mostly seated in the living room, but those who have a close relationship with the host will be invited to sit in the study. There are chairs and tea table in the study, and sometimes even *guqin* table and flower *ji* depending on the host's interests. All these pieces of furniture need to be elegant (as shown in Picture 10.59). This is exactly why Ming-style furniture is especially liked by so many people (as shown in Picture 10.60).

Picture 10.58 Furniture display of a farmer's living room in Zhejiang, cited from Cai Cheng's *Geographical Works*

In the hundred years since the late Qing dynasty, western-style furniture, such as sofas and glass bookcases, has become increasingly popular in China, and new materials and decorative techniques have emerged one after another. Since China's reform and opening up, with the revival of traditional culture and the change of people's aesthetic tastes, traditional furniture has gotten into people's good graces once more. Classical furniture markets have boomed in recent years. They have caused a huge supply and production chain to appear in some counties and cities, among which those specialized in Ming-style furniture seem to be the most promising. It is conceivable that in the next few years, Chinese furniture and Western furniture will coexist and prosper in China, causing our furniture culture to become more colorful. If that is the case, this reemergence will become able to meet people's growing material and spiritual needs.

10 Furniture Making

Picture 10.59 Wang Shixiang's study

Picture 10.60 Ming-style furniture in the Minneapolis Museum, USA, cited from *Ma Weidu Talking about Collection: Furniture*

Chapter 11
Making Calligrapher's Tools

Hua Jueming and Guan Xiaowu

Papermaking is one of China's greatest inventions. Paper, ink stick, writing brush, and inkstone are collectively called the "Four Treasures of Study", which are organically linked together to form a holistic cultural phenomenon and play a great role in the inheritance, dissemination, and development of Chinese civilization. The production of the Four Treasures of Study follows the scientific method, requires exquisite skills, follows a standardized production process, and is carefully designed and produced by craftsmen. All this ensures their good quality and ease of use. These skills still exist in all parts of the country and play an important role in manufacturing and people's daily lives.

11.1 Paper

11.1.1 Definition of Paper and Mechanism of Papermaking

In early days of human civilization, many different kinds of writing materials were used, such as bamboo slips and silk paper in China, papyrus in Egypt (as shown in Picture 11.1), parchment in Europe, pattra in India, and treebark by the Mayan and Aztecs in South America. They cannot be called paper because they are prepared and produced differently, and even perform differently from paper.

Xu Shen of the Han dynasty explains in his book *Discussing Writing and Explaining Characters*: The pronunciation of the characters for paper, 纸 (*zhi*), and

---

H. Jueming · G. Xiaowu (✉)
The Institute for the History of Natural Sciences, Chinese Academy of Sciences, Beijing, China
e-mail: gxiaowu@ihns.ac.cn

H. Jueming
e-mail: huajueming@163.com

© Elephant Press Co., Ltd 2022
H. Jueming et al. (eds.), *Chinese Handicrafts*,
https://doi.org/10.1007/978-981-19-5379-8_11

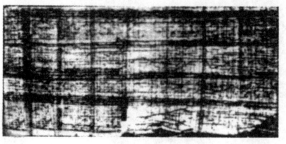

Picture 11.1 Sedge (left) and papyrus (right), cited from *Complete Collection of Traditional Chinese Handcrafts: Papermaking and Printing*

for cotton wad, 絮(*xu*), comes from the characters "糸" (*si*) and "氏" (*zhi*)". This means that paper is a sheet formed by combining wads of hemp with bamboo. Pan Jixing, a contemporary Chinese scholar, once said that traditional paper is usually made of a suspension made mechanically and chemically of plant fiber. This is then made into (or poured into) thin slices through the surface of a porous curtain. After drying, the cellulose will rely on hydrogen bonding with it to enhance its hardness. It is composed of hemp, bark, and bamboo fibers, and is a long chain polymer formed by linking d-glucose ($C_6H_{10}O_5$) with a 1–4-pullulin bond (as shown in Picture 11.2). When the plant fiber is purified and dispersed in water, the hydroxyl groups of cellulose will absorb the water and swell, which in turn cause the oxygen atoms and water molecules in adjacent molecules to form a water bridge. After drying, it will turn into paper by forming hydrogen bonds (as shown in Picture 11.3). A hydrogen bond is a kind of chemical bond, so papermaking is a feat of both physical and chemical engineering. The hydrogen bond association is key to the creation of paper. Paper is a synthesized material rather than a natural material.

Paper is thin and white in color, light and thin in texture, and it is very smooth to write on with ink, with the proper level of hardness and durability. The width of the paper is measured to be easily rolled and folded. Paper is cheap and widely used, and can be produced in large quantities to meet people's various needs. These characteristics of paper are not possessed by other writing materials such as bamboo slips, silk, papyrus, or parchment. Therefore, as soon as paper was invented, its advantages quickly became evident and thus made it the most important writing, painting, and printing material. Even in modern times when science and technology are highly developed, paper is still the modus operandi and probably will be for a long time.

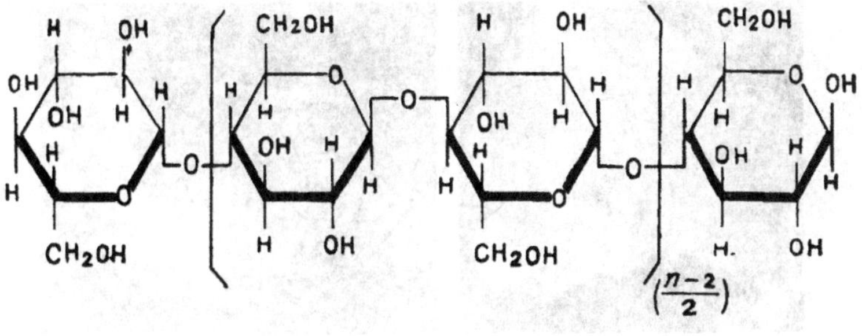

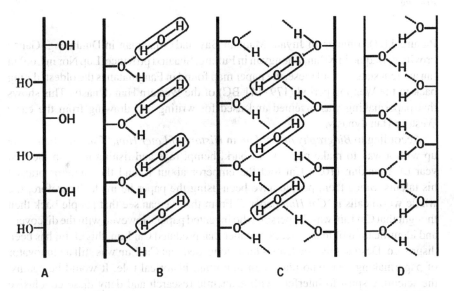

Picture 11.2 Molecular structure of cellulose, cited from *History of Science and Technology in China: Papermaking and Printing*

Picture 11.3 Mechanism of papermaking, cited from *History of Science and Technology in China: Papermaking and Printing*

A. cellulose
B. swelling through water absorption
C. formation of water bridge
D. hydrogen bond association

11.1.2 Origin and Development of Papermaking

Up to now, there are many places where ancient paper dating back to the Western Han dynasty has been discovered, namely, at Fangmatan in Tianshui (as shown in

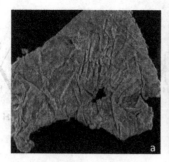 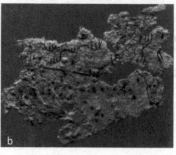

Picture 11.4 Map of Fangmatan, unearthed in Tianshui, Gansu Province dating back to the Western Han dynasty, cited from *Complete Collection of Traditional Chinese Handcrafts: Papermaking and Printing*

Picture 11.4), Jinguan in Juyan, Maquan Bay and Xuanquan in Dunhuang, Gansu province, Baqiao in Xi'an, Zhongyan in Fufeng, Shaanxi province, Lup Nor in Loulan (an ancient state). Of all these, the paper map found in Fangmatan is the oldest, dating back to the Wenjing period (179–141 BC) of the Western Han dynasty. This shows that papermaking was invented and used for writing and drawing from the early Western Han dynasty.

According to *Biography of Cai Lun* in *History of Later Han*, "Cai Lun then came up with a way to make paper from bark, hemp, rags and fishing nets. In the first year of Yuanxing (105), Cai told the emperor about it, and the emperor praised his talents. Since then, people have been using the paper he made. Therefore, the whole world calls it '*Cai Hou* paper'." From this we can see that people back then thought that Cai Lun was the person who invented paper. However, with the discovery and identification of many pieces of paper that predated Cai Lun, this claim has been disproven. Despite this, academics have conceded that Cai Lun was still an innovator of papermaking, one who played an important historical role. It would be against the scientific spirit to interfere with academic research and deny these conclusive historical facts that paper predates Cai Lun.

Cui Yuan (77–142), a scholar in the Eastern Han dynasty, says in his letter to Ge Gong: "I give you ten volumes of *Xu Zi*. I wanted to copy the books with white silks for you, but my family is poor and couldn't afford silks, so I have to replace them with paper". This shows that silk was still used for writing at that time, but paper was used more widely. The extensive use of paper led to the progress in the production of paper, writing brushes, and ink sticks. Zuo Bo paper, a writing brush made by Zhang Zhi, and Wei Dan ink were the representative works of this period. During the Three Kingdoms period, paper production spread all over the north and south of China, and transcriptions became fashionable. According to volume 92 in *Book of Jin*, after Zuo Si wrote *Odes to Three Capitals* in the Western Jin dynasty, "so many people in wealthy families competed with each other to circulate and copy books. Luoyang paper became expensive for this reason". At the end of the Eastern Jin dynasty, the imperial court ordered that all those who used bamboo slips should use yellow paper instead. Adequate and cheap writing materials also promoted the progress of calligraphy and changes in fonts. From the Han dynasty to the Southern and Northern dynasties period, Chinese characters evolved from the official script to the regular script, which gave birth to the running script and cursive script. After Buddhism was introduced in China, believers copied a large number of Buddhist scriptures, which was also an important factor in the development of the paper industry.

In the early days, hemp was used to make paper. In the Tang dynasty, bark paper became the most important kind of paper (as shown in Pictures 11.5 and 11.6). Han Yu honored the use of mulberry paper by calling it "Mr. Mulberry". Later, it was called "mulberry ink" and "mulberry sheets". Under the advocacy of Li Yu, the last ruler of the Southern Tang dynasty (937–975), a bureau was set up to make paper for the royal palace, which became the most famous producer of *chengxintang* paper in history.

Great achievements had been made in the production of letter paper during this period. According to *Notes of Past Famous Paintings* written by Zhang Yanyuan,

Picture 11.5 Five Oxen by Han Huang of Tang (partial), drawn on mulberry paper

Picture 11.6 Feng Chengsu of Tang copied Wang Xizhi's *The Orchid Pavilion* (partial) on mulberry bark paper

a great painter of the Tang dynasty: "A good family should own at least a hundred pieces of *xuan* paper, wax them and store them for copying". This is a reference to the famous waxing method. Yellow waxed letter paper was dyed with phellodendron. It was hard, smooth, and moth-proof. Powdered and waxed letter paper refers to coating the paper with white mineral powder and then waxing and calendering it, so that it has the advantages of both powdered paper and waxed paper. Using alum to size paper can prevent it from wrinkling easily, like it does when using starch to size paper. This method not only meets the requirements of fine painting, but also allows long-term preservation, which is a major technological breakthrough in the production of paper in the Tang dynasty. Sprinkling gold on paper also began at that time, creating varieties such as gold flower, silver flower, gold sprinkling, and silver sprinkling.

With the wide application of paper and the vigorous development of the paper industry, Su Yijian's *Four Treasures of Study* and Mi Fu's *On Ten Types of Paper* were written in the Northern Song dynasty. In order to meet the large demand for raw materials, bamboo paper developed greatly in the Song dynasty and competed with bark paper because of its low price. Bamboo paper accounted for 90 percent of paper output after this transition period. Zhejiang, Sichuan, Fujian, and Jiangxi provinces became the main areas of bamboo paper production. Jianyang, Fujian province was one of the Song dynasty's major printing centers. The printed "*jianben*" used locally sourced and produced bamboo paper. Jade plate paper produced in Yanshan, Jiangxi province, which is close to Fujian province, was designated as tribute paper by the government because of its excellent texture. During the Ming and Qing dynasties, cooking, pounding, and bleaching methods improved, which further enhanced the quality of bamboo paper. The *lianshi* paper in Fujian, *dagong* paper in Hunan, *fangxuan* paper in Sichuan and white paper in Zhejiang were all top grade, pure white bamboo paper.

During the Song and Yuan dynasties, bark paper continued to develop. The length of "*pi* paper" could reach up to 50 feet while its thickness remained uniform. *Xuande* paper had categories such as five-colored powdered paper and porcelain blue paper (as shown in Picture 11.7), which was similar in value to *chengxintang* paper. Yet, *Huizi*, which was the most widely circulated banknote of the Song dynasty, was printed on mulberry paper produced in the north (as shown in Picture 11.8).

The peak of bark paper came with the arrival of the *xuan* paper era. The word "*xuan* paper" first appeared in the Tang dynasty. After a long period of improvement, it became an outstanding representative of handmade paper. *Complete Library in Four Sections* was written on *xuan* paper (as shown in Picture 11.9). Cao Dasan, the ancestor of Cao in Xiaoling, Jingxian county, Anhui province, moved from Nanling to this place during the Song and Yuan dynasties to make paper. He passed down his papermaking methods from generation to generation and played an important historical role in the inheritance and development of *xuan* paper. During the Guangxu period (1875–1908), there were 3,000 households surnamed Cao in Xiaoling, most of whom had papermaking as their main source of income. At present, Cao's descendants still play an important role in the *xuan* paper industry.

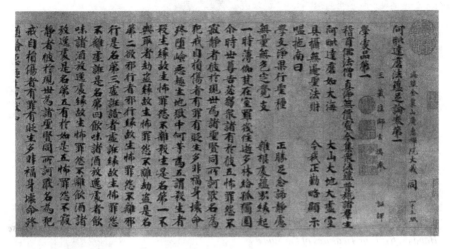

Picture 11.7 Scriptures written on *Jinsushan* Tibetan scripture paper of the Song dynasty, cited from *Complete Collection of Traditional Chinese Handcrafts: Papermaking and Printing*

Picture 11.8 Huizi (paper money) from the Southern Song dynasty, cited from *Chinese Handicrafts: Four Treasures of the Study*

At the end of the Qing dynasty and in the beginning of the Republic of China, Western machine-made paper and papermaking machinery entered China. This had a devastating effect on the handmade paper industry, causing it to shrink drastically. During the Second Sino-Japanese War (1937–1945), however, materials were scarce, and the traditional paper industry once again flourished. For example, there were

Picture 11.9 *Complete Library in Four Sections* written on *xuan* paper, cited from *Chinese Handicrafts: Four Treasures of the Study*

5,000 papermakers in Jiajiang, Sichuan province, who had an annual output of more than 6,000 tons of bamboo paper.

In the early 1950s, the import of pulp was once interrupted, and handmade paper played an important role in alleviating market demand. Some bamboo paper workshops adapted to producing bamboo pulp for machine-made paper factories, which contributed to the sustainable development of the paper industry.

Due to the unbalanced development of China's social economy and common folk activities and traditions, handmade paper still widely exists all over the country. For example, the midfield paper factory in Nanzhang county, in northwest Hubei province was founded by the Chen family who moved from Jiangxi a hundred years ago, and it still produces paper using traditional methods. Another example is the people in Zhifang village, Qufu, Shandong province, who make mulberry paper on the side. It is said that their skills originated from Xinjiang and were brought over from Hongdong, Shanxi, by people surnamed Qiao and Zheng. One of the current practitioners is Kong Xiqiao, who is a 56th generation descendant of Confucius. During China's reform and opening up, traditional culture started to recover. People paid more and more attention to intangible cultural heritage and the demand for handmade paper for calligraphy, painting, and ancient book printing increased. At present, there are 11 pieces and 16 kinds of handmade papermaking skills listed in *List of National Intangible Cultural Heritage*, such as *xuan* paper from Anhui's Jingxian, bark paper from Sichuan's Jiajiang, Zhejiang's Fuyang, Guizhou's Zhenfeng, Yunnan's Danzhai, as well as from Tibetan people and Yunnan's Dai and Naxi people, mulberry paper from Anhui's Qianshan and Yuexi, as well as Xinjiang, *liansi* paper from Jiangxi's Qianshan, and letter paper from Anhui's Chaohu. Xing chunrong from Jingxian county, Anhui province, Yang Zhanyao from Jiajiang county, Sichuan province and Zhuang Fuquan from Fuyang county, Zhejiang province, were named as national inheritors of handmade paper. The development of the traditional paper industry still maintains its vitality. It seems like it will be passed down in an orderly manner and continue its development.

11.1.3 Production of Hemp Paper

Hemp paper is mostly made by recycling worn out items made from hemp or ramie, such as rope heads, rags, or fishing nets. Pan Jixing did research on the hemp papermaking skills used in Zhifang village, Fengxiang, Shaanxi province in 1965. Their production process included 18 processes: (1) material preparation, (2) soaking, (3) cutting, (4) first grinding, (5) elutriation, (6) washing out the plant ash, (7) second grinding, (8) drying, (9) cooking, (10) elutriation, (11) third grinding, (12) washing, (13) slotting, (14) papermaking, (15) squeezing, (16) paper drying, (17) paper uncovering, and (18) packaging.

Hemp waste items must be soaked, cleaned, chopped, crushed with stone, and elutriated in baskets three times for about 10 h. The color of washed hemp is beige, which is mixed with a lime slurry in a grinding tank and ground until it is fully mixed. The filtered slurry is piled on the slate for 10 days in summer to 30 days in winter. It is then rolled into balls and put into a pot for cooking. First, use high heat, then low heat. Turn off the heat after about 2 h. Then, cover the pot and let it simmer for about 12 h.

After the lumped materials are taken out of the pot, they are scattered with wooden sticks and the plant ash is washed out. Then it is ground for 6–8 h and washed into white flocculants. Then, it is put in a trough, stirred with wood to spread the paper into pulp, and distribute it evenly. Craftsmen use their experience to judge and decide whether or not the concentration of pulp is satisfactory.

Papermaking bamboo curtains require a combination of crafts, namely, one to make the curtain and one to make the frame. The curtain is made of thin bamboo silk and silk threads (or horsetail). Select adult bamboo, remove the bamboo nodes, scrape off the skin, cut it open, splice them through the middle, and then draw them into filaments with a diameter of 0.3 to 0.5 mm using a drawing board (as shown in Picture 11.10). Weaving is carried out on a wooden frame, and several pairs of shuttles are hung on the crossbeam. The craftsman presses the bamboo silk onto the frame, turns the shuttle inside and outside, and makes the silk thread cross and wrap around the bamboo silk, thus weaving it into a curtain line by line and painting it. The resulting curtain is uniform and smooth, firm and straight in the direction of the warps, rolled up in the direction of the wefts, and held at both ends by bamboo strips. Curtain holders are mostly made of wood or bamboo, and must be equipped with a side post to hold the curtain taut (as shown in Picture 11.11). Before papermaking, it is necessary to slot it again to make the pulp uniform.

The hemp papermaking skills of Zhifang village in Fengxiang, Shaanxi province, have a long history, similar to the techniques and processes of the Han dynasty (as shown in Picture 11.12). Since no paper additives, like glue, are added, the pulp must

Picture 11.10 Silk drawing, cited from *Complete Collection of Traditional Chinese Handcrafts: Papermaking and Printing*

be stirred until it becomes more uniform, before the papermaking process is started. Skilled paper workers can make 1,200 sheets of paper a day.

This paper mill was discontinued in the late 1960s. For a long time, many scholars thought that this skill had been lost, but, luckily, it was not. When he visited Wutai Mountain in Shanxi province in the 1980s, Song Zhaolin found that Jiangcun produced very thin and tough hemp paper, which was used locally for accounting, pasted on windows and shed roofs, and sold in Hebei province and Inner Mongolia. Its technological process was similar to that in Fengxiang, except that, here, they baked the paper with a fire wall. After this discovery was published in 2008, it attracted the attention of academics. According to the recent investigations, there are still many hemp paper workshops in the counties of Dingxiang and Yuanping near Wutai Mountain in Shanxi Province. This shows that China's traditional paper industry lives on and its distribution is far wider than assumed. This also shows that there is still much work to be done in regard to the research, protection, and inheritance of traditional handicrafts.

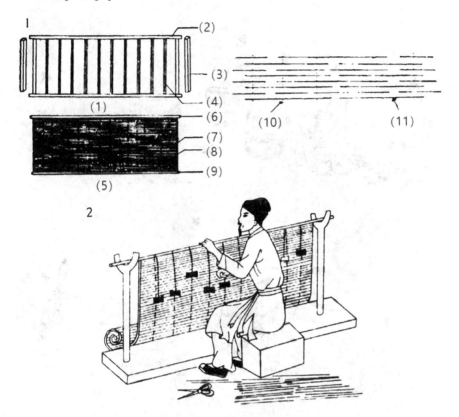

Picture 11.11 1. Movable papermaking curtain, cited from *History of Science and Technology in China: papermaking and Printing*. Left: paper curtain and curtain bed, (1) bed, (2) bed frame, (3) side post (for holding the curtains taut), (4) supporting column, (5) curtain, (6) curtain shaft, (7) bamboo strips, (8) silk threads, (9) edge strips; Right: schematic diagram of a paper curtain, (10) silk threads, (11) bamboo strips 2. Making paper on a movable papermaking curtain

11.1.4 Production of Bark Paper

11.1.4.1 Xuan Paper

Hemp paper is hard and its surface is rough. Mulberry paper is superior to regular bark paper, and is more suitable for writing and painting, so it is called "national paper". There are many kinds of bark paper. We will take a quick look at the production methods of *xuan* paper, mulberry paper, and stellera paper as our prime examples.

The production process of *xuan* paper is usually called "the 72 working procedures", which are divided into three major steps: material preparation, pulp preparation, and papermaking. If subdivided, there can be more than a hundred processes. *Xuan* paper is made from local blue sandalwood trees (as shown in Picture 11.13) and straw growing in sand flats. The best time to cut blue sandalwood trees is from

Picture 11.12 Process of hemp paper production in the Han dynasty, cited from *History of Science and Technology in China: Papermaking and Printing*. 1. washing 2. cutting 3. washing 4. heating up plant ash 5. cooking 6. tamping 7. slotting 8. papermaking 9. drying and uncovering the paper

the Beginning of Winter festival to the Beginning of Spring. The cortexes of 2- to 3-year-old trees are the best. Bundles of branches are stacked on bamboo steaming racks and cooked for 24 h. Then they are taken out, soaked, peeled, and dried in the sun. They are commonly called "fur" at this time.

The tied and bundled "fur" is once again soaked in water, then untied and air dried to soften it. Then it is bundled again, soaked in lime water, then left on the ground to

Picture 11.13 Blue sandalwood tree

cool for 15 to 40 days, after which it is cooked in a pot. The resulting "fur" is placed on a stone slab and trampled with feet to separate the outer bark from the inner bark (the "phloem"). It is then covered with straw for 4 to 10 days, after which it is washed in water, soaked, and bleached for about 12 h. Any impurities, such as the chaff or ash, are picked out and dried in the sun to obtain the bark paper blank.

These blanks are steamed twice with water mixed with alkali, washed, peeled, spread out, and dried in the sun (as shown in Picture 11.14) to obtain green bamboo peel and *liao* peel, respectively. The latter is beaten to remove impurities and made into *xiacao* peel.

After the third alkali steaming and cleaning, the *xiacao* peel is pounded into strips weighing about 1–2.5 kg each with tilt hammers (as shown in Picture 11.15). The strips are then chopped into pieces, mixed with water in a jar, and stepped on to spread them out. They are then put into sacks and stepped on to form streams, stirred with sticks, and elutriated until the water flowing out of the bags is clear, so as to obtain a leather-like slurry, which is how the required "fur" slurry is obtained. Repeated steaming and alkali boiling make its tissues soft and decompose, as well as remove and disperse the fibers. Repeated elutriation and sorting can wash out the plant ash, remove any dirt and impurities until clean.

The preparation of foraged slurry is similar to that of the aforementioned tree "fur" slurry (as shown in Picture 11.16). However, the ratio of "fur" and foraged slurry differ more or less depending on the requirements of each type of paper item, ranging from 8:2 to 4:6. After bleaching, the two kinds of slurry are placed in a

Picture 11.14 The arduous task of spreading out the pulp and drying it

Picture 11.15 1. Soaking 2. Pounding, cited from *Complete Collection of Traditional Chinese Handcrafts: Papermaking and Printing*

trough in their respective ratios and stirred evenly. The pulp material is obtained by filtering out the water.

Add water to the whole material and mix well, then add paper additive. The paper additive of *xuan* paper is made of kiwi stem juice. The function of paper additive is to make the pulp fiber set and make the paper's texture symmetrical. At the same time, it is also used as a lubricant to prevent the paper sticking together. The latter point

Picture 11.16 Alkali cooking of forage, cited from *Complete Collection of Traditional Chinese Handcrafts: Papermaking and Printing*

is particularly important. Thanks to this paper additive, thousands of pieces of wet paper can be stacked together and have the water pressed out, yet still be separated one by one, thus greatly improving the production efficiency.

When making paper, the master controls the curtain and the apprentices lift the curtain. It takes two people to operate the production line for four-*chi xuan* paper (138 × 69 cm) and six people for the two-*zhang xuan* paper (180 × 142 cm). Every 120 sheets require just one batch of pulp and one curtain can be used for 20–30 days. The keys to papermaking: The first is to "pat waves", that is, to shake the pulp with a curtain before the papermaking process in order to make it uniform. The second is to "swing curtains", which, according to *Heavenly Creations*, means that: "The thickness of paper is controlled by craftsmen's techniques. The sheets are thin when swung lightly and thick when swung heavily". Skilled craftsmen can accurately grasp the thickness of paper, so the paper thickness is uniform. This is an artistic handicraft, which can only be realized by those attentive learners under the guidance of their masters and through repeated practice. This serves as a typical example of how intangible cultural heritage is inherited.

The paper is pressed, dehydrated, uncovered, baked (as shown in Picture 11.17), sorted, trimmed, and packaged to become a qualified product. *Xuan* paper has high standards; even its weight must meet the standard. For example, a four-*chi* cotton sheet must weigh 2.4 plus 0.1 kg per 100 sheets. *Xuan* paper that does not meet the standard cannot leave the factory.

It takes 2 years to make *xuan* paper according to the above strict process. At present, some processes have been replaced by machinery, which can shorten the production cycle by one-third.

Xuan paper is famous for its exquisite material, exquisite workmanship, strict requirements, and high quality. Because of its durability, it has the reputation of being "paper that lasts for a thousand years" (as shown in Picture 11.18). Because the ink on *xuan* paper can be thick, light, dry, and wet, it creates the effect of the "ink being divided into five colors". It is especially suitable for expressing the full beauty of Chinese calligraphic work and painting. These characteristics of *xuan*

Picture 11.17 Baking paper, cited from *Complete Collection of Traditional Chinese Handcrafts: Papermaking and Printing*

paper are closely related to local plant resources, water quality, and traditional handicraft production. It cannot be easily replaced by machine-made paper. Previously, people often mistakenly thought that handicrafts were hasty and unscientific, but they evidently were not. It can be seen from the above that handicrafts, especially the top-level crafts, such as the production of *xuan* paper, are quite complex and formidable. They not only conform to scientific principles but also have strict rules and norms. Because they were limited by the times, they did not have the means for a thorough rational analysis and multiple practical tests, which sometimes resulted in craftsmen knowing the "how" but not knowing the "why". Nowadays, *xuan* paper has been awarded the Certificate of Origin and is a well-known national trademark. With the joint efforts of craftsmen, communities, enterprises, and the government, *xuan* paper techniques can be well protected and passed on.

11.1.4.2 Mulberry Paper

Mulberry paper, which was called *han* bark paper in ancient times, occupies an important position in the history of bark paper. Mulberry paper is mostly made of mulberry bark, but sometimes *sanya* trees or buttonwood bark are used. Many villages and towns in Yuexi county and Qianshan county, Anhui province, thrived on the production of mulberry paper in the Qing dynasty, but there are only a few workshops left now. In 2004, the Palace Museum searched for mulberry paper to restore the scenic illusion paintings (*tongjinghua*) in Juanqinzhai. It took years to find paper workers such as Liu Tongyan and Wang Bailin in Anhui province to supply the required paper. After that, mulberry paper-producing areas were also discovered in Xinjiang and Shandong provinces.

Taking Qianshan in Anhui province as an example, during the period from the 3rd Solar Term (known as the Waking of Insects) until the Qingming festival, people chose annual wild mulberry phloem to make paper. Their way of preparing the materials was similar to that of other bark papers. Paper additives made of carambola

Picture 11.18 "Paper lasts for a thousand years. The color of the ink forever distinct", written by Liu Haisu, a well-known contemporary painter

birch were added when the pulp was prepared, which could be buried in wet sand to stay fresh. Then the paper was made, squeezed, uncovered, and baked.

Mulberry paper is also produced in Qian'an, Hebei province. This kind of paper is rich in cotton, strong in tensile force, does not deteriorate easily, and is resistant to insects. It was used by the court in the Ming and Qing dynasties and is known as the "*xuan* paper from the south that moved northward".

11.1.4.3 Stellera Paper

Both the Naxi and Tibetan people use the white phloem of the rootstalks of the stellera (chamaejasme) to make paper (as shown in Pictures 11.19 and 11.20). Dege's Tibetan paper is divided into five grades. The best paper is used for writing official documents, followed by scripture writing, and then for printing scripture (as shown in Picture 11.21), common people's goods and packaging paper. The tools used in papermaking differ slightly from area to area. For example, in Tibet, they may use a pestle in a yak butter barrel (as shown in Picture 11.22). The paper is made by pouring. More specifically, the pulp is poured onto a fixed curtain, stirred, dried, then removed and calendered (as shown in Pictures 11.23, 11.24, and 11.25). Naxi white paper is used for writing scriptures (as shown in Picture 11.26) and printing more mundane things such as *jiama* (paper charms). Its production techniques are similar

Picture 11.19 Stellera, from Dege

Picture 11.20 Removing the skin of a stellera root

to those of Tibetan paper. These two kinds of paper are thick and strong, durable, moth-proof, and associated with folk customs and religious beliefs, which cannot be easily replaced by machine-made paper. This is another example of the strong vitality of traditional crafts because of their intricate connection with indigenous cultures, customs, and beliefs.

The process, production efficiency, and the paper itself of paper pouring and paper-making methods have clear differences. Some scholars believe that these differences are related to the fact that bamboo curtains were not produced nor paper additives used in these areas, which makes them difficult to continuously move, stack, and uncover.

11.1.5 Production of Bamboo Paper

Bamboo, the main material used to make bamboo paper, is widely and abundantly available, causing the production of bamboo paper to have the largest distribution

11 Making Calligrapher's Tools

Picture 11.21 Printing, Dege Parkhang Sutra-Printing House, cited from *Chinese Handicrafts: Four Treasures of the Study*

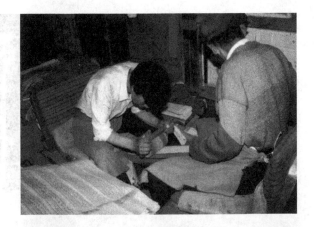

Picture 11.22 Pounding tools (left: wooden barrel; Right: stone mortar)

Picture 11.23 Pouring paper

Picture 11.24 Paper drying

Picture 11.25 Calendering of the Dai people in Menghai, Yunnan Province

and output. According to the record, *Heavenly Creations* contains a description of papermaking, taking Fujian bamboo paper as an example. "*Shaqing*" (Killing the Green) is the name of a volume in *Heavenly Creations*, which refers to when bamboo is "taken from the pool after 100 days of soaking for further processing and the rind and green skins are removed by washing (author's note: this is why it is named 'killing the green')" (as shown in Picture 11.27). The fact that this is a separate volume in this great work of literature is proof of its importance. At present, some paper mills mainly produce low-grade "burnt-offering" paper, while Jiajiang in Sichuan province and Fuyang in Zhejiang province are famous for making high-grade painting and calligraphy paper. The production process in Jiajiang is as follows: cutting bamboo, *shaqing*, beating, pulp lye, first steaming, pestling, second steaming,

Picture 11.26 *Dongba Scripture* written on white paper

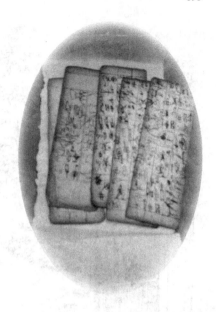

rinsing, fermentation, pounding, bleaching, paper additives, pressing, uncovering, baking, cutting, and packaging.

There are more than 20 kinds of bamboo that can be used for papermaking in Jiajiang (as shown in Picture 11.28), among which henon bamboo and fishscale bamboo are the best. Bamboo shoots that start growing in the second month of the lunar calendar are cut down when the trunk starts growing culms in the fifth and sixth month, which are referred to as the "spring batch". Clumping bamboo shoots start growing in the eighth month of the lunar calendar and are cut down in the tenth month. Henon bamboo and fishscale bamboo can also be cut at that time. The bamboo cut at this time are referred to as the "winter batch".

Remove the top branches of the cut bamboo, leaving the trunk of the bamboo and stacking it in the cellar for retting, which is also known as "killing the green" (as shown in Picture 11.29). It must be retted for 20 days in summer and 40 days in winter until the color of bamboo completely changes from green to yellow. The processes of beating, pulping, cooking, pestling, rinsing, retting, bleaching, papermaking, pressing, uncovering, and baking are similar to those of bark paper (as shown in Pictures 11.30, 11.31, 11.32, and 11.33). However, it takes 5 to 7 days and nights for the first steaming and the second steaming, which is far longer than with bark. Paper additive is made by mixing dried and crushed alum leaves with water, and is only used one day at a time.

Bamboo paper production is very hard, as recorded in *Annals of Jiajiang County*: The intensity of the other work process is no less arduous than papermaking work, nor are its tedious procedures any less tedious than papermaking procedures. In ancient times, the men who spent their days plowing the fields and the women who spent all day weaving still had time to rest. Papermaking, on the other hand, was

Picture 11.27 Cutting bamboo to bleach in the pond, cited from *Heavenly Creations–Shaqing*

made all throughout spring and summer, day and night. Men and women of all ages had to work on it all the time, which was commonly referred to as "causing trouble with their families". Fuyang's mountainous area on the south bank of the Fuchun River, especially around Da and Xiao Yuan creek, are covered in bamboo paper mills. According to *Annals of Fuyang County*: "Among all the different counties and cities in Zhejiang, Fuyang makes the highest quality paper, and of all the places in Fuyang,

11 Making Calligrapher's Tools

Picture 11.28 Bamboo near the Jiajiang River

Picture 11.29 *Shaqing* (killing the green)

Picture 11.30 Pounding

Picture 11.31 Cooking

Picture 11.32 Rinsing

Dayuan's *yuanshu* paper is the best. This is because of its handmade process and the water used, which differentiate it from others".

There is a special process in Fuyang bamboo paper production in which the paper, after cooking, soaking, and washing, is soaked in human urine (as shown in Picture 11.34), then stacked into a canopy, wrapped with green hay, and fermented and degraded in a closed environment for 1 week (summer) to half a month (winter). Until the 1950s, paper additive was still used in this area. Later, it was found that after changing flat silk curtains into round silk curtains, bamboo can be easily separated without the use of paper additive, so people stopped using it.

Picture 11.33 Wood press to remove water

Picture 11.34 Soaking the paper in baby urine (Baby urine refers to the urine a baby boy less than a month old produces in the morning. Soaked in baby urine, the paper produced is of better quality and popular with consumers.)

The quality of paper is related to the quality of the fiber in the source materials. The length–width ratio of hemp and ramie fibers is 1,000–3,000; that of blue sandalwood, mulberry bark, mulberry bark, and stellera fibers is 222–290; that of bamboo fiber is 123–133; and that of grass fiber is only 102–114. Of all kinds of paper, bark paper, and hemp paper are the best, bamboo paper is less, and papyrus can only be used as wrapping paper or toilet paper. Bamboo paper is considered one of the most important kinds of handmade paper, mainly because of its abundance in source material and its low price. In 1939, Zhang Daqian, a well-known painter, went to the Shiziqing Paper Workshop in Wu Cun, Jiajiang, and added hemp fiber to the bamboo pulp to improve the paper's strength, naming it "*dafeng* paper" and praising "*xuan* and *jia* paper as the two treasures of paper". Nowadays, bamboo paper used in painting and calligraphic work is still made in a similar way.

11.1.6 Production of Letter Paper

Letter paper is a high-quality paper product produced by calendering, powdering, adding glue alum, waxing, dyeing, pressing and smoothing, sprinkling gold, and tracing silver. The most famous dyed paper in history is the letter paper designed by Xue Tao, a female poet in the Tang dynasty. It was also known as *Huanhua* paper, named after Xue Tao's residence at Huanhua brook, Chengdu. According to *Heavenly Creations—Shaqing,* the paper is "made of hibiscus skin and aqueous extract of powdered hibiscus flowers is added". In addition, in Yuan Zhen's poem to Xue Tao, it says "two people have fallen in love, but cannot be together. I may not be able to promise you anything, but I always carry the red letter paper you sent me", which shows that this paper was dyed red. Famous kinds of letter paper in the Song dynasty include Jinsushan Tibetan scripture paper, porcelain blue paper, *Juan* paper, pepper paper, *Xuande* paper, *yangnao* paper, jade plate *xuan* paper, and linear paper (as shown in Pictures 11.35 and 11.36). Until recently, many kinds of letter paper were still produced in Jingxian and Jiajiang. For example, in Jiajiang, more than 70 kinds of *hupi xuan* paper and gold sprinkled paper were still produced. In the 1990s, Liu Xihong and Liu Jiang of Chaohu Lake, Anhui province, noticed that the production of paper was declining day by day, endangering the existence of many varieties of paper and even losing some. For this reason, they set up the Duoyingxuan Stationery Research Institute at their own expense and joined forces with Professor Zhang Binglun of the University of Science and Technology of China to do research on the restoration of various kinds of letter paper such as gold traced, gold sprinkled, and powdered and waxed. The manufacturing process of the latter is described below:

11.1.6.1 Material Selection

The production of powdered and waxed paper is very particular, since there can be no impurities nor defects on the paper's surface. The texture of the paper must be tough and stretchy. Jingxian *xuan* paper is the best material to use because of this, but is especially good because of its longevity.

11.1.6.2 Adding Glue Alum

Xuan paper is soaked in alum water and then hung to dry, so that the paper becomes more stretchy and more convenient for powdering.

11.1.6.3 Powdering

Brush powder on the front of the paper with a broad brush comprised of a row of pen-shaped brushes and hang it to dry. If you want to make double-sided powder

11 Making Calligrapher's Tools

Picture 11.35 Scripture written on *yangnao* paper in the Ming dynasty

 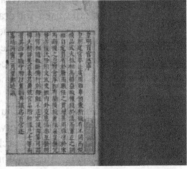

Picture 11.36 "Red forever" paper is used as the title page, and the uncoated pages have been eaten by moth, cited from *Complete Collection of Traditional Chinese Handcrafts: Papermaking and Printing*

wax paper, the back of the paper also needs to be soaked in alum water and brushed with powder.

11.1.6.4 Mounting

The front paper and the back paper are combined into a whole, and then spread out on a board to dry.

11.1.6.5 Waxing

The paper is evenly coated with white wax. The wax layer can make it airtight and thus prolong the paper's lifespan.

11.1.6.6 Calendering

Grinding the paper's surface with fine stones makes the wax more symmetrical and the paper more meticulous, shiny, clear, and waterproof.

11.1.6.7 Gold Tracing

Gold powder and silver powder (commonly known as gold mud and silver mud) are blended with glue and used for drawing auspicious patterns. If the paper is to be framed in a painting scroll, only one side of the paper is powdered.

If paper doesn't need to be framed in a painting scroll, people often sprinkle gold foil on the back of the paper for decoration. For the kind of paper with powder on both sides, people usually only process the front of the paper, and only stick gold foil on the back.

The cooperation between the Liu family's father and son duo at the Duoyingxuan Institute of Stationery and the University of Science and Technology of China is a typical example of the joint research and development of traditional handicrafts by craftsmen and scholars. The resulting paper is rich in variety, excellent in quality, and steeped in culture. They can be used for writing, painting, and making high-end products such as gifts, letterheads, and invitations. They are highly appreciated by customers at home and abroad (as shown in Picture 11.37). Liu Jing is part of a new generation of papermaking craftsmen. After graduating from college, with the encouragement of his father, he resolutely resigned from his job to devote himself to the restoration research and market development of letter paper, and with great success. He was quickly hired as a special researcher by the Chinese National Academy of Arts and invited to give lectures and skill demonstrations at the Central Academy of Fine Arts and at Tsinghua University Academy of Fine Arts.

Picture 11.37 Hand-painted gold powdered and waxed paper, made by the Duoyingxuan Institute of Stationery

The future development of letter paper is looking good. Highly educated young people devoting themselves to handicrafts can only promote the protection, inheritance, development, and revitalization of traditional crafts to a higher level. This new phenomenon in this modern era can only be endorsed.

11.1.7 Papermaking and Its Impact on Chinese Society and the World

Historically, paper has been one of the most important materials, just like iron. Pig iron and paper are both great Chinese inventions. For a long time, China was the only country in the world that used pig iron and paper. In a certain sense, it can be said that Chinese civilization is both a pig iron civilization and a paper civilization. Pan Jixing, a contemporary Chinese scholar, said: "As early as the Tang dynasty (618–907), China entered the era of paper". This statement is very accurate. All ethnic groups all over China write, draw, print, and manufacture all kinds of paper and paper products on a daily basis, such as ancient books, scriptures, albums, scrolls, Spring Festival couplets, couplets hung on columns of a hall, New Year pictures, *jiama* (paper charms), and *fengma* (paper charms) as well as business cards, *gengtie* (a card to be exchanged on engagement which states the couple's personal information), account books, documents, contracts, banknotes, bills, window paper, wallpaper, paper cuttings, paper carvings, paper fans, paper umbrellas, paper kites, paper armors, paper bags, cards, joss paper, paper models, lanterns, fireworks, firecrackers, and silkworm breeding paper, foil paper, burnt-offering paper, toilet paper, and wrapping paper. It is safe to say that paper is omnipresent (as shown in Picture 11.38). As recorded in Jiajiang's *Record of Cai Weng's Monument* (as shown in Picture 11.39):

Picture 11.38 Cards from the Qing dynasty, cited from *Collection of Civilization: the Traditional Techniques of Papermaking and Printing*

"Paper has a wide range of uses, making it convenient not only for our people's daily lives, but everyone in the world can benefit from it".

What Chinese people do and produce every day cannot be done without paper. Cai Lun, the Founder of the paper industry, is therefore highly respected. In the first year of the Yuanchu period (114) in the Eastern Han dynasty, Cai Lun was conferred the title of Longting Marquis, with Longtingpu, Yangxian county, Hanzhong city, Shaanxi province as his fief, by the emperor.

It is said that Zhifang Township in Yangxian county is the place where Cai Lun made paper. Up to now, there are still more than ten villages, with places such as Zhifang street and Huayang Daguping, which make paper using the ancient methods. Cai Hou Temple (as shown in Picture 11.40) was built in this area. In the early years, people set up their own incense festivals for paying homage on Tomb-Sweeping Day. In the 3 days before this incense festival, people didn't drink, eat meat, or have sex. On the day of the pilgrimage, everyone bathed and changed clothes, raised embroidered flags, carried offerings, and beat gongs and drums. After arriving at Cai Hou Temple, with the sound of drums and music, each village presented their offerings in turn, burned incense, paper, candles, and performed the three kowtows and nine prostrations. Nowadays, there are still grand local festivals in spring and in autumn, but these are to worship the paper saint Cai Lun instead. In 2006, the temple was added to the *Sixth Batch of National Key Cultural Relics Protection Units*. Cai Lun's birthplace, Leiyang in Hunan province, has a cenotaph, which is also worshipped regularly. Papermakers in Jiajiang regard steaming bamboo as a very important thing. Before cooking paper on the stove, they must worship the ancestor Cai Lun, slaughter white chickens and eat tofu pudding, in order to pray that the paper produced comes out clean and white. Every year, the 16th day of the third lunar month

Picture 11.39 Record of Cai Weng's Monument

is Cai Lun's birthday and the paper mill has to be closed for one day. The owner of the mill leads all the craftsmen to bring offerings, incense sticks and firecrackers to the temple of Cai Lun. The 18th day of the ninth lunar month is the Cai Lun Festival. Then, it is also necessary to set up incense tables for worship and set up banquets as a treat for paper workers. Cai Lun Temple in Xuwan, Xiaoling, Jingxian county covers an area of more than 300 m². It is where the statue of Cai Hou is enshrined. According to *Record of Rebuilding Cai Hou Temple for Longting Marquis in the Han dynasty*, a stone tablet that was rebuilt in 1935, "If people work together to make *xuan* paper, the handicraft of paper manufacturing will not be lost, which will have a far-reaching impact on future generations". Unfortunately, this temple was destroyed during the Cultural Revolution (1966–1976) and has not been repaired yet. Jingxian, Jiajiang, and Fuyang Counties are famous papermaking towns in China. For example, half of the 51 towns in Fuyang's main industry is the production of paper, accounting for more than 100,000 jobs. Paper culture museums are now built in these papermaking areas and paper culture festivals are held every year. At those times, besides worshipping ancestors, there are also cultural performances such as yangko dances, and bamboo and hemp work songs showcasing paper production. These have become grand festivals for local people (as shown in Pictures 11.41, 11.42, and 11.43). The papermaking technique of mulberry bark paper in Beizhang Village of Xi'an is known as the "ancestor of papermaking". According to local people, this technique began even before Cai Lun. As a folk saying goes, "Cang Jie's words, Lei Gong's bowls, paper produced along the Fenghe River, and bark soaked in river water". Another one also states that "if the family has a daughter, she will not marry someone from the Beizhang village, because the people in the village to keep working in the middle of the night", which is a metaphor for the amount of hard work that is required to make paper. The village used to hold a grand Cai Lun temple fair every New Year's Eve, where men, women, and children went to the market and sang Shaanxi Opera. Every 3 years, the person in charge of the "*piaohang*" would be identified by lottery at the temple fair, and every paper worker could pay to stand for election. The person in charge purchased mulberry skins from *piao* merchants around Qingling. This semi-finished paper was called "*piao*", which was managed, purchased, and stored by the "*piaohang*" (commercial house) as part of a unified supply chain.

According to Pan Jixing's research, there is a chicken and a pig under the statue of Cai Hou in Cai Lun Temple, and the same image can also be seen in the color overprinted woodcut from the Qing dynasty's Qianlong period. According to the

Picture 11.40 Cai Lun's Tomb at Cai Hou Temple

Picture 11.41 Jiajiang Handmade Paper Museum shaded by bamboo, cited from *Jiajiang, the Hometown of Chinese Painting and Calligraphy*

Picture 11.42 Fuyang, ancient Chinese papermaking and printing cultural village, cited from *Complete Collection of Traditional Chinese Handcrafts: Papermaking and Printing*

Picture 11.43 Mass cultural activities of writing and painting on *xuan* paper

old paper worker, it is said that in the Han dynasty, people didn't know that the wet paper could be split into two layers. At one point some people started to notice that chickens and pigs would split one corner of the wet paper with their mouths, which inspired them to make great improvements in the papermaking process, causing them to worship chickens and pigs along with Cai Lun. Although there is no definite evidence for this legend, it shows from one aspect that great inventions often need to be accumulated and improved for many years, while some skills are acquired by chance. People usually show reverence and gratitude for anything that participates in inventions and creations.

Chinese papermaking was introduced to Vietnam in the second century; Korea in the third century; Japan in the early seventh century; and then spread to India, Pakistan, Nepal, Myanmar, Thailand, Cambodia, Malaysia, and Indonesia.

Therefore, the Western Regions in Central Asia and West Asia started using the paper produced in China very early on. In the Battle of Talos in the mid-eighth century, the Arab Empire captured some Chinese paper workers. From then on, Chinese papermaking was introduced to Arab countries. Its earliest paper mill was built and put to use in Samarkand in 751. After that, from the twelfth century, it spread to Spain, Italy, and all parts of Europe. The worldwide spread of papermaking has played an extremely important role in the development of human civilization and culture, so it is highly regarded by Western scholars as one of China's four great inventions. China's traditional papermaking technology and its technological ideas have also contributed to the transformation from handmade paper to machine-made paper. As Pan Jixing pointed out, without such technical integration, there would be no papermaking machinery.

11.2 Writing Brush

Ancient civilizations used a variety of writing tools, such as the ancient Egyptian reed pen and the European quill pen. But only the ancient Chinese writing brush is still widely used today.

11.2.1 Evolution

As early as the late Neolithic Age, people began to draw patterns on pottery with tools like writing brush. Bronze wares of the Shang dynasty were made of clay molds. Scholars claim that complex and smooth patterns can only be drawn on its surface with a writing brush. Up to now, the earliest physical writing brush is the writing brush unearthed from the Chu tombs in Changtaiguan, Xinyang, Henan province. It is made of rabbit hair, with a brush handle length of 18.5 cm and a brush hair length of 2.5 cm. According to *Quiet Girl* in *Book of Songs*: "The gentle and elegant girl is so beautiful and charming. The gift she gave me is a new writing brush with a red brush shaft". It refers to a writing brush painted with vermilion paint. During the Warring States period, the silk paintings consisted of natural and smooth lines, painted with brushes. *Tian Zifang* in *Zhuang Zi* contains the saying "to provide a brush and grind ink", which refers to "brush and ink" as a set. The brush unearthed from the Qin tombs in Shuihudi, Yunmeng county, Hubei province, has a hollow tube inside, which was made to incorporate the writing head and became the rudimentary form used for later writing brushes (as shown in Picture 11.44).

With the development of the economy and culture throughout the Han dynasty, the writing brush and its production developed along with it and became an indispensable

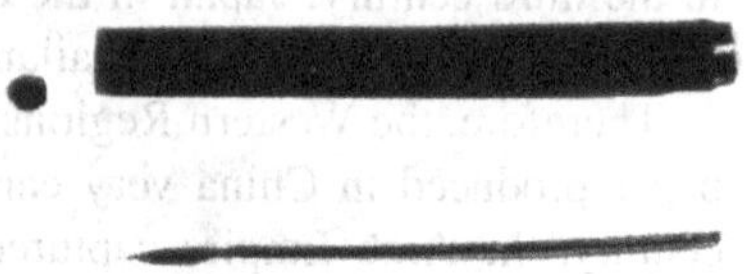

Picture 11.44 Writing brush of the Warring States period

tool. At that time, the tail of the writing brush was cone shaped. Because the scribes often used their writing brush to put their hair up, people sometimes called them "*zanbili*" (scribe of hairpin brush). According to *Ode to Brush Pen* written by Cai Yong in the Eastern Han dynasty, the hair of the writing brush is made of rabbit hair collected in winter, with bamboo as the brush tube, tied with silk thread, and then glued together with raw lacquer. According to *Miscellaneous Records of the Western Capital*, the writing brush used by the emperor had to be inlaid with jewelry and matched with a valuable case to store it in, all very expensive. From this, it can be seen that people began to pay attention to the appearance and decoration of the writing brush. During the Wei, Jin, Southern, and Northern dynasties, brush-making technology developed further. Besides rabbit hair, wool and mice whiskers were also used to make writing brushes. *The Orchid Pavilion* written by Wang Xizhi, a famous calligrapher of the Eastern Jin dynasty, is written with a brush made of mice whiskers, which was praised by people as being "charming, vigorous and unique".

In the Tang dynasty, Xuanzhou, Anhui province, became a brush-making hub. According to Bai Juyi's poem *Writing Brush made of Brownish Rabbit Hair*: "There are wild rabbits south of the Yangtze River. Because rabbits often eat bamboo in the mountains and drink spring water, black and hard rabbit hair grows on the back of rabbits. In order to collect rabbit hair to make writing brushes, the people of Xuancheng spared no effort to pluck thousands of hairs from rabbits before selecting a usable fur. Every year, when Xuancheng has produced writing brushes, also known as *xuanbi* or Xuan brushes, the price of those made of brownish rabbit hair are the same as those made of gold". In the Tang dynasty, Zhuge and Chen were the most famous brush craftsmen. According to Tao Gu's *Record of the Pure and Uncommon* from the Song dynasty, the brushes made by Zhuge from Xuancheng were so expensive that they were also called "treasure brushes". Liu Gongquan (778–865), a famous calligrapher, once participated in the research and production of "*xuanbi*", and this cooperation between brush-makers and calligraphers has been maintained until now.

High tables and chairs were popular in the Song dynasty. This caused the writing posture to change from the hanging elbow to the lifted wrist, which resulted in new requirements for the materials and shape of writing brushes. The brush shaft became shorter, and the hairs of the writing brush were mixed with soft hair (as shown in Picture 11.45). The most famous brush craftsman was Zhuge Gao of Xuancheng, whose brush-making techniques are considered far and wide to have been the best. Huang Tingjian, a famous Song calligrapher, received a Zhuge brush and wrote a poem that reads: "In the *jiju* brushes produced in Xuancheng, the brush maker Zhuge uses mice whiskers for the hairs of the writing brush. I'm glad I got this brush from somewhere else. No matter how much you spend, you can't buy it in the market". Driven by the success of *Xuanbi*, the brush-making industry in Shexian, Yixian, Yangzhou, and Hangzhou also started to flourish. Wang Boli's brush, Chengxintang's paper, Li Tinggui's ink, and Yangdouling's inkstone are called the "Xin'an's Four Treasures of the Study". After the Song dynasty moved its capital southward, the success of *xuanbi* gradually declined and Huzhou became the center of the brush-making industry, especially Shanlian town, which flourished in the Yuan, Ming, and Qing dynasties. At the beginning of the Yuan dynasty, Feng Yingke and Shen Rixin were the most skilled craftsmen to make *Hubi*, which is a kind of writing brush. *Hubi* craftsmen at that time also made *Lanrui* writing brushes, which were made of soft, lustrous, and elastic goat hair, making them suitable for writing and painting. In the Ming and Qing dynasties, the production of *Hubi* reached its peak, and shops were set up in Beijing, Tianjin, Suzhou, and Shanghai, among which the most famous shops were Daiyuexuan and He Lianqing in Beijing, Beisongquan in Suzhou, and Yang Zhenhua and Li Dinghe in Shanghai. During this period, people paid great attention to the decoration on the writing brushes: using gold, silver, jade, ivory, rosewood, bamboo, and pterocarpus for the shafts and decorating them through engraving, inlaying, painting, or other techniques. This made brushes both practical items and art treasures (as shown in Picture 11.46).

In modern times, writing brushes have gradually been replaced by the pen and fountain pen, yet they are still commonly used and produced. Up to now, there are still hundreds of workshops and more than 1,000 people working at Shanlian town, Zhejiang province, who produce 3 million *Hubi* a year, of which one-third are exported overseas. *Xuanbi* and *Xiangbi* take up a considerable portion of production. The writing brushes industry in Wengang, Jiangxi province, has continuously developed in recent years, occupying more than half of the market share and even being awarded the title of "Writing Brush Capital of China".

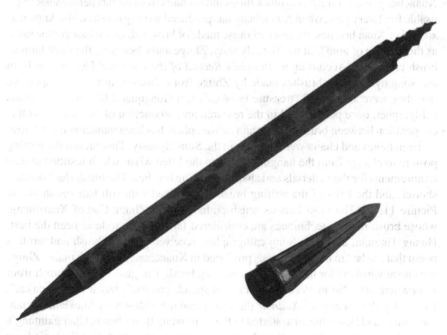

Picture 11.45 *Jiju* writing brush (replica) of the Tang dynasty

11 Making Calligrapher's Tools

Picture 11.46 *Hubi*, cited from *Chinese Handicrafts: Four Treasures of the Study*

11.2.2 Making Techniques

Taking *Hubi* as an example, its production process is also known as "the 72 steps".

11.2.2.1 Material Selection

The most valuable part of a writing brush is its hair. The writing brush industry has strict requirements on the origin of brush hairs, the collection season, and which part of the animal fur is collected. The goat hair used in *Hubi* is collected from the Hangzhou-Jiaxing-Huzhou area, the rabbit hair is collected from the mountainous area of southern Anhui, and the tail hair of weasels is collected in Northeast China. The writing brush tube is supplied by bamboo-producing areas in mountainous areas. Depending on the length, thickness, and sharpness, the brush hair can be divided into dozens of varieties to match different kinds of writing brushes.

11.2.2.2 Basin

This is the most complicated and critical process in the production of *Hubi*, because it contains many sub-processes. The goat hair brush, for example, involves 15 processes such as mixing, linking, arranging, closing, and rounding when being processed in the basin. The mixed hair brush has 22 processes, with additional steps such as soaking, arranging, matching, making, and stirring in the basin. The wolf hair brush has 13 processes in the basin, such as pulling, binding, and lifting. The main job of the basin craftsman is to straighten out the brush hairs in the basin, remove miscellaneous hairs, fluff and unusable hairs, and arrange these brush hairs into semi-finished products (as shown in Pictures 11.47, 11.48, and 11.49). For example, the craftsman in charge of the goat hair basin should not only remove the unusable hairs, but also tidy the hairs with sharp tips to make them "uniform and neat". The former means that the lower end of the translucent part of the pen should be clear and flush, while the latter

Picture 11.47 *Hubi* production site at Shanlian

means that the translucent "*heizi*" (sunspot) part should be uniform in length. This requirement for hairs with sharp tips best reflects the unique style of *Hubi*, but is also the key point of its usage. The brighter the "*heizi*" of the hair tip, the better the quality of the brush hairs. The words written with this brush will be full and neat, and the brush hairs will fold together into a sharp point when lifted. Take goat hair as an example. There are very few hairs on a goat that can be used for making writing brushes. Only three *liang* (150 g) hairs of a goat can be made into the writing brush, and there are only six *qian* (30 g) hairs with sharp tips. It is true to say that only "one hair out of a million" is suitable. It is for this reason that *Hubi* is lauded as "the best writing brush in the world". Goat hair can be degreased by soaking it in lime water. However, to prevent the lime water from possibly damaging the hair, it can also be exposed to the sun and cooled at night. The color of the hair will also become whiter. This is another unique feature of producing the *Hubi*, "people get darker and darker in the sun, while goat hair gets whiter and whiter".

11.2.2.3 Ligation

Ligating is also called bundling hair. The semi-finished brush hair is ligated and molten rosin dripped on the root to make it difficult for the hair to fall off (as shown in Picture 11.50). When binding, it is necessary for the bottom of the tip of the brush to be flat and firmly tied to prevent hair loss and an uneven tip.

Picture 11.48 Removing miscellaneous hairs

Picture 11.49 Making a brush tip

Picture 11.50 Ligation

11.2.2.4 Sorting

Remove any materials with insect marks, faded colors, uneven thickness, and classify and select the materials with the same colors, thickness, and length according to the grade requirements of each of the various writing brushes.

11.2.2.5 Making a Brush Cap

Dig the end of the brush shaft into a hollow, requiring the head to be flat and the navel to be neat, and install the tip of the brush into it, and the brush cap will be made.

11.2.2.6 Inlaying

For some types of writing brushes, the head and tail of the brush shaft must be inlaid with horn or glass to make it more beautiful.

11.2.2.7 Trimming

This is another key process in making *Hubi*. Semi-finished writing brush workblanks must be inspected and trimmed before they can become qualified products. Therefore, it is necessary to carefully eliminate any residual hair that would have adverse effects on the quality and appearance of the writing brushes (as shown in Picture 11.51), so as to reach the grade standard of "sharp, neat, round and durable", which are considered the "four virtues" of the writing brush. "Sharp" means that the writing brush is pointed and shaped like a cone, which does not bifurcate and is beneficial to point, hook, and skim when writing. "Neat" means that the writing brush hair is neat, the top is not jagged after being pressed down, the ink absorption is full, and the ink pay-off is uniform. "Round" means that the tip of the writing brush is very symmetrical, and it is very smooth and convenient to write. "Durable" means that the brush is strong, elastic, retractable, and durable. The trimming process includes picking and cutting, and selecting and twisting. Selecting and twisting is to twist the tip of the writing brush out of shape by hand. Professionals say that "30% depends on selection and 70% depends on twisting", which shows the importance of twisting by hand. Apprentices must thoroughly practice these basic skills. Only when the twisting gestures are perfect, are they allowed to carry out practical operations.

11.2.2.8 Inscribing

The craftsmen in charge of inscribing are familiar with the various forms and shapes, resulting in the inscribed characters being arranged neatly and evenly, beautifully and generously. It is precisely because people have strict requirements on material selection, meticulous production, and pay attention to the harmony and perfection of

Picture 11.51 Trimming

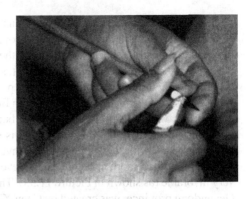

the brush tip, tube shape, texture, color, inscribing, and decoration that they can be seen as unique within China's brush-making industry and even rank among the best.

Xuanbi has a long history. In the Qing dynasty, the brushes made by the crafting families Liu, Cheng, and Cui in Taiping county, Anhui province, were the best, with an output of nearly one million. It is also very strict in selecting its materials. The goat hairs of their writing brushes are selected from the male goats born in Qidong and Nantong, Jiangsu province, and the goats are slaughtered in January after the Winter Solstice. The hairs at the front end of the goat's back and root of the neck are the first-class hairs that can be used to make writing brushes. A goat only has about 100 mg of that hair, called "*xiguang* front", the hair on the chest is called "*baijia* front", and the hair at the top of the hips is called "*huangjian* front". All types of front hairs have their own uses (as shown in Picture 11.52). The hair of the *Zihao* writing brush is the hair of wild rabbits, which live in the shade of mountainous areas in southern Anhui. It is said that only 50 g of *zihao* hairs can be obtained from every 1,000 mountain rabbits, so the price of *Zihao* brushes is like the price of gold.

Picture 11.52 Different qualities of hairs

In addition, *Xiangbi* from Changsha, Hunan province, and *Hengbi* from Hengshui, Hebei province, are also famous. The nibs of *Xiangbi* are mixed but not layered, and the caps are mostly copper, which makes them more durable. In the Qing dynasty, *Hengbi* were supplied to the imperial court, so their market price doubled. Emperor Guangxu once ordered the government to purchase all writing brushes from Hengshui. Li Fushou Writing Brush Workshop in Beijing also used to be the representative of wolf hair writing brushes in the north, but not anymore. However, Daiyuexuan, an important production branch of *Hubi*, is still famous as a time-honored brand. In recent years, with the support of the government, it has expanded its workshop area and recruited more apprentices. Its development and market trend even seems very favorable (as shown in Picture 11.53). The *Maolong* writing brush from Baisha, Guangdong province, was created by Chen Xianzhang (1428–1500), a scholar in the Ming dynasty. He selected plant fibers and, after a special kind of treatment, used it for its nib. This method is still in use. The production techniques of these two writing brushes have also been added to *List of First Batch of National Intangible Cultural Heritage*.

People in the writing brush industry regard Meng Tian (the general of Qin state) as their ancestor. They built statues and temples for him and worship him regularly. The writing brush factory located at Shanlian has a specially made "*Fuyu* brush" to commemorate Meng Tian. The brush is 112 cm long, the maximum diameter of the nib is 24 cm, and the weight is 13 kg. It is the largest *Hubi* known so far. Chinese people have always respected writing brushes. Scholars love writing brushes. They love them so much that they can't even bear to discard the residual writing brushes, so

Picture 11.53 Li Shuyuan, the second-generation descendant of Daiyuexuan, inscribing on the brush shaft

Picture 11.54 A painting named Bi Ding Sheng Guan (Be Sure to Be Promoted)

they respectfully bury them in the ground, which is called "*bizhong*" (brush tombs). When Li Bai was young, he dreamed that a beautiful flower grew on the brush he commonly used. Since then, Li Bai has been famous all over the world for his quick writing and rich poetry. Folk legends also speak highly of brushes, as in *The Magical Brush of Ma Liang* (as shown in Picture 11.54).

11.3 Ink Stick

In China, literati are called "ink men", calligraphic works are called "ink treasures", engraved block printing techniques are called "ink blocks", and imperial examination papers are called "ink paper". The ink stick, as one of the "Four Treasures of the Study", is closely related to Chinese cultural heritage and Chinese people's daily lives.

11.3.1 Evolution

The earliest known ink balls were unearthed at the Shuihudi ancient tombs in Yunmeng county, Hubei province, and dates back to the fourth- to third-century BC. More than 4,000 pieces of ink from the Western Han dynasty were unearthed in Wangdu, Hebei province; Hunyuan, Shanxi province; Jiangling, Hubei province (as shown in Picture 11.55); and Xianggang Hill, Guangzhou, Guangdong province. According to *Collection of Laws and Regulations of the Han dynasty* written by Ying Shao, in the Eastern Han dynasty, "officials will reward secondary officials with one big piece of Yumi ink and one small piece of Yumi ink every month". Yumi is the old name of Qianyang, Shaanxi province today, where the ink produced was famous for its high quality. Later generations still use Yumi to refer to ink, just as people still use the word *"chu"* instead of the name of paper. By the Wei, Jin, Southern and Northern dynasties period, people were using graphite molds to hold ink sticks. The production method of ink is recorded in detail in *Writing Brushes and Ink Sticks* in *Essential Techniques for the Peasantry* written by Jia Sixie of the Northern Wei dynasty: "Mash the prepared lampblack, then sieve it in a jar with fine silk [...] get a kilo of ink, prepare 250 g of fine glue, and then soak it in *chen* bark juice. This *chen* refers to the bark of ash trees found around Jiangnan area. Its bark is green when put into water, which removes grease quickly, conveniently and effectively, and is compatible with the color of ink. Prepare five eggs, remove the yolks and leave only the egg whites. Prepare 50 g of cinnabar and 50 g of musk, and mix them together according to the requirements of the formula. Put the prepared material into an iron mortar. The more times it is ground, the better. It should not be ground less than 30,000 times. The time of making ink is between February and September every year. The weather in this period is neither hot nor cold, which is the best time to make ink". As can be seen from the above, the main processes of ink making in that period included collecting lampblack, making a selection, making colloid, blending, pounding, and drying the ink. These processes had been nearly perfected, laying a solid foundation for later generations to make ink. In the process of making ink, various ingredients are added to prevent corrosion, increase its fragrance, increase its strength, and ensure a good color.

A long time ago, the main producing areas of ink were mostly around Yumi and Fufeng in Shaanxi province. By the Tang dynasty, production was mainly transferred to Yishui and Luzhou in Hebei province. At the end of the Tang dynasty, in order to avoid the influence of war, Xi Chao and his son, ink workers from Yishui, moved to Shexian county, Anhui province to make ink. According to the fifth volume in *Four Treasures of the Study* by Su Yijian, "The ink they made was as strong as jade, and the lines made by it were exquisite", which was deeply appreciated by Li Yu, the

Picture 11.55 Ink, writing brush, inkstone, and wood box from the Western Han dynasty

last emperor of the Southern Tang dynasty; hence, they were given the emperor's surname (Li). Since then, the ink created by the craftsman surnamed Li has become famous all over the country and has been passed down from generation to generation. Huizhou ink has been the most famous for more than a thousand years.

Famous Huizhou ink artists in the Song dynasty were surnamed Geng in Shezhou and surnamed Sheng in Xuanzhou. Monographs such as *History of Ink* and *Ink Spectrum* (as shown in Picture 11.56) were also published. In the Ming dynasty, Huizhou became the national ink-making center, with many ink shops and ink-making experts. People used pine smoke and lampblack to make ink. There were two ink-making schools, namely, the *She-style* and the *Ti-style*. The former (*She-style*) was represented by Luo Xiaohua, and according to *Annals of Shexian county*, his tung oil smoke ink is "as strong as stone, with exquisite lines and priceless value". The latter (*Ti-style*) was founded by Wang Zhongshan and Shao Gezhi, who pioneered a complete set of ink sticks, which was famous for its exquisite ink style and was imitated by other schools (as shown in Picture 11.57). After the mid-Ming dynasty, the Huizhou ink industry developed outward and its distribution area greatly expanded.

There were four masters of Huizhou ink in the Qing dynasty: Cao Sugong, Wang Jinsheng, Wang Jie'an, and Hu Kaiwen. "Hu Kaiwen" was the name of the shop opened by Hu Tianzhu in Jixi county, Anhui province. The name of the store was named after the plaque "*Tian Kai Wen Yun*" found in Jiangnan Imperial Examination Centre in Nanjing. He set up an ink factory in Yixian county, Anhui province, refined pine smoke with high-quality pine, improved the production formula, raised the production standards, and produced famous inks such as "*Cangpeishi*" and "*Wujin*" with excellent ink quality. His son Hu Yude's "Cotton Map" and set of ink sticks "Zodiac" were fine ink mold products, and the "*Cangpeishi*" ink was made by mixing 500 g of pine smoke; 150 g of pearls; 50 g of jade chips; and 50 g of borneol, and raw lacquer, and then pounding it 100,000 times. Hu Yude had nine sons, all of whom were engaged in ink-making. They used Hu Kaiwen Ink Shop in Xiucheng, Anhui province, as the head office and set up branches in Anqing, Guanhu, Wuhan, Yangzhou, and Shanghai, making Hu Kaiwen ink industry famous all over the country. In 1915, Hu Kaiwen's "*Cangpeishi*" ink won a gold medal at Panama Pacific International Exposition and won honor for Chinese ink. At present, the ink

Picture 11.56 Fire painting in volume 1 of *Ink Spectrum*, cited from Chinese Handicrafts: Four Treasures of the Study

Picture 11.57 Wurui Lianhun colorful ink sticks, made by Hu Kaiwen Ink Shop in the Qing dynasty, cited from *Appreciation and Collection of Four Treasures of the Study*

factories in Jixi, Tunxi, and Shexian are all named after "Hu Kaiwen", and they are all listed in *List of National Intangible Cultural Heritage*. Cao Sugong Ink Factory in Shanghai is another school of Huizhou ink, which has made great progress in recent years.

11.3.2 Preparation of Pine Smoke and Lampblack

Ink comes in many colors, such as black, red, blue, and yellow. Black ink is the most commonly produced and used. Pine smoke and lampblack are the main source materials for making ink.

11.3.2.1 Preparation of Pine Smoke

Pine smoke refers to the soot produced by the incomplete combustion of pine wood. It was discovered a very long time ago. For example, it was recorded in *A Song in Slow Time* written by Cao Zhi (192–232) that "ink is made of pine smoke and writing brushes are made of rabbit hair".

There are many ways to make pine smoke. Let's take *Heavenly Creations*' entry as an example, which cites its technological process as follows:

Removal of Turpentine. Punch holes in the roots of the pine and burn them with an oil lamp to make the turpentine flow out. If the turpentine is not cleaned, there will be small pieces of condensation when using the ink.

Burning. Bamboo awnings are woven with bamboo strips, which are connected one after another and are more than 30 m high. The inside and outside of the bamboo awning and the interface are wrapped tightly with paper and bamboo mats, so it does not leak smoke. The bamboo awning is provided with a flue and a smoke outlet. Pine chips are slowly burned at one end of the awning and attached to the inner wall of the awning through the flue (as shown in Picture 11.58).

Collection. After the bamboo awning has cooled, pine smoke is collected with goose feathers. The "light smoke" particles found at the end of the bamboo awning are the finest, which can be used to make fine ink for painting and calligraphic work, the "mixed smoke" in the middle is used as general ink for painting and calligraphic work, and the "coarse smoke" found closest is used as printing ink.

11.3.2.2 Preparation of Lampblack

Lampblack is made of animal fat, vegetable oil, and mineral oil. According to *Method of Making Ink* written by Shen Jisun in the Ming dynasty, the production process of lampblack is divided into soaking, burning, and collecting. Soaking oil refers to

Picture 11.58 Production process of pine smoke

soaking traditional Chinese medicine such as hematoxylin, Chinese goldthread, the bark of Himalayan coralbean, almonds, arnebia euchroma, sandalwood, gardenia, dahurian angelica, and strychnine seeds in tung oil and sesame oil (or vegetable oil, soybean oil). It is heated before it is burned, cooled, filtered, and then added to oil. The time of burning smoke is usually in late autumn and early winter, and is carried out in a clear kiln room to ensure that the ink is pure and does not easily spoil or become moldy.

11.3.3 Production Techniques of Huizhou Ink

Taking the production techniques used by Hu Kaiwen Ink Factory in Jixi, Anhui province, as the example, the main processes are as follows.

11.3.3.1 Refining Smoke

The tung oil is heated and atomized, and then burned to obtain soot. The efficiency of this method is very high, and the quality of the obtained oil fumes is also good. In order to ensure a steady source of raw materials, the factory specially plants more than 500 acres of tung trees. Refined smoke should be stored for 3 years to avoid the ink sticks being too dry when used. Pine smoke should be soaked in water for several days before use, and then washed with a screen to remove impurities.

11.3.3.2 Making Glue

The glue is made by boiling animal skins and bones, which can bond soot into pieces. People used to make glue from antlers and swim bladders, but now they make glue from animal skins and bones, which are of inferior quality and are used as ordinary ink sticks.

11.3.3.3 Blending

The ratio of lampblack, glue, and auxiliary materials has a great influence on the ink quality. Therefore, the secret of making Huizhou ink by hand is usually kept and done by only one person in a workshop. The preliminary mixed material is called ink *ke*, which must be kept warm in a heated *kang* furnace.

11.3.3.4 Pestling

Ink *ke* is repeatedly ground and crushed with a square pestle. After being crushed, the ink material is fine and uniform, and the fine and highly dispersed soot is mixed with glue, which can improve the quality of the ink.

11.3.3.5 Stamping

Ink molds are also called *motuo* (as shown in Picture 11.59), which is recorded in *Method of Making Ink* as follows: "The mold is made up of seven pieces of board, with four enclosed as a frame. Two pieces engraved with patterns and characters are fixed in the frame, one on the top and the other on the bottom. Another piece of wood is used to hoop them. When the ink stick is made, remove the hoop and take the stick out". The wooden board with patterns and characters carved in the ink mold is called "ink printing", which is carved with solid briar.

11.3.3.6 Drying

The drying time is at least half a year. During this period, it must be turned over to prevent deformation, and some should be wrapped in paper and hung to dry.

Picture 11.59 Ink stick mold

Picture 11.60 Lighting room, in Hu Kaiwen Ink Factory in Tunxi, cited from *Chinese Handicrafts: Four Treasures of the Study*

11.3.3.7 Polishing

There are different polishing methods. The scraping method adopted by Fang Yulu in the Ming dynasty is done by "first grinding the shape with tools, then adding equisetum, then adding animal fat, soaking it in raw lacquer, and then adding fragrant medicine". This causes the produced ink to look like it contains water and resulting in a very bright color.

11.3.3.8 Tracing

According to the patterns and characters on the ink's surface, it can be filled with gold and silver powder or other pigments.

The ink sticks are packed in boxes and packaged for sale. Good ink sticks should be paired with a good box.

Most workshops of Huizhou ink have the same ink-making process, but some may have unique techniques. For example, the ancient methods to make soot and the various tools, such as rushes and oil tanks, which are still used at Hu Kaiwen Ink Factory in Tunxi are very similar to those described in the Ming dynasty's *Method of Making Ink* (as shown in Picture 11.60). The factory also has more than 7,000 pieces of ink molds made by many famous artisans since the Ming and Qing dynasties. Some varieties of ink sticks are refined from pure tung oil fumes, gold and silver foil and musk. They look dark and moist, and don't fade when put on paper.

People in ethnic minority areas make their own ink for printing and writing. For example, Dege Parkhang Sutra-Printing House in Sichuan uses the bark of the local rhododendrons for ink making. They smolder it in a ground stove for 20–30 days, cool it for 3–5 days, and then grind it into ink. Writing ink is made of magnolia-leaf willow with oxhide gelatin.

After the Ming and Qing dynasties, after having been continuously passed on and developed, the techniques of making ink, as well as the quality, graphics, and packaging of ink, have been improved, and the excellent ink has been upgraded to actual non-practical works of art that are valued by collectors.

By convention, the ink sticks used for writing and painting must be ground. In ancient times, many literati yearned for a life in which they wrote and painted with beautiful young women helping them lay paper and grind ink sticks. However, when a large amount of ink is needed, it is most convenient to use directly prepared Chinese ink instead of grinding ink sticks then and there. According to the records in *Writing of a Scholar* by Lu Qian, the ink used for printing is paste-like, which is mixed with coarse lampblack, glue, and wine. It must be stored for three or four summers to remove the odor. The longer it is stored, the better its quality. Temporarily ground ink melts easily, causing the words written to smudge easily. Long-stored ink paste can be mixed with water for use. Beijing's *Yidege Ink* was created by Xie Songdai in 1865, which is one of the innovations in ink-making technology. This kind of ink is very convenient to use, the color and shade of the written words are consistent, and the words will not run when mounted, which makes it a very popular choice. Nowadays, people use industrial carbon ink instead of pine smoke and oil smoke to make ink, and add phenol as preservative. However, for the ink used in high-grade paintings and calligraphic works, spices such as borneol should still be added.

11.4 Inkstone

11.4.1 Evolution

Stone inkstones unearthed from Yangshao Ancient Cultural Relic Site in Baoji and Lintong, Shaanxi province have a history of 5,000 years. They were used to grind natural pigments such as ochre and cinnabar (as shown in Picture 11.61). Most of the stone inkstones from the Warring States period to the Western Han dynasty are round cakes. The three-legged inkstone unearthed in Nanle, Henan province from the third year of the Yanxi period (160) in the Eastern Han dynasty, has inscriptions on it. The inkstone cover is integrated with the inkstone body, and the cover button is decorated with six relief carved dragons, which is also made of inkstones.

The inkstones in the Wei, Jin, Southern, and Northern dynasties period are stone inkstones and porcelain inkstones that can be used to directly grind ink on, since they are, in fact, special tools for grinding ink (as shown in Picture 11.62).

This is an epoch-making progress in the history of inkstone making and is related to the maturity of ink stick production technology at the time.

In the Tang dynasty, many kinds of inkstones, such as *Duanshi*, *She* inkstones, *Hongsi* stones, and *Chengni* inkstones, were used, among which dustpan-shaped inkstones, turtle-shaped inkstones, and porcelain *Biyong* inkstones were common (as shown in Picture 11.63). The shapes of inkstones in the Song dynasty were more diverse. It gradually became fashionable for literati to appreciate and collect inkstones, which created an era in which both the practical value and artistic value of inkstones were appraised. Monographs such as *Inkstone Spectrum* written by Ouyang Xiu, *Inkstone History* written by Mi Fu and *Inkstone Record* written by Cai Xiang were published one after the other. *Tao* inkstones became famous to be given as a

Picture 11.61 Stone inkstone and water cup of Yangshao culture, unearthed at Jiangzhai, Lintong, Shaanxi Province, cited from *Appreciation of Famous Chinese Inkstones*

Picture 11.62 Inkstone of the Northern Wei dynasty, cited from *Appreciation of Chinese Famous Inkstones*

Picture 11.63 *Biyong* inkstone of the Tang dynasty, cited from *Appreciation of Famous Chinese Inkstones*

Picture 11.64 *Taohe chaoshou* inkstone of the Song dynasty, cited from *Appreciation of Chinese Famous Inkstones*

tribute (as shown in Picture 11.64), and later the production technology of *Chengni* inkstones became increasingly mature.

The inkstones in the Ming dynasty still maintained a dignified style, and, at the same time, people created a variety of inkstone shapes and styles. The trend of collecting inkstones was prevalent. Excellent inkstones were not only excellent in the quality of the stone but also in the way it was engraved. They often integrated painting, calligraphic work, poetry, and seal cutting, and were both practical and high-grade works of art. The skill of making inkstone in the Qing dynasty was the most brilliant. During the Kangxi period (1662–1722), the green and beautiful *Songhua* inkstones became a tribute, and the lacquered sand inkstones by Lu Kuisheng in Jiangnan area also became famous for a while. In this period, inkstone engraving was usually complicated, branching into the Zhejiang School, Guangzuo School, and Gongzuo School. Some scholars cooperated with inkstone workers to make inkstones that showed off the elegant style and tastes of the literati. After developing and being compared for a long time, the *Duan* inkstone (as shown in Picture 11.65), *She* inkstone, *Tao* inkstone, and *Chengni* inkstone have become recognized as the four most famous inkstones; however, there are also famous inkstones such as the *Zijin* inkstone in Anhui, *Yishui* inkstone in Hebei, *Jinxing* inkstone in Jiangxi, and *Sizhou* inkstone in Guizhou. At present, the above four famous inkstones, as well as the *Zijin* inkstone, *Yishui* inkstone, and *Jinxing* inkstone, have been listed in *List of National Intangible Cultural Heritage*, which causes them to become better protected and developed with the concerted efforts of artisans, communities, and the government. As a result, driven by interests, some manufacturers have recently made large inkstones weighing more than ten tons using super-large stones, which are decorated with cumbersome engravings. That is the abnormal development of

Picture 11.65 *Duan* inkstone in the Qing dynasty

inkstone production, which only highlights the lack of innovative ability of inkstone makers.

11.4.2 Making Techniques

There are many varieties of inkstones, and first-class inkstone materials have to perform perfectly, so that the developed ink is delicate and does not hurt the hairs of the brush. It needs a very low rate of water absorption and be of high quality. In addition, the inkstone should be dense and clear in texture, bright in color, and good in visual effect.

11.4.2.1 Making of *Duan* Inkstone

Let's take the *Duan* inkstone as an example, which is valued by people due to its noble quality.

The stone mined at Fuke Mountain in Duanxi, Guangdong province is the best for making *Duan* inkstones (as shown in Picture 11.66). Because it is not earthquake-resistant, the stone layer sometimes extends to the bottom of the river, and stones with a thickness of only 20–30 cm can be used as *Duan* inkstone material. Therefore, it has been mined by hand since ancient times. Quarriers have to squeeze in a very narrow space, and then look for inkstone material based on their expertise. They will knock from the edge of the stone layer, shake the stone open, and then chisel out the usable parts. Quarrying is extremely hard, as Li He, a famous Tang poet, mentioned in his *Stone Inkstone Song*: "Duanzhou masonry's craftsmanship is really superb. After sharpening their knives, going up the mountain to the quarry is like stepping on the sky to cut clouds, which looks very divine. The inkstone is ground evenly, and water is injected, just like the lips with lip glaze. The pattern in the inkstone is beautiful". The ancients said that "a kilo of stone is worth a thousand dollars", which still rings true. In addition to its excellent quality and low water absorption, the stone patterns are extremely rich and colorful. The famous patterns are fish brain, azure, banana leaf white, blue and white, water ripple, gold thread, silver thread, stone eyes, insect spots, and cinnabar spots. Fish brain is the most delicate and pure. It is praised as "white as clouds and loose as catkins". It is only found in a very small number of stone, and this part is often used as the center when making the inkstone. With its dark blue and gray color, azure as mentioned by the ancients is: "It's like the color of the sky just after autumn rain, which is beautiful". Water ripples can only be seen in older stones, and the looming ripples display a unique aesthetic pattern. Stone eyes come in various colors, such as emerald green, beige, pink green, yellow and white, and their shapes are equally diverse, such as finch eyes, cat eyes, myna eyes, and ivory eyes. These stone eyes are extremely precious (as shown in Picture 11.67).

Picture 11.66 Fuke Mountain in Duanxi, located in Zhaoqing city, Guangdong province, cited from *Chinese Handicrafts: Four Treasures of the Study*

Picture 11.67 Stone eye, produced in Zhaoqing city, Guangdong province, cited from *Chinese Handicrafts: Four Treasures of the Study*

The production process of the *Duan* inkstone mainly includes material preparation, design, carving, box matching, polishing, ink dipping, and waxing.

The inkstone must be observed with the naked eye, and the stone skin, bottom plate, and any defects must be removed. A clever miner does not only have the unique ability to identify the perfect stone, but can also accurately identify good quality stone and detect its internal patterns. After the stone is selected, it is processed into inkstone with hammer and chisel.

The design should be based on the internal structure of the stone, so as to show off its excellence and turn defects into beauty as much as possible, thus increasing its artistic value. Workers usually consider the theme, shape and technique carefully, and usually use the best parts to make ink hearts.

Carving techniques include shallow knife carving (relief carving), deep knife carving (high-relief carving), fine engraving, and shallow engraving. The more precious the inkstone, the less it is carved. Carving tools include mallet, hammer, chisel (square mouth, round mouth, hook line), drill, saw, and talc. Carvers should pay attention to the overall coordination in their operations. The inkstone box plays a role in protecting and decorating the inkstone, and the materials used to produce it are exquisite, which sets off the inkstone's simplicity, dignity and finesse. After the box is ready, carefully polish it with natural talc by first grinding it coarsely to remove the gouging marks and knife paths, and then continue to grind it finely until it is smooth. Polished stone inkstone should be waxed. In order to make the color look better, it is sometimes necessary to soak the inkstone in ink, and then take it out to let the ink fade for a day or two.

At present, there are more than 200 *Duan* inkstone workshops in Zhaoqing, Guangdong province, with nearly 10,000 employees. Every year, an inkstone culture festival is held, and a *Duan* inkstone museum has been built. The future development of the *Duan* inkstone is looking good. In view of the shortage of stone resources, measures have been taken to strictly control the amount of mining and the collected inkstones are rationally used.

11.4.2.2 Production of *Chengni* Inkstone

The *Chengni* inkstone developed from the pottery inkstone and tile inkstone. The *Chengni* inkstone was considered the best in the Tang dynasty, and it was famous in Jiangzhou, Shanxi province; Guozhou, Henan province; and Qingzhou, Shandong province. This skill was perfected in the Song dynasty. According to Su Yijian's *Four Treasures of the Study*: "The method of making the *Chengni* inkstone is as follows: Put the stone into the water with mud, wash it, and store it in an urn. Then prepare another urn of clear water for later use, and compress and filter the mud, so as to filter out the clear water and dry it. Add red lead, knead it with the mud-like dough, and beat it with a stick to make it firm. Carve it in the shape of an inkstone with a bamboo knife, and the size can be random. After having dried slightly, start carving with a sharp knife. Place it in an open space, expose it to the sun, then add rice bran

Picture 11.68 Yunhai Tengjiao Inkstone

and cow dung, stir it, and cook it for another night. Then add ink, wax, rice vinegar and steam it for five to seven hours".

This skill was forgotten until the late Qing dynasty. In the 1980s, Lin Yongmao, Lin Tao, and his son were able to successfully reproduce it in 1991 after years of trial. The newly developed Jiangzhou *Chengni* inkstone inherits the tradition yet is also innovative, with more diverse shapes. The workers also burned yellow, red, green, and cyan inkstones, and the excellent products such as the "Haitian Yuri Inkstone" and the "Luoshu Inkstone" won many awards. The "Yunhai Tengjiao Inkstone" (as shown in Picture 11.68) and "Peace Inkstone" have been collected by UNESCO.

The selection of mud materials is the most important process in the production of *Chengni* inkstones. *Chengni* inkstones use alluvial mud material from Fenhe Bay, which is of moderate size, good plasticity, strong toughness, and smoothness.

The mud is mixed with water, filtered with a silk basket, and then put into a silk bag for press filtration (as shown in Picture 11.69). Then the mud is taken out and piled in a cool place to dry. It takes 1 year to remove the dryness. After this, it is kneaded to make it dense and the flux and color developer added to make a workblank.

The inkstone workblank should be carefully designed, and carved and sanded after drying.

Kiln firing is another key process in the production of *Chengni* inkstones (as shown in Picture 11.70). The burning temperature is related to the composition of clay, so people must strictly control the temperature to get a good quality inkstone. If the temperature is too high and the inkstone surface is too smooth, it will affect the performance of the ink. If the temperature is too low, the porosity of the inkstone increases, and the quality of the inkstone will become very poor. Therefore, it can be seen that a temperature that is too high or too low is not conducive to the production of ink. After leaving the kiln, the workblank should be sanded and polished with

Picture 11.69 Filtering with bag

oil stone and sandpaper to remove the defects caused by firing and increase the smoothness of the inkstone body. The Four Treasures of the Study have always been closely linked with Chinese cultural undertakings and Chinese people's daily lives. All these are brought about by the hard work of paper workers, brush workers, ink workers, and inkstone workers. Thousands of sheets of paper can be made per day, yet only one hair used as a writing brush can be selected from hundreds of hairs. Ink is filtered ten thousand times and inkstones are chiseled from the bottom of the river. The production of the Four Treasures of the Study symbolizes the cohesion of heart and strength, and the synergy of wisdom and skills. This precious skill is related to the integrity of China's cultural sovereignty, helps to strengthen China's cultural identity, safeguards the national spirit, and should be protected and passed down. Under the influence of the ideology of the "extreme left", our traditional culture was destroyed, which led to wrong educational ideas and a low national quality. For example, China had a fine tradition of learning Chinese characters from an early age.

Calligraphy is an art that Chinese people are proud of and it has a profound cultural significance. However, the calligraphy classes in primary schools were cancelled during the Cultural Revolution (1966–1976) and has not been resumed despite repeated appeals. In the meantime, primary schools in Japan and South Korea have always adhered to this tradition of learning characters written on *xuan* paper. 70% of *xuan* paper produced in Jingxian county is sold to Japan and South Korea. This situation is obviously not conducive to the cultural construction and the revitalization of the Chinese national handicraft industry. Therefore, relevant departments should attach great importance to it and take effective measures to rectify it sooner rather than later.

11 Making Calligrapher's Tools

Picture 11.70 Kiln firing

Chapter 12
Printing

Guan Xiaowu and Fang Xiaoyan

Printing technology was originated in China. It was first used over 1,000 years ago, as far back as at least the early Tang dynasty. This technique may first have been used to print Buddhist scriptures and images of Buddha to fulfill the wish of spreading the teachings of Buddha. Later, it was used for the engraving and printing of written works from daily almanacs, cosmological theories pertaining *yin* and *yang*, divination and dream interpretations, *feng shui*, poetry and prose, up to the Taoist and Confucian classics. Printing is far more efficient than manual transcribing and tracing when a large number of texts and pictures need to be copied. It carved a new path for preserving documents, disseminating information, popularizing education, and promoting knowledge that benefited the world. Printing inevitably played an important role in the development and continuation of China's traditional culture. However, it has also been spread to every corner of the world, which in turn contributed substantially to the spread and development of all civilizations. Therefore, printing, along with the invention of gunpowder and the compass, is referred to as the "Three Great Inventions" by Westerners, since they each caused changes on a global scale.

The earliest printing technique that served as the source of all subsequent printing techniques is engraved block printing. In order to meet various needs, printing plate materials, printing processes, and methods were continuously innovated, causing successive printing inventions such as movable type printing, chromatic printing, watercolor block printing, *gonghua* printing, wax block printing, magnetic block printing, and clay block printing. They can be classified under three categories based on how the printing plates were made and which printing techniques were used, namely, (1) engraved block printing, (2) movable type printing, and (3) composite

---

G. Xiaowu
The Institute for the History of Natural Sciences, Chinese Academy of Sciences, Beijing, China
e-mail: gxiaowu@ihns.ac.cn

F. Xiaoyan (✉)
School of Humanities, University of Chinese Academy of Sciences, Beijing, China
e-mail: fangxy@gucas.ac.cn

© Elephant Press Co., Ltd 2022
H. Jueming et al. (eds.), *Chinese Handicrafts*,
https://doi.org/10.1007/978-981-19-5379-8_12

printing. Despite their own characteristics, wax block printing, magnetic block printing, clay block printing, and graphic printing are all engraved block printing techniques. The different movable type printers are divided based on the materials they are made of, such as clay, wood, or metal. The metal movable type can be further specified as the copper or tin type. Chromatic printing can be achieved using multiple crafts that may belong to different techniques. For example, when printing something with colors it can be done using several blocks each with a different color, thus using a composite printing technique, or it can be done using multiple colors on a single block at once, thus using a block printing technique. Composite printing techniques are used for printing paper money and wooden New Year pictures, watercolor block printing, and *gonghua* printing. In this chapter, we will be looking at the processes of engraved block printing, movable type printing, chromatic printing, watercolor block printing, and *gonghua* printing.

12.1 Engraved Block Printing

12.1.1 Origin and Development

Engraved block printing is a technique that integrates various materials and processes. It was not invented by one person overnight. Its appearance originated from several existing techniques and its popularization was largely due to the prevalence of Buddhism.

In China, seals have been commonly used for a long time. They have existed since the Shang dynasty. Seals were made of bronze, jade, pottery, ivory, horn, or wood and used to make an imprint on the sealing clay of bamboo slips and letters, and on silk with ink in order to show someone's name (proof of authenticity) and official position (proof of authority). The practice of stamping seals with Chinese characters on paper appeared following the popularization of paper as a writing medium. In the late Sui and early Tang dynasties, it was common to transcribe books on paper, and the techniques for copying characters and images from steles and bronze wares with paper and ink were quite mature. After the Wei and Jin dynasties, wooden seals for stamping prints of Buddha appeared on a large scale and were used for copying Buddhist images and scriptures. One example is the stamped Buddha images from the Southern Qi dynasty on the back of Volume 10 of the Buddhist scripture *Za A-pi-tan Xin Lun* (Samyuktabhidharma-hrdaya-sastra) kept at the National Library of China (as shown in Pictures 12.1 and 12.2).

After tremendous development in the Northern and Southern dynasties, Buddhism gained even more popularity during the Sui and Tang dynasties. In China, it spread far and wide, with a large number of followers who all wanted their own copy of the Buddhist images and scriptures. In the late Sui and early Tang dynasties, transcribing and stamping were no longer enough to meet the demand, so they came up with block printing. In fact, the reverse engraving techniques and the proper printing materials that were necessary for block printing had already been used to create stamped

Picture 12.1 The written edition of *Za A-pi-tan Xin Lun* in the Eastern Jin dynasty (partial)

Picture 12.2 The stamped Buddha images on the back of the *Za A-pi-tan Xin Lun* (partial)

prints of Buddhist images and scriptures. In other words, to a certain extent, the very foundation for block printing, namely the combination of the wood printing blocks, ink and materials to print on, had already been laid. The shift from seal stamping to engraved block printing was a major breakthrough, which opened a colorful new chapter in the history of printing that, from then on, would evolve constantly.

Unfortunately, there are not many written references or unearthed objects regarding the early use of printing. *Yunxian Miscellaneous Notes,* written by Feng Zhi in the late Tang dynasty, states that from 645 to 664 AD, during the Zhenguan period (627–649) of the Tang dynasty, Xuan Zang printed the image of Samantabhadra, the Bodhisattva of Practice, on a large scale using *Huifeng* paper at a speed of 5 bales per year once he returned from India. According to the estimates of Pan Jixing (1931–2020), an expert in the history of Chinese science and technology, the annual printing volume of those Samantabhadra images was an incredible 200–250 thousand copies. The information written in *Yunxian Misscellaneous Notes* is controversial as its information is contested. However, according to scholars such as Tsien Tsuen-hsuin (1910–2015), the information is not unfounded.

Most of the existing printed materials from the Tang dynasty are related to Buddhist scriptures. The *Karma Purifying Mantra*, also known as the *Rasmi Vimala Visuddha Prabha Dharani* (as shown in Picture 12.3), found in 1966 in the Sakyamuni Pagoda of Fogong Temple in Gyeongju, South Korea, is estimated to have been printed in China during, or after, the last year of Wu Zetian's reign (684–704) in the Tang dynasty. The *Lotus Sutra* (Saddharma Pundarika Sutra), unearthed in Turpan, Xinjiang in 1906, which was published during the first half of the Second Zhou dynasty (690–699), is the earliest printed volume found in China so far. From 1906 to 1908, M.A. Stein, a Hungarian archaeologist, discovered the printed version of the *Diamond Sutra* (Vajracchedika Prajnaparamita Sutra) in a stone library in the Mogao Caves, southeast of Dunhuang in Gansu province (as shown in Picture 12.4). At the end of the sutra, there is an inscription that reads: "On April 15, in the ninth year of Xiantong period, printed by Wang Jie with respect and in commemoration of my parents". This signifies that the *Sutra* was printed in 868 and is the earliest printed book that is marked with a clear date. Other printed materials found in the library include the *Usnisa Vijaya Dharani Sutra* (*Puriying All Evil Paths Sutra*), the almanac printed in the fourth year of the Qianfu period during the reign of Emperor Xizong (877) in the Tang dynasty, and the remnant pages of the almanac printed in the second year of the Zhonghe period during the reign of Emperor Xizong (882). Of the pages printed in 877, those printed by nongovernmental workshops seem to be of higher quality.

During the Tang dynasty, mainly in its capital Chang'an and the city of Chengdu, they did not only print Buddhist scriptures and almanacs but also books on topics such as astrology, dream interpretation, *feng shui* and medicine, as well as dictionaries. Records of that time combined with these printed volumes prove that engraved block printing technology was relatively mature in the Tang dynasty and steadily growing in popularity.

During the Five Dynasties and Ten Kingdoms period, the use of engrave block printing spread to many more areas in China. It was widely used in governmental offices, in workshops, and in private homes. The central area of the printing industry encompassed Luoyang, Kaifeng, Sichuan, Jinling (capital of the Southern Tang dynasty), and Hangzhou (capital of the Wuyue kingdom). During the Tang dynasty, a large number of Buddhist materials were printed. The most famous was the *Tathagatagarbha Sutra* (as shown in Picture 12.5) printed by King Qian Chu of Wuyue,

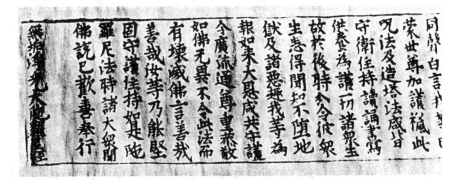

Picture 12.3 The *Karma Purifying Mantra* found in 1966 in the Sakyamuni Pagoda of Fogong Temple in Gyeongju, South Korea (partial)

Picture 12.4 The title page drawing of the *Diamond Sutra*, printed in 868, unearthed in Dunhuang, kept in the British Museum

who reigned from 929 to 988, which was then distributed among pagodas in various parts of the Wuyue kingdom. There are three printed copies, respectively, from 956, 965, and 975, all of which have similar inscriptions with the characters used in the "84,000 teachings" also known as the "dharma". A large number of Buddhist images have also been engraved and printed in the northern region. One example is the single-page print of the *Guanyin* Bodhisattva (as shown in Picture 12.6), printed in 947, which shows the name of the engraver of the printing blocks "Lei Yanmei",

Picture 12.5 *Tathagatagarbha Sutra* printed by King Qian of Wuyue kingdom (partial) in 975, kept in the Far Eastern Library of the University of Chicago, cited from *Paper and Printing*, written by Tsien Tsuen-hsuin

who was probably the first engraver in ancient times to leave their name in a printed book.

In addition to Buddhist scriptures and almanacs, many Taoist and Confucian classics, literary anthologies, and historical reviews were engraved and printed during the Five Dynasties and Ten Kingdoms period. Du Guangting, a Taoist priest, wrote the *Review on Tao Te Ching*, based on the *Lao Zi* with the annotations of Emperor Xuanzong of Tang, and produced the first printed version of the Taoist classics, requiring over 460 printing blocks, at his own expense. Based on the *Kaicheng Stone Classics* written in the Tang dynasty, grand chancellor Feng Dao and his colleague Li Yu presided over the engraving of printing blocks for the Confucian classics such as *Book of Changes*, *Book of Songs*, *Book of Documents*, *Commentary of Zuo*, *The Commentary of Gongyang*, *The Commentary of Guliang*, *Ceremonies and Rites*, *Rites of Zhou* and *Book of Rites*. The project took four dynasties to complete: starting from the third year of the Changxing period (932) in the Later Tang dynasty and finishing in the third year of the Guangshun period (953) in the Later Zhou dynasty, producing printing blocks for the 130 volumes of the "Nine Classics". During the project, one volume of *Characters of Five Classics* and one of *Characters of Nine Classics* were published by the Imperial Academy (*Guozijian*) in the third year of the Kaiyun period (946) in the Later Jin dynasty, which provided standard fonts for the style and characters of the woodblocks used for printing.

During the Song dynasty, engraved block printing techniques became more sophisticated. All over the country, government offices, workshops, private homes, and even temples were providing block printing services. The government issued a number of decrees regulating the printing industry. For books of great value, publishers were entitled to apply to the government for protection of their rights and interests in defense against piracy, demonstrating a clear awareness of copyright protection. The printing industry also covered a more extensive range of subjects including Confucianism, history, philosophy, Buddhism, and Taoism. The rulers of the Song dynasty personally incentivized the compilation and printing of books, causing books to get published on an unprecedented scale. For example, Zhao Guangyi, Emperor Taizong of Song, had 2,500 volumes of books compiled and

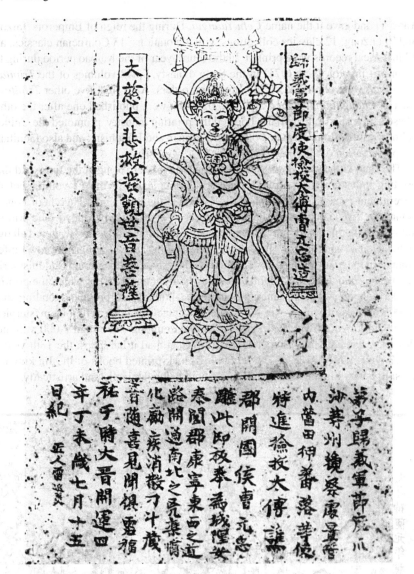

Picture 12.6 *The Greatly Merciful Greatly Compassionate Rescuer from Suffering*, the Bodhisattva Avalokiteshvara, carved by Cao Yuanzhong, cited from *Block Printing in Tang and Song dynasties*, edited by Su Bai (1922–2018)

printed including *Readings of the Taiping Era*, *Extensive Records of the Taiping Era* and *Finest Blossoms in the Garden of Literature* under the imperial direction of Li Fang in the second year of the Taiping Xingguo period (977). In the second year of the Jingde period (1005), Emperor Zhenzong of Song had thousands of volumes of *Records of Relations Between Rulers and Officials in Past Dynasties* compiled by

Yang Yi and gave it the name *Cefu Yuangui*. During the reign of Emperors Taizong and Zhenzong, 12 classics collections with annotations, 13 Confucian classics, and 17 historical records were printed. In the fourth year of the Kaibao period during the reign of Emperor Taizu (971) in the Song dynasty, 5,048 volumes of the *Tripitaka in Kaibao* were published on the emperor's orders. After that, five other *Tripitakas* such as *Chongning Tripitaka* and *Zifu Tripitaka* were published one after the other. These actions had substantially boosted the printing industry, promoted the Chinese culture, politics, and religion, both Confucianism and Buddhism, and also facilitated the preservation of all types of books.

The Liao kingdom, which was opposing the Northern Song at the time, had their printing industry set up around Yanjing. The remnants of the 12 volumes of the government-issued *Liao Tripitaka*, discovered in the Sakyamuni Pagoda of Fogong Temple in Yingxian county, Shanxi province, were exquisitely printed. The printing industry in the Jin kingdom was also prosperous, with printing sites scattered all over Hebei, Henan, and Shanxi. The most famous printing site was Pingshui (now Linfen, Shanxi) in Pingyang prefecture of Shanxi. The most important printed Buddhist scripture is the *Zhaocheng Jin Tripitaka* (as shown in Picture 12.7) of which the printing blocks were engraved at Tianning Temple, Jiezhou, Shanxi province, totaling some 7,000 volumes of which over 4,000 are still intact. Another religious masterpiece of the Jin kingdom was the *Jindao Tripitaka* counting more than 7,000 volumes. However, Kublai Khan, Emperor Shizu of Yuan, had it burned in the 18th year of the Zhiyuan period (1281). The remaining block-printed books of the Jin kingdom attest to a very high level of proficiency, both technologically and artistically.

Picture 12.7 Title page painting in the *Zhaocheng Jin Tripitaka* engraved in the Jin dynasty, cited from *Hundred Pictures of Ancient Chinese Prints*, edited by Zhou Wu

The rulers of the Yuan dynasty had already been encouraging governmental block printing before they gained sovereignty over the central plains. In the eighth year during the reign of Emperor Taizong (1236), an editorial office and a Classics Office was set up in Pingyang to edit and print classics and historical records. Jianyang in Fujian province and Pingshui in Shanxi province were the two places where the block printing industry prospered the most. Dadu, present-day Beijing, also became a printing hub after the *Classics* office was moved there from Pingyang by Kublai Khan. In addition, provinces such as Zhejiang, Jiangxi, Jiangnan, Jiangdong, and Huguang also generated many printed books. Back then, printing organizations could be divided into governmental, private, and workshop services. Governmental services included central and local government organizations. The representative work of the local government services is the *Comprehensive Examination of Literature* written by Ma Duanlin (as shown in Picture 12.8) with blocks engraved by the West Lake Academy. This piece of work boasts beautiful fonts and exquisite engraving skills. Private printing shops also produced many high-quality works, some even finer than those printed in the Song dynasty.

The abolishment of the printing tax in the Ming dynasty greatly stimulated the further development of the printing industry. Books on Confucianism, Buddhism, Taoism, popular novels, operas, music, craft technology, navigation, shipbuilding, and even those related to western science and technology were printed and published in large quantities in order to meet social demand. The official printing services of the central government were mainly provided by the Directorate of Ceremonial (*Si Li Jian*). The Ministry of Rites (*Li Bu*), Ministry of War (*Bing Bu*), Ministry of Industry (*Gong Bu*), Board of Public Censors (*Du Cha Yuan*), Bureau of Astronomy (*Qin Tian Jian*), Imperial Academy (*Guo Zi Jian*), and Imperial Academy of Medicine (*Tai Yi Yuan*) occasionally printed books as well. There were three sutra printing factories subordinate to the Directorate of Ceremonials with about 1,200 craftsmen specializing in engraving printing blocks. In the first year of the Yongle period (1403), blocks for the 6,331 volumes of the *Hongwu Tripitaka* were engraved. Later, an additional 6,771 volumes of the *Bei Tripitaka* were engraved at the Beijing printing factory, first in the fifth year of the Zhengtong period during the reign of Emperor Yingzong (1440) and once more during the Wanli period (1573–1620) in the Ming dynasty. This exemplifies the prosperity of the printing industry back then. Private block printing was common as well, and on quite a large scale.

The Qing dynasty continued the constant development of the block printing industry and produced a substantial amount of printed works. Beijing, as the political center, played an important role in the publishing industry. Cities including Nanjing, Suzhou, Hangzhou, and Yangzhou also maintained momentum in printing. The rulers of the Qing dynasty, just like their Song predecessors, devoted considerable energy to compiling and publishing books. For example, Emperor Shunzhi had the laws and regulations of the Qing dynasty compiled and printed. A large number of books were engraved in blocks and printed by the *Hall of Martial Valor* (*Wuying Dian*) during the reign of Emperor Kangxi (1662–1722). The 7,240-volume *Dragon Tripitaka* was commissioned to be engraved in blocks, which took from the reign of Emperor Yongzheng (1723–1735) until the third year of Emperor Qianlong's reign (1738) to

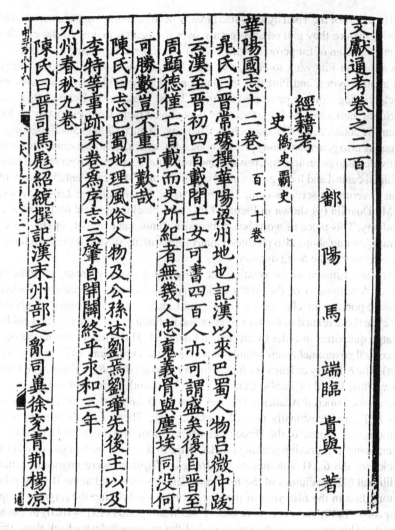

Picture 12.8 The *Comprehensive Examination of Literature* written by Ma Duanlin with blocks engraved by the West Lake Academy in the Yuan dynasty (partial)

finish. The printing blocks of *Notes and Examinations of Thirteen Classics* and the *Twenty-Four Histories* were also engraved during Emperor Qianlong's reign, and thousands of volumes of the *Full Literature of the Tang Dynasty* were printed during the reign of Emperor Jiaqing.

12.1.2 Materials, Tools, and Procedures

Engraved block printing essentially entails applying special printing ink to woodblocks with the print's mirror image engraved, and then printing the texts on the printing medium. The most important printing materials are woodblocks, paper, and printing ink. The woodblocks should be fine-grained, smooth and easy to plane, the calluses uniform in size, have even water absorption and a moderate hardness. Pear wood and jujube wood are the most common materials for engraving woodblocks, while other kinds of wood such as apple wood, apricot wood, boxwood, ginkgo wood, acacia wood, Chinese fir and banyan wood can also be used.

The raw materials for creating printing paper include hemp, bark, rattan, and bamboo. The *Karma Purifying Mantra* in the Tang dynasty, now kept in Gyeongju Museum of South Korea, was printed on mulberry paper, while *Mengqiu*, the ancient children's literacy textbooks from the Liao dynasty found in the Wooden Pagoda in Yingxian county, Shanxi province, was printed on white hemp paper. No matter what kind of paper is used for printing, its surface must be smooth and flat with a moderate level of water absorption.

Black pigment is generally used in engraved block printing, while other pigments can also be used if needed. Manuscripts are usually printed in blue pigment for proofreading before formal printing. After that, black pigment is used to print on a large scale. Black pigment is made of soot, while blue pigment is made of indigo.

Three types of tools are used to facilitate woodblock printing: (1) writing tools, (2) engraving tools, and (3) printing tools. The most commonly used writing tool at the time was simply a brush. While engraving tools could have 20–30 different kinds, each with a different shape and function (as shown in Pictures 12.9 and 12.10). Printing tools include those not used for writing or engraving, such as the *zongba* (a brush made from palm leaves) and *cazi* (an eraser made from plant leaves and wood) (as shown in Pictures 12.11 and 12.12).

Different printing houses have different printing processes. We will be using the main procedures of the Guangling Ancient Books Engraving Society in Jiangsu province as our example.

12.1.2.1 Woodblock Engraving

Woodblock engraving consists of three basic steps: (1) drafting, (2) proofreading, and (3) engraving.

When drafting (as shown in Picture 12.13), the transcriber copies the content that needs to be printed on special transcription paper. Generally, each book would be transcribed by only one person to ensure the consistency of the font. It is a requirement for the handwriting to be similar to that of the original. Whenever special fonts are used, they should also be transcribed with accurate strokes and vivid characters.

Drafting is followed by proofreading. If there is any mistake, the paper is removed and substituted with a new one. If there are too many mistakes, it needs to be transcribed again to make sure everything is correct.

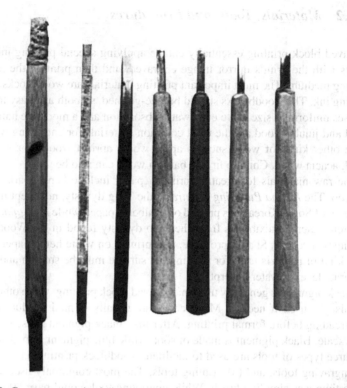

Picture 12.9 Common engraving knives

Picture 12.10 Common engraving tools—curved chisels and flat chisels

Picture 12.11 Zongba

Picture 12.12 *Cazi* made from palm leaves and the outer skin of bamboo shoots

When engraving, the transcription paper needs to be accurately positioned and firmly glued onto the woodblock in reverse. After the paper is dry, engravers use carving knives to remove the unnecessary parts, leaving only the strokes of the characters and useful lines (as shown in Pictures 12.14 and 12.15). Finally, the parts on the woodblock that are not needed are cut away, before being proofread and corrected once more.

12.1.2.2 Printing

A printing table is composed of two countertops with a specified gap left between them: One countertop for mounting woodblocks and another for paper (as shown in Picture 12.16). The specific positions and methods for using and arranging the woodblocks and paper on the printing table are decided by the printing houses themselves, depending on their preferred printing process. Once mounted, the woodblocks are

Picture 12.13 Sample writing

Picture 12.14 Woodblock engraving

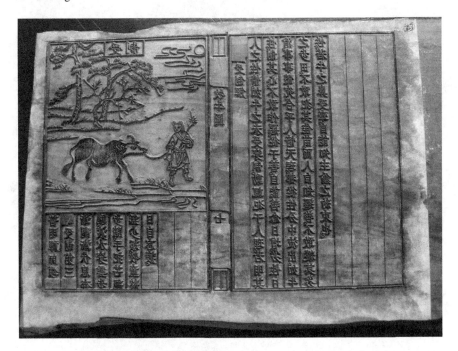

Picture 12.15 Engraved woodblocks (with the unnecessary parts still remaining)

brushed with clear water. After the water has been absorbed, the colors are brushed onto it. Generally, a draft will be printed with blue pigment for proofreading before the formal book is printed with black ink. After brushing the ink onto the woodblocks, the workers place the paper onto the woodblocks and rub it flat with a *cazi*. Then, they lift it up and place it in the gap between the two countertops, so that the printed paper can dry.

12.1.2.3 Bookbinding

The binding styles of traditional books have roughly gone through the stages of using (1) bamboo or wooden slips, (2) scrolls, (3) folded books, (4) whirlwind binding, (5) butterfly binding, (6) wrapped back binding, and finally (7) thread binding in the Ming dynasty.

Before the invention of papermaking, bamboo or wooden slips, and occasionally silk, were used as the first and main media for written work in China (as shown in Picture 12.17). The bamboo or wood first needed to be scraped into slips before they could be used as a medium. During the Shang and Zhou dynasties, characters would be engraved or painted on bamboo or wooden slips. Later generations wrote on the slips with brushes that had been dipped in ink and revised mistakes they made with knives. Once written, the final step was to use hemp thread or silk tape to string the

Picture 12.16 Printing

slips together and wind it up into a bundle from left to right with one of the slips functioning as an axis. According to historical records, the length of different slips varied. These bamboo tablets are the earliest form of Chinese books. They were written from top to bottom and arranged in vertical rows from right to left. The same way of writing in modern times originated from having to use bamboo slips strung together as a medium in ancient times. Similarly, the modern unit of measurement for books is the character "册" (*ce*), which depicts two slips of bamboo together, and the one for articles is "篇" (*pian*), which depicts a flattened piece of bamboo.

Based on the way bamboo slips were strung together and then rolled up, scrolls (as shown in Picture 12.18) came into being after silk books were produced and improved further once paper books appeared. A large number of ancient scrolls were found in Cave No. 17 of the Mogao Caves in Dunhuang, which is known as the "Sutra Cave", providing evidence of the prevalence of binding paper books in the form of scrolls before the Tang dynasty and the Five Dynasties period. A scroll is made of multiple sheets of paper, connected at the edges, in one long sequential sheet and a wooden roller, made of several wooden strips tied together, around which the paper is rolled. A scroll also has a backing of protective and decorative paper or silk added before the first page, which is called *biao*, *baotou* or *yuchi*. A cord is fastened around the middle of the *biao* to prevent the scroll from unrolling. Every five or ten scrolls are wrapped in cloth or put in bags for preservation and classification. A label with the title and volume number is stuck on the roller's head to easily distinguish the scroll's contents. However, scrolls take a lot of effort to create as well as to read, having to roll and unroll it every time.

Picture 12.17 *Lao Zi* in bamboo slips (partial), cited from *Guodian Chu Slips* compiled by the Jingmen Museum

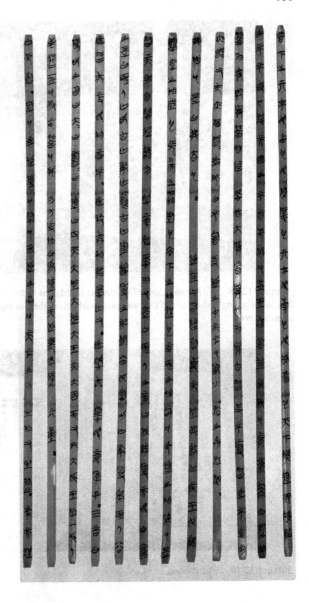

Folded books (as shown in Picture 12.19) solved this problem. They were inspired by the "Sanskrit" binding method of the *Palm-leaf manuscript* from India. The pages of the books are folded into rectangles alternately forward and backward, and hard paper covers are added in the front and back for protection. They were more convenient to open and read than scrolls. Folded Sutras have been found that prove they were still used after the Song dynasty. However, the more pages are added to these folded books, the more easily they scatter and tear.

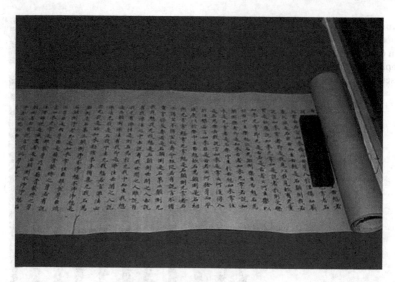

Picture 12.18 The *Nirvana Sutra* (Mahāyāna Mahāparinirvāṇa Sutra) of the Sui dynasty (partial), the longest single-sheet scroll that was produced before the Sui and Tang dynasties, now kept in the China Books

Picture 12.19 Folded book

Whirlwind binding (as shown in Picture 12.20) was the improved version of folded books. One binding method used in whirlwind binding was to stick one half of a piece of paper to the first page of the book and the other half to the last page. In other words, it uses one whole piece of paper to hold the book in order to avoid the pages from scattering. This improvement made it much easier and less time-consuming to read a book, almost as fast as a whirlwind, hence the name. Another binding method used in whirlwind binding was to stick all the pages on a whole piece of paper in sequence and then roll them up into the shape of a scroll, this was also known as dragon scale binding.

Butterfly binding (as shown in Picture 12.21) was first used in the Song dynasty. It refers to the binding method to fold the pages in half with the text facing inward, stick the back of the pages on the front and back paper, and use hard paper as the cover. This binding method was conducive to the protection of books. However, butterfly-bound books were inconvenient to read because one had to turn over two blank pages for every new page.

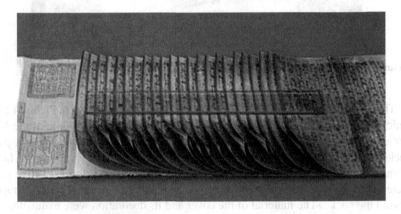

Picture 12.20 Whirlwind binding

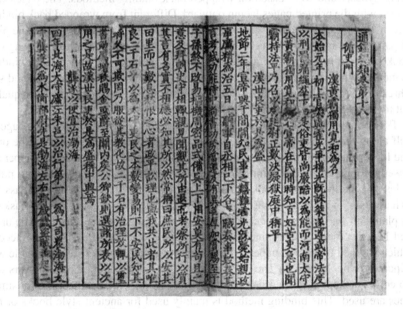

Picture 12.21 Butterfly binding

Picture 12.22 Wrapped back binding

This issue was solved by using back-wrapped binding (as shown in Picture 12.22), which was used from the late Southern Song dynasty until the middle of the Ming dynasty. *Yongle Encyclopedia* used this binding method. This method folds the pages back-to-back, with the printed pages facing outward and the empty backs facing inward, and gluing the margins of the pages onto the book ridge. Next, the pages were held by paper spills and a book cover was adhered to the spine or "wrapped around the back". The material of the cover and its decoration were exquisite.

Thread-bound books (as shown in Picture 12.23) appeared in the middle of the Ming dynasty and were based on wrapped back binding methods. The pages were folded with the printed pages facing outward. Different from wrapped back binding, thread binding uses two and a half pages of cover paper instead of one whole piece of paper to wrap the back of the book. The last step is to punch holes through the book and bind it with thread. Generally, four or six holes are punched for binding. Sometimes the two corners of the book will even be wrapped with thread and silk. Thread-bound books are simple and unsophisticated in style yet strong in quality, but the process of making them is complicated and time-consuming.

For contemporary thread-bound books, the method is usually to fold the pages in half along the central line with the printed page outward, neatly arrange the outer and lower margins of the paper, bind it with paper spills, and then cut the upper and lower margins of the book until they are equal in height. The last step is to punch holes near the spine of the book and bind it with thread. This can be further classified as hardcover and paperback thread-bound books. The former uses cloth, damask silk or plain silk for the book cover and the binding method is complicated. The upper and lower corners of the spine need to be wrapped, and folded flaps are needed, which is glued to the end paper before being wrapped or packaged as a whole with slipcases or book holders. The latter is fairly easy using common paper covers with no corner-wrapping, folded flaps or cover decoration. No slipcases or only simple ones are used. This binding method is mainly used for ancient-style books or rare editions.

Picture 12.23 Thread binding

12.1.3 Major Printing Centers and Institutions

12.1.3.1 Major Printing Centers in Jianyang and Sibao, Liancheng in Fujian Province

The woodblock printing industry in Jianyang, Fujian province came into being during the Five Dynasties period and gained momentum in the Song dynasties. In the Southern Song, it had become one of the three major woodblock engraving and printing centers of China (which included Sichuan, Zhejiang, and Fujian) and its prosperity continued until the Yuan, Ming, and early Qing dynasties. Masha (now Masha town) and Chonghua (now Shufang town) served as its birthplace. There were 37 (some said over 50) woodblock engraving and printing workshops in Jianyang in the Song dynasty and 220 in the Ming dynasty. The book *Shulin Qinghua*, written by Ye Deqing (1864–1927) in the Qing dynasty, recorded that there used to be 32 private workshops of its kind, of which 10 were in Masha and Chonghua. During the Yuan and Ming dynasties, the raging fires of war caused the Jianyang workshops and publishing houses to incur heavy losses and gradually fall into decline. In the 27th year of Guangxu period (1901), the Chonghua workshops and publishing houses were ravaged by fire. The buildings were destroyed, and the engraved woodblocks and books all went up in flames. Hence, the prosperity of the printing industry in Jianyang ended abruptly.

The engraved block printing industry in Jianyang was unique and initiated new techniques and traditions such as copyright protection and the use of detailed outlines in history books. They were also the first to combine original text and annotations in their engraved woodblocks, to print bamboo paper, to use butterfly-fold binding, to include bookmarks (displaying the title or chapter), and to integrate text with pictures. They printed the Confucian classics, historical records, philosophical writings, and anthologies. Besides, books for daily use also took a large proportion of their printing material. According to the *Chinese Ancient Rare Books Bibliography*, 171 Confucian classics, 480 historical records, 505 philosophical writings, 304 anthologies, and 8 collections were engraved and printed in Jiangyang. These have all been listed as rare ancient books on a national level, indicating the important role Jianyang played in protecting and teaching classical books. In the Southern Song dynasty, Jianyang-printed books spread to Korea and Japan, contributing substantially to cultural exchanges. In 2005, the Jianyang woodblock printing technique was listed in *List of the Intangible Cultural Heritage* in Fujian province.

The engraved block printing in Sibao town, Liancheng county, Fujian province came into being in the Song dynasty, continued its development in the Ming dynasty, and reached its peak in the Qing dynasty. Sibao at that time referred to the four (*si*) counties including Changting, Liancheng, Qingliu, and Ninghua, which covered more than a dozen villages in western Fujian province. The name of Sibao literally translates as "Four Protection" and implies the intention to defend the four counties. Nowadays, Sibao refers to the Sibao township under the jurisdiction of Changting county during the Ming and Qing dynasties. The Ma and Zou families from Mawu and Wuge village were the two main printing households. During their most flourishing period, over 70% of these two families, about 1,200 people, were involved in printing, spread across more than 100 publishing houses, but forming one integrated industrial chain of production, supply and marketing. In the Ming and Qing dynasties, Sibao, Beijing, Hankou, and Huwan were celebrated as China's four block printing epicenters. During the Qianlong, Jiaqing, and Daoguang periods in the Qing dynasty, Sibao replaced Jianyang as southern China's printing epicenter. Its publications were sold nationwide and even spread overseas. During the Taiping Rebellion (1851–1864), the civil war extended to Sibao and greatly impacted the local printing industry, but it soon recovered. After the techniques of lithographic printing and letterpress printing were introduced to China, the block printing industry in Sibao gradually declined and finally came to an end in 1942. The ancient woodblocks became coveted collectables in the eyes of antique dealers, therefore causing many private woodblocks to be claimed and lost (as shown in Picture 12.24). Not even the faintest trace of the former glory of the Sibao printing industry is visible today, to the great disappointment of Chinese and foreign experts who went there for research purposes. In 2008, the woodblock printing technique of Liancheng's Sibao was listed in *Extended List of the First Batch of Intangible Cultural Heritage*.

12.1.3.2 The Guangling Ancient Books Engraving Society in Yangzhou, Jiangsu Province

The Guangling Ancient Books Engraving Society (as shown in Picture 12.25) was founded in 1960. It is famous for collecting and preserving over 300,000 engraved woodblocks from the Ming and Qing dynasties (as shown in Picture 12.26) and a complete set of thread-binding tools for block printed materials. The Society has

Picture 12.24 Woodblocks and printing works from Sibao

successively published more than 600 collections of thread-bound books such as *Collected Commentaries on Elegies of Chu State, A Collection of Transmitted Plays Published by the Warm-Crimson Studio, Collection of Chinese Buddhist Temple Gazetteers, Great Overview on Brush Note Style Novellas, Commentaries to the Classics from the Hall of the Free Mind*, the Four Great Classical Novels of China and many Chinese modern literature masterpieces. Many new woodblocks have also been engraved in the Society, such as the 100,000-word manuscript *Litang Daoting Lu* written by Jiao Xun, a scholar in the Qing dynasty, which took over 20 years. At present, it has an annual output of 500,000 thread-bound books. In 2005, its block printing technique was listed in *List of the First Batch of National Intangible Cultural Heritage*.

12.1.3.3 Jinling Scriptural Press in Nanjing (as Shown in Picture 12.27)

Jinling Scriptural Press, stemming from the Jinling Scripture Office, was established in the fifth year of the Qing dynasty's Tongzhi period (1866) by 10 like-minded individuals, which included Yang Renshan from Shidai (now Shitai county), Anhui province. After the Press was set up, Yang Renshan, with the help of Nanjo Bunyu, a Japanese Buddhist scholar, recovered about 300 important ancient books and records that had been lost from China ever since the late Tang dynasty and the Five Dynasties period. These works, such as *Zhong Annotation, Bai Annotation, Cheng Weishi Annotation, Yinming Ruzhengli Annotation, Treatise on the Huayan Samadhi* and *Wuliangshou Jing Youpotishe Yuansheng Ji Zhu*, were newly engraved, printed, and circulated, and became famed for their fine quality.

Picture 12.25 Guangling Ancient Books Engraving Society in Yangzhou

Picture 12.26 Engraved woodblocks in Yangzhou

The Buddhist images printed in the Press were also exquisite. There are still 18 large-scale Buddhist image woodblocks left, such as the *Image of the Magnificent World of Ultimate Joy in the West*, the *Merciful Guanyin Image*, and *Lingshan Religious Ceremony*. They were meticulously engraved in accordance with the textual materials, these included the *Statue Measurement Sutra* by famous engravers from Nanjing, such as Pan Wenfa under the invitation of Yang Renshan during the

Picture 12.27 Jinling Scriptural Press

Tongzhi and Guangxu period (1862–1908) in the Qing dynasty. *Lingshan Religious Ceremony* is a masterpiece of Buddhist image engraving with the 97 characters of distinct expressions well-arranged and neatly structured.

During the Second Sino-Japanese War (1937–1945), Nanjing was captured by the Japanese army, and the woodblocks and buildings of Jinling Scriptural Press were seriously damaged. After 1949, the Press gradually resumed its business thanks to the attention and maintenance of the Buddhist community and collected more than 50,000 pieces of scripture woodblocks from the former scriptural press in Beijing and Tianjin. The total quantity of woodblocks had once reached some 150,000 pieces. It now houses 125,318 pieces of Buddhist scripture woodblocks and serves as a Buddhist scripture publishing organization that not only collects ancient scriptures and scripture woodblocks but also engages in engraved block printing and the circulation of scriptures.

The engraving and printing techniques of Jinling Scriptural Press have been successfully passed down through oral tradition for seven generations already. Its printing process is essentially the same as that of Guangling Ancient Books Engraving Society. However, they usually only use one single color to print so that it is not necessary to use a printing table with two countertops, and the paper does not need to be fixed. Choosing scriptures to be engraved is a rigid process and selecting the right paper and ink as well as designing the layout of the blocks needs to be just right. The Press is rigorous in its collation and prefers a larger font size for the reader's convenience. These quality-printed books are commonly called Jinling Edition. So far, they have manually printed more than 300 types of Buddhist classics, such as the 80 volumes of the *Avatamsaka Sutra* and the 1,347 volumes of the *Complete Works*

Translated by Master Xuan Zang. In 2005, the engraved block printing technique of the Jinling Scriptural Press was listed in *List of the First Batch of National Intangible Cultural Heritage*.

12.1.3.4 Dege Parkhang Sutra-Printing House (as Shown in Picture 12.28)

The Dege Parkhang, also known as the Dege Sutra-Printing House or the Bakong Scripture Printing Press and Monastery, was founded in 1729 in the Cultural Street of Dege county (Gengqing town). It was approved as a *National Priority Cultural Relic Protection Site* by the State Council in 1996. In 2005, the block printing technique of the Dege Parkhang was listed in *List of the First Batch of National Intangible Cultural Heritage*. There are three major Tibetan sutra-printing houses, respectively in Lhasa, Tibet, in Labrang, Gansu and in Dege, Sichuan. The Dege Parkhang ranks first among the three for its most complete collection of Tibetan cultural classics, its highest printing quality and its protection of architectural murals, engravings, and other cultural relics. As early as 1703, *Tusi* officials in Dege started providing funds for the engraving and printing of sutra. Except for the suspension from 1958 to 1979, the history of printing sutras in Dege has lasted for nearly 300 years. During that period, Tibetan paper was used to print sutras and about 500,000 sheets of paper were used every year. Since its establishment, the Parkhang has printed more than 830 ancient books and carved 2.9 million printing woodblocks (as shown in Picture 12.29).

Picture 12.28 The sutra printing in the Dege Parkhang

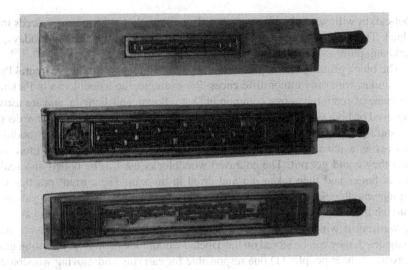

Picture 12.29 Sutra printing blocks

They only used the Tibetan paper they produced themselves to print sutras. After the production of Tibetan paper was forcefully stopped, the Press resorted to handmade paper from Ya'an and Jiajiang in Sichuan. Later, once Tibetan paper resumed production, their output fell short of what they needed, so they used both kinds of paper together.

Before printing, the sheets of paper are cut into strips slightly larger than that of the woodblock and soaked in water for later use. Another method is to soak several sheets of paper in water, evenly distribute them on the sutra printing paper, and clamp them together with planks so that the papers can soak gradually and evenly. The sutra printing process consists of papermaking, woodblock engraving, and printing. In Tibet they used the root of chamaejasme and the pour method to make paper. The resulting paper is light, tough, moth-resistant and absorbs ink very well. In 1958, the production of Tibetan paper was shut down. In 2000, the Dege Parkhang invited an 80-year-old man to teach young people the process of making Tibetan paper, thus bringing this traditional technique back to life.

The woodblocks of the Parkhang were made of red birch, which keeps its shape for hundreds of years after being smoked, soaked in manure, boiled in water, dried, and planed. The woodblocks are deeply engraved with beautiful calligraphic work and are suitable for repeated printing.

At Dege Parkhang, the sutras are only printed in two colors: (1) red, using cinnabar and (2) black, using soot ink. The latter is used for printing ordinary books, and the former for valuable classics. In the early days, soot ink was made inside the Parkhang itself, and could be divided in two types depending on what was used to create them, namely, writing ink and printing ink. Printing ink, made with bark of a rhododendron tree, is delicate, soft, and bright in color. When using it, workers only need to add a bit of water instead of glue. Writing ink is made by grinding the leaves of local foot-catkin willows and adding ox glue. To write with it, a bit of water needs to be added. The ink has a bright color and will not fade for a long time. Nowadays, the Parkhang purchases inks from other regions.

The block printing techniques of the Drge Parkhang and Jinling Scriptural Press are similar, with only minor differences. For example, the woodblocks in Parkhang are made of red birch, a tree common in Dege, Baiyu, and Jiangda, and are usually engraved on both sides. It is said that, in order to encourage engravers to carve deeper, the sutra-printing houses in Tibet used to sprinkle gold sand into the woodblock grooves as a reward, which meant that the deeper they engraved, the more gold sand they could get out. The engraved woodblocks used to be boiled and soaked in yak butter and then taken out and dried in the sun. The current practice is to put the woodblocks in the sun and using a brush to repeatedly brush the yak butter onto the blocks (as shown in Picture 12.30), and then removing the extra yak butter by washing it with water boiled with the root of a local plant called *suba*. After being dried, they can be stored on the Tibetan woodblock rack. Every printing group consists of three people: (1) one responsible for carrying and moving woodblocks, (2) one for brushing ink onto the blocks, and (3) another for printing (as shown in Picture 12.31). To avoid mistakes in the printing sequence, workers should abide by the fixed procedures when carrying and returning woodblocks to the rack. The pigment remaining on the woodblocks should be washed away before being returned. The already-scrubbed woodblocks are first put on a wooden frame to dry in the shade, then brushed again with yak butter and finally put back on the rack after drying (as shown in Picture 12.32).

Picture 12.30 Brushing yak butter on the woodblocks

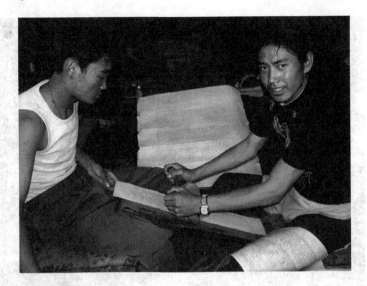

Picture 12.31 Printing

The sutras printed by the Dege Parkhang enjoy a great reputation and are widely circulated in Tibet and among related academic circles at home and abroad. The problems faced by the printing technique in Derge include its high cost and complicated process, which causes difficulties in its inheritance. Furthermore, the production process of Tibetan ink has not yet been restored.

12.2 Movable Type Printing

Movable type printing gradually evolved from engraved block printing. Historical records show that movable type printing was invented by a civilian called Bi Sheng in the Qingli period during Emperor Renzong's reign (1041–1048) in the Song dynasty. After Bi Sheng invented the ceramic movable type, other materials, including wood and bronze, were also used to produce movable types. Bi Sheng is thus revered as the founding father of printing. Compared to block printing, movable type printing uses less raw material, shortens the printing cycle, and thus reduces the overall cost. Ceramic movable type printing applies a similar idea to that of the stamping process used in the Shang and Zhou dynasties when casting bronze wares.

Picture 12.32 Woodblocks on the rack

12.2.1 Ceramic Movable Type Printing

12.2.1.1 The Invention and Spread

The fact that Bi Sheng invented ceramic movable type printing is unequivocally recorded in volume 18 named *Skills* in *Dream Pool Essays* written by Shen Kuo (1031–1095) (as shown in Picture 12.33). Volume 52 of *Imperial Facts Encyclopedia* written by Jiang Shaoyu in 1145 also recorded similar historical events. It is unclear which books Bi Sheng printed, but the techniques for producing clay types and ceramic movable type printing are widely circulated. In the fourth year of the Shaoxi period under Emperor Guangzong's reign (1193) in the Song dynasty, Zhou Bida, a famous scholar and politician, wrote in his correspondence with his friend Cheng Yuancheng: "I recently learned of the technique used by Shen Cunzhong (Shen Kuo). Today I used ceramic types in bronze blocks and printed 28 chapters of my *Notes of the Jade Hall*". This technique also spread to the Western Xia Empire for the printing of Buddhist scriptures, which at the time was at war with the Northern Song.

The earliest work printed with a ceramic movable type is the *Vimalakirti Nirdesa-Sutra* from the Western Xia Empire. Based on the inscription following the name of the Sutra's writer and the documents unearthed together with it, which were printed, respectively, in 1224, 1225, and 1226, it is speculated that the Sutra may have been printed during Emperor Renzong's reign (1141–1193).

During the Yuan dynasty, Yao Shu, a counselor working for Kublai Khan, once asked his student named Yang Gu to print Zhu Xi's *Elementary Learning* and *Reflections on Things at Hand* and Lv Zuqian's *Classics and History in Donglai* with Shen Kuo's movable type printing technique. When examining the technique Yang Gu used, it can be inferred that the ceramic movable type printing technique in the Yuan dynasty was quite like that invented by Bi Sheng.

Historical records regarding the use of movable type printing in the Ming dynasty are yet to be found.

Picture 12.33 *Dream Pool Essays* written by Shen Kuo in the Song dynasty (partial) containing records of Bi Sheng's invention of ceramic movable type printing [the version printed in Dongshan Academy in 1305, the 9th year of the Dade period of the Yuan dynasty], cited from *Pictures of Ancient Chinese Civilization*

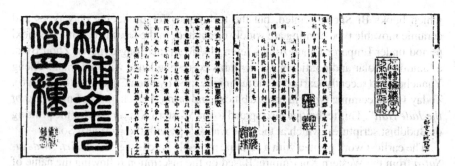

Picture 12.34 *Analysis on Proofreading of Sphragistics and Four Examples* written and printed by Li Yao with ceramic movable types

In the Qing dynasty, craftsmen engaging in ceramic movable type printing included Li Yao from Suzhou, Jiangsu, and Zhai Jinsheng from Jingxian county, Anhui, whose printed works can still be seen today.

Li Yao, who lived in Hangzhou at the time, printed the *Biography in the Southern Ming Dynasty*, written by Wen Ruilin in the Qing dynasty, with ceramic movable types twice, respectively in the ninth (1829) and tenth year (1830) of the Daoguang period. In 1830, he also reprinted the 18 volumes of *Biography in the Southern Ming Dynasty* and in 1832, he again printed his self-written *Analysis on Proofreading of Sphragistics and Four Examples* (as shown in Picture 12.34), all with ceramic movable types.

Zhai Jinsheng, after reading *Dream Pool Essays*, "found the ceramic movable type quite interesting", so he "didn't bother to exhaust the wisdom to study this technique", and finally "produced hundreds of thousands of ceramic movable types after 30 years of hard work" to carry out the printing. In 1844, the 24th year of the Daoguang period, Zhai Jinsheng printed the *Preliminary Compilation of Ceramic Movable Type Trial Printing* (as shown in Picture 12.35) with the assistance of his descendants using white *lianshi* paper. In the 27th year of the Daoguang period (1847), he used small fonts to print the poetry anthology *First Collection of Xianping Bookstore*, totaling 5 books covering 18 volumes, written by Huang Juezi, one of his friends from Yihuang county, Jiangxi. In the winter of the same year, his brother Zhai Tingzhen used the same movable types and printed *First Collection of Elegant Poems in Xiuye Hall* in three books covering four volumes written by himself. In 1857, the seventh year of the Xianfeng period, 82-year-old Zhai Jinsheng had the *Genealogy of the Zhai Family in the East of Jinghe River* printed by his grandson, which had been compiled by Zhai Zhenchuan during the Jiajing period (1522–1566) in the Ming dynasty.

The copies of *Preliminary Compilation of Ceramic Movable Type Trial Printing* and *Genealogy of the Zhai Family in the East of Jinghe River* are still part of many libraries' collections, including the National Library of China. The National Library of China also houses the 12 volumes of *Preliminary Compilation of Ceramic Movable Type Trial Printing* and the 5 books (18 volumes) of *First Collection of Xianping Bookstore*. However, the *First Collection of Elegant Poems in Xiuye Hall* is stored at

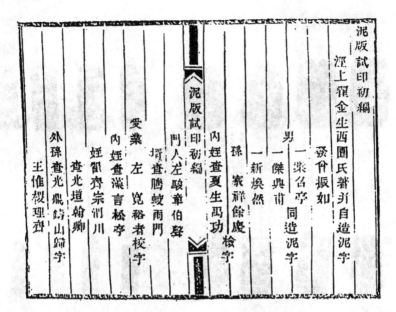

Picture 12.35 *Preliminary Compilation of Ceramic Movable Type Trial Printing* printed by Zhai Jinsheng

Jilin Normal University. Thousands of ceramic movable types made by Zhai's family (as shown in Picture 12.36) are now in the possession of organizations such as the National Museum of China or private collectors.

12.2.1.2 Ceramic Movable Type Printing Technique

Some people are skeptical about ceramic movable type printing, thinking that clay is too fragile to be used for printing, or believing that instead of ordinary clay, what Bi Sheng used was actually *liu i ni* (six-one lute), which was a cement-like substance originally synthesized from seven components and used to seal the top of a furnace. Some foreign scholars believe that Bi Sheng used metal types rather than ceramic ones. The unearthed materials printed with ceramic movable types and movable types made by Zhai Jinsheng indicate that ceramic movable types indeed existed and were actually "as hard and tough as bones and horns". Besides, the characters and pictures in the printed materials are all still very clear. This evidence discredits the above-mentioned skepticism. Zhang Binglun (1938–2006), a Chinese historian of science, simulated and recreated the ceramic movable types invented and used by Bi Sheng and Zhai Jinsheng by carrying out experiments with other scholars. They were able to clarify many technical issues related to the production and printing of movable types.

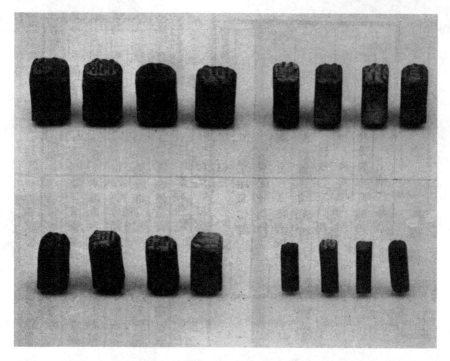

Picture 12.36 Four different kinds of the ceramic movable type made by Zhai Jinsheng

Based on *Dream Pool Essays* and *Imperial Facts Encyclopedia*, Zhang Binglun and others summed up the ceramic movable type creation and printing processes as follows:

Clay processing. Select good quality clay and put it under the blazing sun for a long time. Crush the clay and remove foreign matter to purify and smoothen it. It then needs to dry to a moderate degree before it is suitable for carving.

Movable type making. Batter the clay and model it into clay blocks of the same size. Dry them in a cool place. When the blocks are moderately hard, write inverted characters on them with a brush and then carve them in relief. Dry the carved clay movable types in a cool and ventilated place, and then fire them to get ceramic types.

Typesetting. Take an iron frame and set it on the iron plate of a similar size. Put the mixture of rosin, wax, and paper ash into the frame to fix the types closely together. Choose movable types according to the content of the text and arrange them in the frame from left to right and from top to bottom. If there are uncarved characters, fetch a clay block immediately, carve the character, fire it, and place it into the frame. When the frame is full of types, heat the iron plate on a low fire. When the types in the iron frame become soft, move the frame onto the table and press the blocks with a flat plate to flatten the surface. After the iron plate cools down, the ceramic movable types are bonded together and have become one flat movable type block. Usually, two iron plates and two iron frames are used alternately to improve efficiency.

Printing. After the printing block is made, it can be used for printing by brushing it with ink and pressing the paper onto it. After printing, heat the iron plate on a low fire once more. When the bonding material softens, the movable types can be taken out for a second use.

Storing. Ceramic movable types that have been carved or removed from the frame will be classified according to their tone and rhyme, and skillfully placed in wooden frames, so that they can be found easily.

According to the research of Zhang Binglun and others, the texts produced by Zhai Jinsheng were all printed by ceramic movable types with characters in intaglio, which were then pressed by wooden blocks carved in relief. Zhai Jinsheng added two extra steps when making ceramic types compared to Bi Sheng. The first extra step was to make wooden movable type blocks with characters cut in intaglio, and the second step is to first make clay movable types in relief by pressing the wooden types in intaglio into clay and fire them into ceramic blocks, then make another set of blocks in intaglio by pressing the ceramic blocks in relief into clay and fire those into ceramic blocks as well (as shown in Pictures 12.37, 12.38 and 12.39). Other procedures are the same as Bi Sheng's technique.

Picture 12.37 Ceramic movable type block

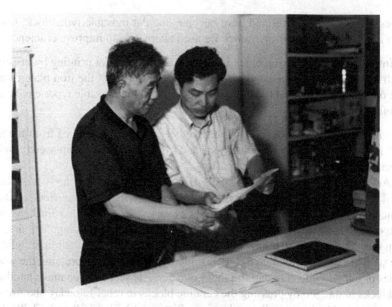

Picture 12.38 Zhang Binglun (left) and Liu Yun (right) proofreading the first printed manuscript when recreating the ceramic movable type printing technique

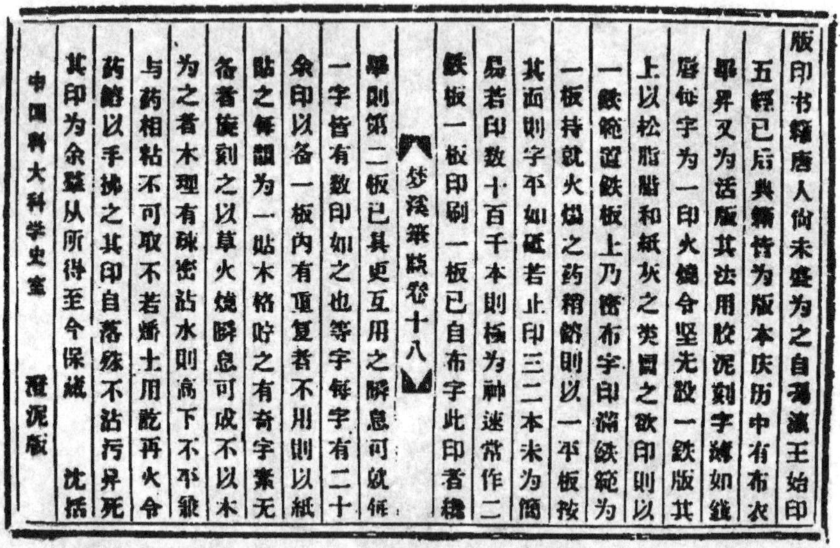

版印书籍唐人尚未盛为之自冯瀛王始印
五经已后典籍皆为版本庆历中有布衣
毕昇又为活版其法用胶泥刻字薄如钱
唇每字为一印火烧令坚先设一铁版其
上以松脂腊和纸灰之类冒之欲印则以
一铁范置铁板上乃密布字印满铁范为
一板持就火炀之药稍镕则以一平板按
其面则字平如砥若止印三二本未为简
易若印数十百千本则极为神速常作二
铁板一板印刷一板已自布字此印者才

【梦溪笔谈卷十八】

毕则第二板已具更互用之瞬息可就每
一字皆有数印如之也等字每字有二十
余印以备一板内有重复者不用则以纸
帖之每韵为一帖木格贮之有奇字素无
备者旋刻之以草火烧瞬息可成不以木
为之者木理有疏密沾水则高下不平兼
与药相粘不可取不若燔土用讫再火令
药镕以手拂之其印自落殊不沾污昇死
其印为余群从所得至今保藏 沈括

中国科大科学史室 澄泥版

Picture 12.39 The paragraphs explaining how Bi Sheng invented ceramic movable type printing in *Dream Pool Essays*, printed by Zhang Binglun and others with self-made ceramic movable types

12.2.2 Wooden Movable Type Printing

12.2.2.1 Origin

Of all the movable type printers in ancient China, the wooden movable type was the most widely used. According to the *Dream Pool Essays*, it may have already been used in the Northern Song dynasty, stating: "The reasons for not using wood to make movable types are that the wood has a different grain, it becomes uneven if it gets wet and it cannot be separated again once glued together". It is possible that wooden movable type printing was first invented, used, and spread during the Song dynasties. However, it is similar to the engraved block printing in its use of materials, paper, and ink as well as its printing processes. This makes it difficult to distinguish the two techniques solely from their printed products. Therefore, works identified as having been printed with a wooden movable type have long been contested.

Wooden movable types were widely used in the Yuan dynasty. In the first year of the Yuanzhen period (1295), Wang Zhen, then governor of Jingde county, Xuanzhou prefecture, asked craftsmen to carve many wooden movable types based on his own design and printed *Annals of Jingde County* in the second year of the Dade period (1298). Later, in October of the sixth year of the Yanyou period (1319), Ma Chengde from Guangping was appointed as the chief of Zhizhou prefecture. During his term, he "had some 100,000 wooden movable types carved" and used them to print books such as *Exposition of the Great Learning* in the second year of the Zhizhi period (1322). Unfortunately, the above-printed copies have not been preserved. The book *On Imperial Testing Policies* printed in the Yuan dynasty and kept in the National Library of China "has large gaps on the margin of the pages with uneven lines and ink color". Zhang Xiumin (1908–2006), a Chinese bibliognost, believes that this book is the earliest existing copy printed with a wooden movable type in China.

Occasionally, Uyghur wooden movable types are unearthed as well. In 1908, Paul Pelliot, a famous French scholar, unearthed a collection of Uyghur wooden movable types in Dunhuang, which were at first mistaken as Mongolian movable types. In 2005, Memtimin Hoshur came across 960 pieces of Uyghur wooden movable types that had been lost overseas in the cultural relics warehouse of the Guimet Museum in Paris. The research group chaired by researcher Peng Jinzhang of Dunhuang Research Academy, which had already discovered 6 Uyghur wooden movable types, unearthed another 48 (as shown in Picture 12.40) in the northern cave of Dunhuang's Mogao Grottoes. Therefore, there are now a total of 1,014 Uyghur wooden movable types preserved all over the world. The Uyghurs played a particularly important role in the spread of Chinese printing technology to the west.

The wooden movable type printing techniques did not greatly improve during the Ming dynasty, but they enjoyed a wider application and popularity. Hu Yinglin, a scholar of the Ming dynasty, wrote: "Nowadays, those eager to print can use movable types. However, this technique originated from the Song dynasty, and today there are no craftsmen who know how to make ceramic movable types, so people have no choice but to resort to wood to produce movable types." Wei Guosong of the

Picture 12.40 The Uyghur wooden movable types unearthed in the northern cave of Dunhuang's Mogao Grottoes and stored at the Dunhuang Research Academy, cited from *Discovery and Early Spread of Chinese Movable Printing Technology* written by Shi Jinbo

Qing dynasty wrote: "The use of the movable type started in the Song dynasty, but the use of wood as the main material for producing them started in the Ming dynasty." Gong Xianceng, another scholar of the Qing dynasty, also wrote: "In the Ming dynasty, it was quite fashionable for people to use wooden movable types for printing." According to Zhang Xiumin, "of the more than 100 different books that were copied and printed with wooden movable types in the Ming dynasty, most were printed during the Wanli period (1573–1620). Not many copies were printed before the Hongzhi period (1488–1505). The names of the places were also printed on some copies. Apart from the previously mentioned cities of Chengdu, Jianyang, and Nanjing, the names of provinces are also printed, including Jiangsu, Zhejiang, Fujian, Jiangxi, and Yunnan". Wooden movable types were also used for printing genealogical books in the Ming dynasty, which further promoted the development and popularization of this printing technique.

It reached its peak in the Qing dynasty. It was widely used in provinces such as Hebei, Shandong, Henan, Jiangsu, Zhejiang, Anhui, Jiangxi, Hubei, Hunan, Sichuan, Fujian, Guangdong, Shaanxi, and Gansu. There were specialized printing houses such as the Movable Type Printing House and Ju Zhen Tang, which used wooden movable types to print Confucian classics, historical records, philosophical writings, and anthologies. The profession of the genealogist also appeared. These genealogists were engaged in printing genealogical books, which popularized the trend of printing genealogies with family lines tracing back to the late Ming dynasty.

During the Qianlong period (1736–1796), Wuying Palace was the largest producer of wooden movable types and printed books. In total, they printed 134 different kinds of books, all bearing the name "Collected Gems Edition". About 250,000 wooden movable types were made here as well as tools such as wooden slots for placing types, character plates, grids, block boxes, large cabinets, and wooden stools, costing a total of about 2,339 *tael*.

After the printing was finished, the project manager Jin Jian laid out all the materials in a comprehensive and systematic way and wrote a book named *Process of Printing in Wuying Dian Using Movable Types*. During the Jiaqing period (1796–1820), several books were printed with the same set of types. During the reign of emperors Daoguang, Guangxu, and Xuantong (1821–1912), books printed by nongovernmental workshops with movable types circulated far and wide. The most influential book is the 120-chapter long *A Dream of Red Mansions*, which was printed with wooden movable types for the first time, presided over by Cheng Weiyuan from Huizhou. It also greatly promoted the spread of this literary masterpiece.

In the middle of the Qing dynasty, books appeared that were printed with wooden movable types and overprinted with color. According to "selected texts in the Tang and Song dynasties" in *Concise Title Catalogue of Siku Quanshu* written by Shao Yichen, "This year, Xie Lanchi from Jiangxi province also used a five-color overprint technique when placing the blocks for printing". Du Zexun, Professor from Ancient Books Organization and Research of Shandong University, possesses a five-color overprinted copy of *Imperial Anthology of Tang and Song Poetry* (two residual volumes) printed with wooden movable types (as shown in Picture 12.41). Yellow is the color last printed and the other colors are vermilion, black, blue and green. This copy confirms the existence of the five-color wooden movable type overprint technique in the mid-Qing dynasty.

In the middle and late Qing dynasty, with the introduction and popularization of lithographic printing and letterpress printing from western countries, the wooden movable type printing market gradually shifted to the countryside of Jiangsu, Zhejiang, Anhui, Jiangxi, Hunan, Hubei, Sichuan, and Fujian province, where it was mainly used for printing genealogical books.

12.2.2.2 Wooden Movable Type Printing Technique

Besides a few minor differences, the techniques of wooden movable type printing and ceramic movable type printing are basically the same. The processes of movable type printing recorded in the *Book of Agriculture* written by Wang Zhen (1271–1368) are (1) carving characters that have been categorized based on a number corresponding to a rhyming pattern, (2) sawing the plate into small movable types with a smooth surface, (3) making revolving typecases, (4) picking the types from the typecase according to the text to be printed, (5) typesetting, (6) printing, (7) breaking up the movable types, and (8) returning them. The wooden movable type printing technique recorded in *Process of Printing in Wuying Dian Using Movable Type* written by Jin Jian in the Qing dynasty is more detailed, rigorous, and scientific than Wang Zhen's description and shows unprecedented logic and preciseness regarding the process, technical aspects, and production schedule. It mainly has the following differences:

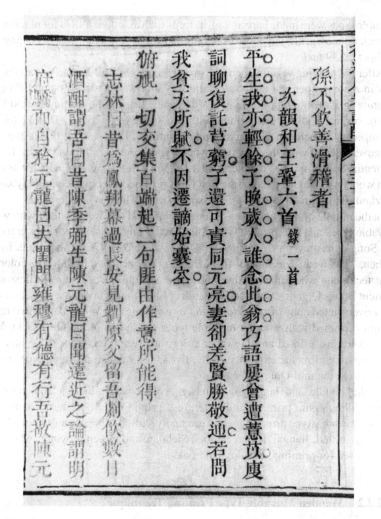

Picture 12.41 Five-color wooden movable type overprint copy of *Imperial Anthology of Tang and Song Poetry* (partial), collected by Du Zexun and photo by Li Yatian

First, according to Wang Zhen's technique, the whole wooden plate needs to be sawed into single movable types after the characters have been carved. But in Wuying Palace, the movable types were sawed into separate pieces and placed on a wooden bed before engravers carved the characters on them (as shown in Picture 12.42).

Second, Wang Zhen arranged the movable types according to a rhyming pattern and stored them in revolving typecases (as shown in Picture 12.43). A worker would have to find the types needed from two typecases, which is very time-consuming. But in Wuying Palace, wooden types were stored in wooden cabinet drawers (as shown in Picture 12.44) and arranged in the order of dictionary radicals and strokes.

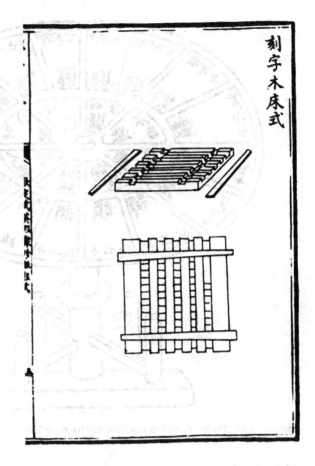

Picture 12.42 The wooden bed for carving characters, cited from *Process of Printing in Wuying Dian Using Movable Type* written by Jin Jian in the Qing dynasty (approximately in 1773)

Dedicated workers were responsible for putting the required characters on trays, thus improving work efficiency.

Third, Wang Zhen arranged the frames, lines, and characters on the same page all together and printed a whole page at a time. In Wuying Palace, the frames, empty backs, and lines were first printed with block printing. Then movable type printing was used for printing characters. This way, the frames on each page became neater and the lines clearer.

Fourth, wooden movable types deform and become uneven after using them several times. In response, Wang Zhen used bamboo clips as the backing plate, while Wuying Palace used folded paper. In Wuying Palace, different books were typeset at the same time (as shown in Picture 12.45) and movable types were swiftly dismantled for other prints. When there were insufficient types for a character, different books were printed alternatingly to solve the problem.

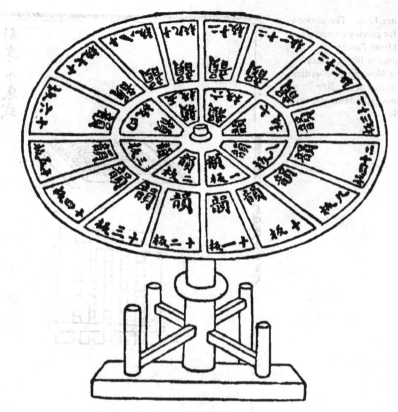

Picture 12.43 The revolving typecase invented by Wang Zhen

12.2.2.3 Examples

Wooden movable types still exist in Ruian, Yiwu and other places in Zhejiang province (as shown in Pictures 12.46 and 12.47) for printing genealogy. At the beginning of the Yuan dynasty, Wang Famao, who lived in seclusion in Changtaili, Anxi, Fujian province, compiled genealogies and printed them with wooden movable types. During the Zhengde period (1506–1521) in the Ming dynasty, his descendants moved from Fujian to Puwei and Xiangyuan in Pingyang, Zhejiang province, and then to Dongyuan village, Ruian city in 1736, the first year of the Qianlong period in the Qing dynasty. Wooden movable type printing began to take root in Dongyuan and is still used in that region today. High-quality pear wood is needed for making wooden movable types. It can only be used to make types after having dried naturally in the wind and sun. At present, there are dozens of people in Dongyuan village who have skillfully mastered the complete technique of wooden movable type printing, most of whom print genealogical books throughout the different counties and districts of Wenzhou city. The traditional printing techniques of wooden movable types have

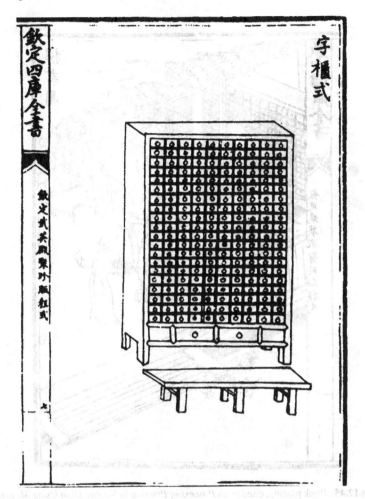

Picture 12.44 Cabinet of Characters, cited from *Process of Printing in Wuying Dian Using Movable Type*, written by Jin Jian in the Qing dynasty

been mostly retained in the printing of genealogy in Dongyuan village and Guangming village, Fotang town, Yiwu city. The main tools they use are carving knives, revolving typecases, and palm leave brushes. The printing material is *xuan* paper and for binding the books they use threads twisted from hemp paper. Ruian city has used special funds to set up the Chinese Wooden-Type Printing Cultural Village and the Wooden Movable Type Printing Exhibition Hall. In 2008, the movable type printing of Ruian, Zhejiang was listed in *List of Second Batch of National Intangible Cultural Heritage*.

There is also a family in Wentang village, Qimen county, Huangshan city, Anhui province, that still prints genealogical books with wooden movable types (as shown in Picture 12.48). It is reported that the family now only preserves one format of

Picture 12.45 Book placing, cited from *Process of Printing in Wuying Dian Using Movable Type*, written by Jin Jian in the Qing dynasty

movable types, totaling more than 36,200 types, which was inherited from their ancestors for typesetting and printing genealogy.

Genealogies dating back to the Sui and Tang dynasties and even beyond have been completely lost. But the tradition of editing and printing genealogical books made a comeback in the Yuan and Ming dynasties and became popular in the Qing dynasty. Genealogy is a publication that needs continuous revision. There is a saying that it needs "a minor revision every 15 years and a major revision every 30 years". Chinese people attach great importance to genealogy, which largely explains why the tradition of printing genealogical books with movable types still exists today.

Picture 12.46 Wooden movable type printing plate arranged by Wang Chaohui from Dongyuan village, Pingyangkeng town, Ruian city, Zhejiang province

Picture 12.47 Newly revised version of the genealogical book printed by Wang Chaohui (partial)

12.2.3 Bronze Movable Type Printing

Judging from existing literature, Chinese printing techniques spread to North Korea from a very early time. Not only engraved block printing, but movable types were

Picture 12.48 Villagers printing genealogical books with wooden movable types in Wentang village, Qimen county, Huangshan city, Anhui province

used in North Korea as well, among which the bronze movable type was the most extensively used. It was in the fifteenth century that North Korea began to cast bronze movable types on a large scale, even earlier than China.

King Taejong of Joseon of North Korea, set up the "casting house" in 1403 and cast hundreds of thousands of bronze movable types for printing. Bronze movable types were cast using the same method as Chinese copper coins and seals: Carving characters on wooden blocks, pressing the blocks into fine sand to create reversed character molds, then pouring liquid copper into the molds to make movable types. Most Chinese scholars now believe that in China bronze movable type printing began in cities of Wuxi, Changzhou and Suzhou, Jiangsu province in the first year of the Hongzhi period (1488) in the Ming dynasty. Representative craftsmen of this printing technique include Hua Sui, Hua Cheng, Hua Jian, and An Guo from Wuxi city.

12.2.3.1 Application

Dating back to the Ming dynasty, Hua Sui's Huitong House was the first to use bronze movable types, and printed the most copies with these types, totaling about 19 different books. Hua Sui (1439–1513) was born in Wuxi, Jiangsu province. Records show that "[Hua Sui] enjoyed reading historical records when he was young and loved proofreading once he grew up. After careful examination of historical information, he transcribed them into bound books, then made bronze movable types to revise and print all these precious books. He once said: 'I now fully understand them'". Shao Bao (1460–1527) also wrote in *Biography of Huitong* in *Rong Chuntang Set*: "[Hua Sui] then made bronze movable types to pass on the historical records, saying: 'I now fully understand them'. This is the reason why he named his publishing house 'Huitong House', meaning 'full understanding'". Hua Sui started printing books with bronze movable types in the third year during Emperor Hongzhi's reign (1490). The first book collection he printed were the 50 volumes of *Memorials of the Ministers Collected in the Song Dynasty*, which is the earliest work printed with bronze movable types in China. Other books he printed include *Splendid Flower Valley, The Fifth Collection of Rongzhai Essays, Hundred Rivers Reach the Sea of Learning, Phonetic Explanation of Nine Classics, Outline of the Finest Blossoms in the Garden of Literature, Phonetic Explanation of Spring and Autumn Annals, Essentials of Categorized Matters Like Joint Jade Circles Quoted From Old and New Literature (Volume One)*, and *Revision on Phonetic Notation of Book of Changes*. These were all printed before the 13th year of the Hongzhi period (1500). These books are extremely precious. They are to China what the incunables are to Europe.

Hua Cheng (1438–1514), a scholar recommended by local government officials in the eighth year of the Chenghua period (1472), once worked as the vice dean of the Court of Imperial Entertainments in Beijing. *Annals of Wuxi County* states: Hua Cheng, whose literary name was Rude, served as the vice dean of the Court of Imperial Entertainments under the recommendation of the government. He was so good at identifying precious ancient utensils, calligraphic works, and paintings, that he built the Shanggu Study to better appreciate all these collections. He also loved to collect books and he created precise movable types. Every time he got his hands on a secret and precious book, he would spring into action and have it ready for printing in a few days' time. Zhu Yunming, a Chinese calligrapher, poet, writer, and scholar-official of the Ming dynasty, said: "Guanglu (referring to Hua Cheng) is over 70 years old, but he is still hungry for knowledge. He is even more curious than youngsters. He made movable types, chose content suitable for learning, and printed them into a collection that benefits all people. Since the explanation of movable type printing in Shen Mengxi's (Shen Kuo) *Dream Pool Essays*, it has been widely used by curious people in the Wu regions. But the quality of the printed works and their influence varied. In 1502, the 15th year of the Hongzhi period, Hua Cheng printed the *Collected Works of Weinan* and the *Sequel Manuscript of Jiannan Poems* of Lu You, a Chinese historian and poet.

Hua Jian, the nephew of Hua Sui, also printed a large number of books with bronze movable types during the Zhengde period (1506–1521). Among them, several collections were handed down to the present, such as the 60 volumes of *Changqing Selection of Yuan Zhen*, printed in the eighth year of the Zhengde period (1513), the 200 volumes of *A Classificatory Compilation of Literary Writings*, printed in the tenth year of the Zhengde period (1515), the 17 volumes of *Luxuriant Dew of the Spring and Autumn Annals*, the 10 volumes of *Collected Writings of Cai Yong* and the one volume of an unauthorized biography printed in the 11th year of the Zhengde period (1516). Most of the books printed by Hua Jian are marked with the label or publisher's words in seal characters stating "锡山兰雪堂华坚允刚活字铜板印行" (Bronze movable type print by Hua Jian [styled Yungang] Lanxue Studio in Xishan) or "兰雪堂华坚活字铜板印" (Bronze movable type print by Hua Jian at Lanxue Studio). Most printed copies are also marked with "bronze movable type print". Books printed at the Lanxue Studio have two lines of characters in every vertical column, thus known as "Lanxue Studio's double-line edition", which are rare and highly praised by bibliophiles. However, the *Collected Writings of Cai Yong* was poorly printed. In addition, the books *Historical Records* and *Talks on Salt and Iron* were known as "Hua Family Bronze Movable Type Prints" since they were also printed with bronze movable types, but it is unclear which family actually printed them.

In the Ming dynasty, people in Changzhou, Suzhou and Jinling also engaged in printing books with bronze movable types. For example, the 50 poetry anthologies of the Tang dynasty were probably printed in Jiangsu during the Zhengde period (1506–1521). In addition, the bronze movable type prints from Zhicheng (Jianning) and Jianyang in Fujian province also boast admirable quality, although these printing houses were for-profit and some of them shared sets of bronze movable types. For example, the copy of *Mo Zi* printed in blue ink (as shown in Picture 12.49) is especially exquisite.

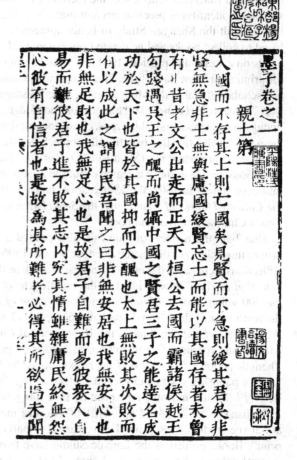

Picture 12.49 *Mo Zi* printed in blue ink with bronze movable types in 1552 (partial), cited from Qian Cunxun's *Paper and Printing*

Bronze movable type printing matured in the Qing dynasty with a wider application than that of the Ming dynasty. Judging from existing literature, bronze movable types were used in the Imperial Household Department during the Kangxi period (1662–1722). Bao Shichen (1775–1855), a calligrapher and reformist scholar, once said: "During the Kangxi period, the Imperial Household Department cast hundreds of thousands of fine bronze movable types for printing." Volume 94 of *Supplementary to the History of the Qing Palace* also states: "During the Kangxi period, the *Complete Collection of Illustrations and Writings from the Earliest to Current Times* was compiled and bronze movable types were cast to print it in a more efficient way. This printed work served as a model for generations to come". It is said that books on astronomy, mathematics and music such as *Textual Criticism on Calendar* and *True Doctrines of Music* were printed with bronze movable types during this period, and the *Essential Principles of Mathematics* in the first year of the Yongzheng period (1723). Research has shown that *Textual Criticism on Calendar* was printed in 1713, the 52th year of the Kangxi period. That same year, Chen Menglei, a scholar from Fuzhou, printed the 9 volumes of *Collection of Poems of Songhe Shan Fang* and the 20 volumes of his own *Collected Works* with bronze movable types borrowed from the Imperial Household Department.

Based on the different fonts of Chen's poem and writing collections and the *Complete Collection of Illustrations and Writings from the Earliest to Current Times*, it is thought that the Qing government made more than one set of bronze movable types.

The *Complete Collection of Illustrations and Writings from the Earliest to Current Times* (as shown in Picture 12.50) is the book collection that the Qing government printed the most of with bronze movable types. The printing project started from the third or fourth year of the Yongzheng period (1725–1726) and finished in the sixth year (1728). The book collection was divided into 5,000 volumes with 20 additional, separately edited catalogs. Every 10 volumes were packed into one case. In other words, this collection consists of 5,020 volumes placed in 502 cases. It was the largest encyclopedia in the world at the time, even four times the size of the famous 11th edition of *Encyclopedia Britannica*. The collection was printed with movable types in two different font sizes on fine and soft *kaihua* paper and *shilian* paper, and ornately and elegantly bound. Every half page contains 9 lines, with each line containing 20 characters in a Song font. There are two border lines around every page and the top, center, and bottom of the back are left blank. The illustrations in the book were printed with engraved wooden blocks. Only 64 prints and a sample print of this book collection were made because it was too voluminous. The Belvedere of Literary Profundity, Hall of Imperial Supremacy, Palace of Heavenly Purity and six other palaces that housed the *Complete Library in the Four Branches of Literature* each kept one copy of this collection. Other copies were given to officials as a reward. Now there are only 12 copies in existence both at home and abroad, due to the small amount of copies printed. Palace Museum and National Library of China each houses a complete set of the collection. This book is of immense value to the research of traditional Chinese science and culture. Printing such a great work with bronze movable types is not only an amazing feat in the history of bronze movable type

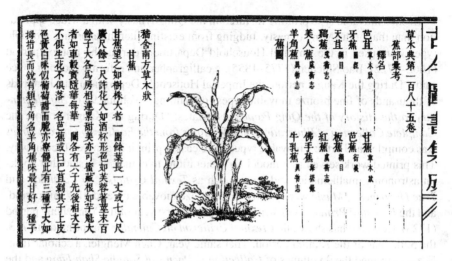

Picture 12.50 *Complete Collection of Illustrations and Writings from the Earliest to Current Times* printed with bronze movable types in the sixth year of the Yongzheng period (1728) (partial). Photocopies provided by the Zhonghua Book Company and Bashu Publishing House

printing, but also an unprecedented feat in the global history of printing. Its complete set of bronze movable types was stored at the Wuying Palace and managed by special staff once the *Complete Collection of Illustrations and Writings from the Earliest to Current Times* was printed. Unfortunately, in the ninth year of the Qianlong period (1744), all the remaining bronze movable types and plates were destroyed and recast into copper coins.

The earliest non-governmental publishing house that used bronze movable types was the Chuili Pavilion in the Qing dynasty. In the 25 year of the Kangxi period (1686), the Pavilion printed the four volumes of *A Selection of Metric Poem and Fu from Finest Blossoms in the Garden of Literature,* which was edited by Qian Lucan and revised by Liu Shihong. The text was printed clearly and delicately in regular script, the so-called "soft characters" (*ruan zi*) or modern style of writing, thus providing a pleasing visual effect. It is the earliest bronze movable type print found in the Qing dynasty, about 40 years before the *Complete Collection of Illustrations and Writings from the Earliest to Current Times.* All that is known of the owner of the Chuili Pavilion is that he probably came from Jiangsu province.

The Ninth Revision of the Genealogy of Xu's Family in Piling, printed in the eighth year of the Xianfeng period (1858), is the only bronze movable type book found in Changzhou city. This 30-volume long genealogy, currently kept at the Oriental Library in Japan, is the only genealogical book known to have been printed with bronze movable types. However, it is still unclear which family in Changzhou ordered this set of bronze movable types.

In Fuzhou city, the most famous person engaged in bronze movable type printing was Lin Chunqi. He started donating money for casting bronze movable types at the age of 18. After 21 years, in the 26th year of the Daoguang period (1846), the project was completed, having cost a total of 200,000 taels of silver. Over 200,000 bronze movable types were engraved by Lin, almost twice as many as those engraved by the Imperial Household Department during the Kangxi period (1662–1722) and more than any set of bronze movable types in North Korea. Lin's set of types is known as the "Futian Shuhai". The cause, process, and technique of making this group of bronze movable types were meticulously recorded in the volume Bronze Plate Narration in *On Phonology*. Lin used this set of bronze movable types to print books on phonology, medicine and war, such as *Five Books on Phonology* by Gu Yanwu (1613–1682). Judging from the printed copies that still exist (as shown in Picture 12.51), this set of types was neatly written, finely engraved and of high printing quality, indicating the maturity of the engraving and printing technique of bronze movable types in the Qing dynasty.

In Hangzhou, Wu Zhongjun (1798–1853) printed three volumes of the *Collected Essays of Miaoxiang Pavilion* and one volume of *Collected Poetry* written by his grandfather Sun Yungui from Changzhou in the second year of the Xianfeng period (1852). The bronze movable types used were not actually owned by Wu Zhongjun. He had gathered and borrowed them when he was working as an official in Hangzhou. The book collection called *Secret Books on Water and Land Attack and Defense Strategy* was also printed with bronze movable types in the third year of the Xianfeng period (1853). This book collection was compiled and printed by Lin Gui, a Manchu who worked as an official in Zhejiang province. Four series from this collection

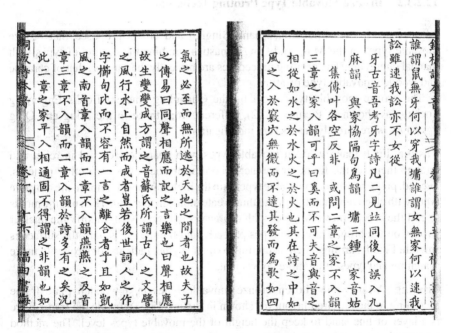

Picture 12.51 *Original Phonics* (partial) printed by Lin Chunqi with the "Futian Shuhai" bronze movable types in the Qing dynasty, now kept in the Nanjing Library

are currently kept in the Peking University Library, namely, *Handy Essentials for Medical Prescriptions for Use in the Army, Art of War from Zhuge Liang* re-edited by Liu Bowen (2 internal volumes and 3 external volumes), *Hundred Battles Strategy from Liu Bowen* (10 volumes) and *Art of War from Shi Shangong* (2 volumes). The abovementioned books account for half of the collection. Based on the characters "省城西湖街正文堂承刊印" in the last volume, which literally translates to "printed by the Zhengwen Hall of Xihu Street in the provincial capital", these books were printed by Hangzhou Publishing House with funds provided by Lin Gui. Some people think the book collection was printed with wooden movable types, but Zhang Xiumin (1908–2006), a Chinese bibliognost, holds that it was printed with bronze movable types because the font is the same as that of the "Futian Shuhai" bronze movable types used by Lin Chunqi.

In addition, Wu Long'a, a Manchu general who served as the chief military officer of Taiwan town, was also engaged in bronze movable type printing in the Qing dynasty and printed *Notes on Sacred Edict of Emperor Kangxi* with exquisite characters and images. *Taiping Heavenly Sun* printed in the 12th year of the Taiping Heavenly Kingdom (1862) with the characters "钦遵旨准刷印，铜板颁行" on its cover, which translates to "the imperial edict was obeyed to print with bronze movable types for issue", may have been printed with bronze movable types as well.

12.2.3.2 Bronze Movable Type Printing Technique

After research and simulation, Zhang Binglun and others proposed that bronze movable types in the Ming and Qing dynasties combine the casting of types with manual engraving. The printing procedures are basically the same as that of ceramic and wooden movable type printing.

The bronze movable types kept by the Guangling Ancient Books Engraving Society (as shown in Picture 12.52), different from those in the Ming and Qing dynasties, were directly cast before modification.

Casting. First, make wooden movable types with characters carved in relief and arrange them on the frame or cabinet according to their rhyme or other patterns. Second, press the wooden movable types into the molding sand to imprint the movable types with characters in intaglio and thus create molds in the sand. Then use the molds to cast the bronze movable types with characters in relief. Next, modify the unclear strokes on the bronze movable types properly and file the types until the surface is smooth. Finally, arrange them according to their rhyme, radicals, or strokes to make searching and typesetting more convenient.

Printing. Arrange the chosen bronze movable types in a wooden plate frame according to the required layout (as shown in Picture 12.53). Inside the frame there is a layer of fine sand to keep the height of the movable types level. The method for selecting movable types is the same as those for other kinds of movable types. After typesetting, tightly lock the bronze movable type plate in place to prevent any from coming loose. Print a sample first to avoid any mistakes in the official prints

Picture 12.52 Materials and tools for casting bronze movable types in the Guangling Ancient Books Engraving Society

(as shown in Picture 12.54). If there are any errors, omissions, or parts that need to be revised, another sample needs to be printed first for proofreading. This is to make sure that every detail is correct. The methods of printing and binding are the same as those for block printing and wooden movable type printing.

12.3 Chromatic Printing, Watercolor Block Printing, and Gonghua Printing

12.3.1 Chromatic Printing

According to *Facts of the Song Dynasty*, *Jiaozi*, a kind of paper money used in the Northern Song dynasty, was printed and distributed in Sichuan in the fourth year of the Dazhong Xiangfu period during the reign of Emperor Zhenzong (1011) in the Northern Song dynasty was "printed with paper of the same color and included seals with images of wooden houses or people. Each issuing company will print special characters and hide unique private marks on *Jiaozi*. Red and black ink are both used in printing." This is the earliest record of two-color chromatic printing. According to *Catalog of Paper Money*, written by Fei Ju in the Yuan dynasty, in the fourth year of the Chongning period during the reign of Emperor Huizong (1105) in the Song

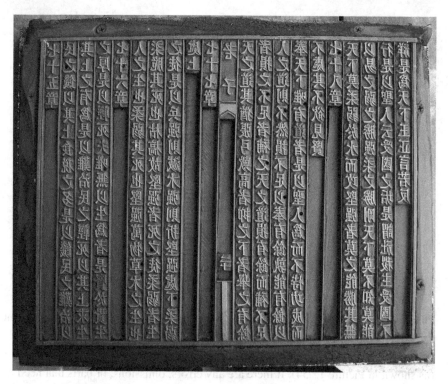

Picture 12.53 Bronze movable type plate

dynasty banknotes were overprinted in three colors using four black, one blue and one red printing plates, which may be the earliest example of three-color chromatic printing.

In July 1983, eight *guanzi* printing plates, used to print the paper money used in the Southern Song dynasty, were discovered in a waste warehouse in Dongzhi county, Anhui province. These are the most complete Song dynasty banknote plates found in China so far. The chromatic printing method used for this kind of paper money was recorded in Volume III of the *History of Liao, Jin and Song Dynasties*: "How to print *guanzi*: The top black seal resembles the character "西", the middle red seal resembles the character "目", and the bottom part is printed with black seals on both sides, making the whole image resemble the character "贾". From this, we can tell that chromatic printing techniques for banknotes had already reached a high level of precision and maturity in the Southern Song dynasty.

In the sixth year of the Zhiyuan period during Emperor Shundi's reign (1340) in the Yuan dynasty, *An Annotation of the Diamond Sutra* (as shown in Picture 12.55) was printed by Zifu Monastery, Zhongxing Road (now Jiangling county, Hubei province) in red and black using chromatic printing techniques. It is the earliest masterpiece found so far that used overprinting techniques so skillfully. It also shows that the application of chromatic printing had expanded from banknotes to books.

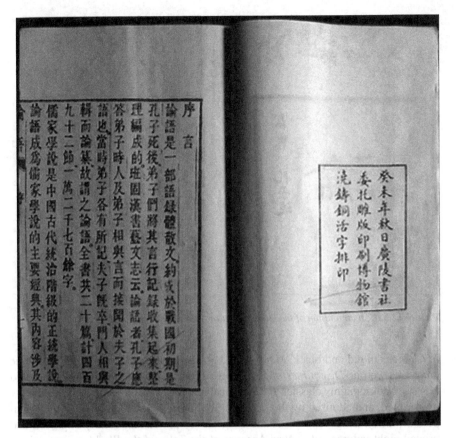

Picture 12.54 Bronze movable type print at the Guangling Ancient Books Engraving Society

Chromatic printing benefited from an all-round development in the Ming dynasty. In 1602, the *Ten Volumes on Lady's Demeanor* were printed and published in Shexian county, Anhui province. This work combines the original four volumes of *A Collection of Stories of Model Women* and six other volumes, such as, *Instructions of the Ancestor of the August Ming*, *Ladies' Classic of Filial Piety*, *Analects on Women* and *Lessons for Women*. This work was overprinted with commentary and annotations in red.

During the Wanli period (1573–1620) in the Ming dynasty, chromatic printing was widely used in private workshops and publishing houses in Wuxing, Jiangsu province and hundreds of classic works, illustrated novels, Chinese operas and medical books were overprinted in two to five colors. The most famous publishing houses were those of the Min and Ling families. Min Qiji from Wuxing, together with his family, overprinted *Commentary of Zuo* in red and black for the first time in the 44th year of the Wanli period (1616). Since then, the Min family has successively overprinted *Lao Zi*, *Zhuang Zi*, *Lie Zi*, *Songs of Chu* and *Commentaries of Dongpo on the Book*

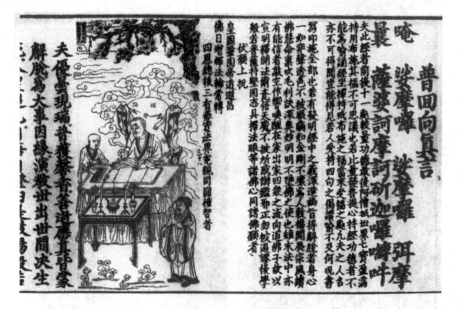

Picture 12.55 *An annotation of the Diamond Sutra* overprinted in red and black (partial)

of Changes in red and black, *Meng Zi* and *Strategies of the Warring States* in three colors, as well as *Discourses of the States* in four colors. A total of over hundred kinds of books, mainly classics and historical records, were elegantly and exquisitely overprinted, and admired by the people at that time. In the ninth year of the Wanli period (1581), Ling Yingchu of the Ling family in Wuxian county engraved and printed eight volumes of *A New Account of the Tales of the World* in four colors: blue, red, yellow and black.

In the Ming dynasty, chromatic printing was also used to print paintings. The color-printed *History of Flowers*, originally owned by Zheng Zhenduo (1898–1958), is now kept in the National Library of China. It initially consisted of four volumes, each corresponding to a season, but now only the volumes for summer, autumn, and winter remain. Each page of *History of Flowers* contains a picture of a colorful flower followed by a brief introduction and its planting method. Even though each volume also came with several ancient poems, it is mainly a flower field guide. The first print of this book may have been produced a little before 1600 with a chromatic printing technique where they apply multiple colors on the same printing block.

The 12 volumes and 18 books of *Cheng's Models of Ink Sticks* were printed in 1605, which used 520 ink patterns created by Cheng Junfang, a famous ink maker from Shexian county. This book series was compiled by Cheng Junfang with pictures painted by Ding Yunpeng, an eminent painter, and woodblocks engraved by a famous woodcarver of the Huang family in Huizhou. The paintings were exquisite, and the engraving skillfully done. However, most noteworthy was that the 50 colorful pictures in the book were mostly overprinted with four or five colors. These overprinting skills were far beyond those of other color prints of the same period. Therefore, this copy showcases the highly developed chromatic printing techniques of the Ming dynasty.

Awakening Zhuang Zi (as shown in Picture 12.56) is the seventh painting in the *Album of Erotic Paintings* (later renamed *Fengliu Juechang*) printed in 1606. This picture was overprinted with six blocks in different colors, namely, dark black, light black, light blue, light red, dark red, and dark yellow. The whole painting is not only logical in layout, appropriate in proportion, fine in engraving and vivid in how it portrays the characters, but it is also efficient in the use of blocks, balanced in the use of colors, and accurate in how it is overprinted. It can be regarded as the epitome of chromatic printing before the invention of watercolor block printing.

However, chromatic printing faces insurmountable difficulties when it is used to print Chinese calligraphic works and paintings, especially freehand brushwork paintings, which are typically smooth and contain wild brush strokes. First, the standard printing paper used in chromatic printing and the unsized paper used in freehand brushwork differ in water absorption and paper quality. The unique characteristics of an ink brush can only be realized on unsized paper. Second, only a limited number of water-based pigments can be absorbed by the woodblocks used in chromatic printing, far less than that absorbed by brushes. So, it is difficult to reproduce the wild style of freehand brushwork on dry paper.

Third, chromatic printing uses several woodblocks with different colors, so the transition of colors on the same woodblock and between different woodblocks

Picture 12.56 *Awakening Zhuang Zi*, cited from *Erotic Color Prints of the Ming Period* written by Robert Hans van Gulik from the Netherlands

cannot be taken into account. Therefore, it is impossible to print freehand brushwork that includes different shades and natural tone transitions using chromatic printing. Fourth, chromatic printing uses strong colors and pursues high contrasts, so the printed works are always wildly different from the original in color.

12.3.2 Watercolor Block Printing

Watercolor block printing was invented to overcome the shortcomings of chromatic printing because it can reproduce Chinese paintings and calligraphic works more vividly, especially freehand brushwork paintings. It was invented by Hu Zhengyan, a native of Xiuning, Anhui province, who was born in the 12th year of the Wanli period (1584) in the Ming dynasty and lived in Nanjing. After a long time of practice with engraving and printing, Hu worked together with other craftsmen to improve chromatic printing by changing the printing medium to wet unsized paper and using wet woodblocks with multiple colors on a single block. The paper, ink, and pigment used were all the same with those used in the original paintings and the woodblock color allocation and printing sequence were all delicately designed based on the original colors and strokes. This way they were able to vividly reproduce freehand flower-and-bird paintings through printing, thus opening a new avenue for printing Chinese painting and calligraphic works, especially freehand brushwork paintings. With this, traditional Chinese color printing reached its peak.

The *Ten Bamboo Studio Painting Spectrum* compiled, engraved, and printed by Hu Zhengyan in the Tianqi period (1621–1627) in the Ming dynasty, *Luoxuan Biangu's Letter Paper Album* engraved by Wu Faxiang in Jinling in the sixth year of the Tianqi period (1626), *A Complete Collection of Ten Bamboo Studio Decorative Letter Papers*

Picture 12.57 Sample page of *A Complete Collection of Ten Bamboo Studio Decorative Letter Papers* printed by Hu Zhengyan with watercolor block printing, now kept in the National Library of China

(as shown in Picture 12.57) engraved by Hu Zhengyan and *Master Yin's Guide to Letter Papers* engraved in the second year of the Shunzhi period (1645) are all representative prints of the Ming and Qing dynasties.

In the 1930s and 1940s, Lu Xun (1881–1936) and Zheng Zhenduo jointly published 200 copies of *Beiping Letter Papers* and 300 copies of *A Complete Collection of Ten Bamboo Studio Decorative Letter Papers* in the Studio of Glorious Treasures in Beijing. In 1950, the Studio of Glorious Treasures became an enterprise affiliated with the National Press and Publication Administration and breathed new life into watercolor block printing. To be better understood, this technique was renamed "woodblock water printing". In 1952, *A Complete Collection of Ten Bamboo Studio Decorative Letter Papers* was reprinted again (as shown in Picture 12.58) using the properly inherited and developed traditional techniques of watercolor block printing and *gonghua* printing. Since then, the Studio of Glorious Treasures' woodblock water printing technique has continuously improved. This technique was previously used only to print poetry and paintings with paper less than a foot long, but later it was also used to print larger paintings. After further research and development, woodblock water printing could be used to print traditional Chinese silk-scroll paintings. The most exquisite work printed this way is the silk painting *The Night Revels of Han Xizai* (as shown in Picture 12.59). This 3-m-long scroll depicts 49 vivid characters, each with different expressions. To reproduce it, a total of 1,667 sets of woodblocks were engraved, paper and pigment same as the original painting was used, and over 6,000 printing sessions were needed to produce a single scroll. This work can be seen as the epitome of woodblock water printing. In 2005, the woodblock water printing technique of the Studio of Glorious Treasures was listed in *List of the First Batch of National Intangible Cultural Heritage*.

Shanghai's Duo Yun Xuan is another important Chinese organization engaged in woodblock water printing. In 1981, *Luoxuan Biangu's Letter Paper Album* (as shown in Picture 12.60) compiled by Wu Faxiang in the late Ming dynasty was reprinted by Duo Yun Xuan. The perfect use of watercolor block printing and *gonghua* printing to create this exquisite copy contributed substantially to the inheritance and development of these two traditional techniques. In 2008, the woodblock water printing of Duo Yun Xuan in Shanghai was listed in *List of the First Batch of National Intangible Cultural Heritage*.

Below, we will look at the watercolor block printing procedures more closely.

12.3.2.1 Drafting

The first step is to select paintings. The paintings should not only be famous masterpieces but also conform to people's aesthetics. The characteristics of watercolor block printing should also be taken into consideration to avoid the process becoming too cumbersome.

The second step is the woodblock arrangement, also known as "*zhaitao*". This means designing several woodblocks of different sizes according to the original painting's color and subtle changes in the shades of the brush strokes. Generally,

Picture 12.58 *A Complete Collection of Ten Bamboo Studio Decorative Letter Papers* reissued by the Studio of Glorious Treasures

Picture 12.59 *The Night Revels of Han Xizai* printed in the Studio of Glorious Treasures with woodblock water printing (partial)

Picture 12.60 *Luoxuan Biangu's Letter Paper Album* printed by Duo Yun Xuan in Shanghai with woodblock water printing (sample page)

each woodblock was brushed with one color, but sometimes multiple colors were applied to make the color transition more natural or to reproduce the effect of a multicolored stroke.

The next step is tracing, which is to cover the painting with translucent Gampi paper and tracing the outline with a pen used for writing regular script in small characters. Several independent drafts should be traced based on the original painting and the woodblock design.

Picture 12.61 Woodblock engraving

12.3.2.2 Woodblock Making

The tools and techniques used in selecting, processing, engraving, and cleaning woodblocks, (as shown in Pictures 12.61 and 12.62) and modifying woodblocks for watercolor block printing are the same as those used for regular block printing.

12.3.2.3 Printing

Printing consists of selecting paper, mounting and wetting the paper, mounting and aligning the woodblocks, mixing and brushing on the color, and the actual printing.

The same (or similar) kind of paper or silk as the original painting should be selected to obtain the best reproduction result. Sometimes the medium needs to be polished and other times it needs to be distressed. To distress means to dye the paper or treat it in ways to make it look used. To polish means to smoothen the surface of the paper by calendering.

Paper needs to be mounted in such a way that it facilitates wetting and drying, is convenient for woodblock alignment and printing, and prevents contamination caused by fresh prints sticking together. The purpose of wetting the paper is to control the moisture content of the paper so that, after brushing and printing have been delicately completed, the printed work reproduces the verve of the original paintings, be it in a dignified, soft, bright, or vivid way.

In addition to the paper, each woodblock must also be mounted in the correct position to ensure the accuracy of overprinting.

After the woodblocks are aligned, the worker holds the block with the left hand to prevent it from moving, and turns the paper marked with the woodblock position

Picture 12.62 Partially engraved woodblock

back to the right side of the printing table with the right hand. Then glue, which is a baked mixture of rosin and beeswax or black plaster oil, is kneaded into small balls and placed on every corner of the woodblock to stick it onto the printing table. The next step is to turn the paper over to doublecheck the position. If there is any gap, the worker will make small adjustments before the glue has completely dried. Then tap it from the side with a wooden mallet to separate the woodblocks from the glue.

Mixing the colors can only be undertaken by experienced craftsmen. What is key is adjusting the content of pigment, water, and glue to their perfect proportion. They need to make adjustments and compare the result with the original painting while mixing, brushing, and printing as to obtain the most similar color constantly and simultaneously.

When brushing the color onto the woodblocks, the worker first takes a little bit of pigment with a writing brush and puts it on a large porcelain plate, then holds a *zongba* (brown brush) with the left hand and makes circular motions on the plate to evenly distribute the color on the brush. Next, the pigment is brushed evenly onto the woodblock in circular motions. Finally, the writing brush is used to paint the details. Sometimes, multiple colors are applied on a single woodblock to reproduce the effect of one stroke with multiple colors in the original painting. Such a technique helps achieve the artistic effect used to show colors transitioning or changing naturally, like on leaves or tree trunks. This technique also saves woodblocks. It was such an innovative way of watercolor block printing that it played a vital role in the progress of color printing in later generations.

When printing, one end of the paper is held with the left hand and the grip is slightly relaxed after the paper is aligned with the woodblock. Then, the paper is pressed from right to left onto the woodblock with a small rake. The final step is to further rub or press the paper with the rake, volarpads, or fingernails, depending on the printer's understanding of the original work.

Printing is the last step in the quality control of watercolor block printing. Printers should be ingenious and aspire to fully embody the verve of the original works.

12.3.2.4 Binding

After printing, the paper is lifted with the left hand and put it in the gap between the two countertops with both hands so that the printed paper can dry without sticking to other prints and thus avoid color contamination. After finishing the printing of one woodblock, all the dried paper is transferred back from the gap to the right countertop, the former woodblock is replaced with another one, and the printing process is repeated. During the printing, the paper should be moisturized by spraying it with water when needed. After all the printing is finished, the paper bail is loosened, the printed paper taken down and the defective pieces removed. Finally, the painting is ready to be mounted or bound. Picture 12.63 shows partial of *Immortals' Birthday Celebration* of Ren Bonian printed with woodblock water printing at Duo Yun Xuan.

12.3.3 Gonghua Printing

There is an inseparable relationship between the techniques for *gonghua* printing and using graphite molds. First, a graphite mold and a *gonghua* woodblock are both engraved from fine and solid wood. Second, the relief effect on the paper embossed by a *gonghua* woodblock is similar to that of ink printed with a graphite mold. Third, the treatment of the bottom of the incised carving lines is exactly the same for graphite molds and *gonghua* woodblocks. Fourth, some graphite molds can be used as *gonghua* woodblocks to make "*gonghua*" in convex relief with distinct layers. Same as watercolor block printing, *gonghua* printing was first mentioned in *A Complete Collection of Ten Bamboo Studio Decorative Letter Papers* in the late Ming and early Qing dynasties. It was even invented by Hu Zhengyan as well. The earliest *gonghua*-printed works were collected in *A Complete Collection of Ten Bamboo*

Picture 12.63 *Immortals' Birthday Celebration* of Ren Bonian printed with woodblock water printing in Duo Yun Xuan (partial)

Studio Decorative Letter Papers, Luoxuan Biangu's Letter Paper Album, and *Master Yin's Guide to Letter Papers.*

There are white *gonghua* and color *gonghua*. White *gonghua* (as shown in Picture 12.64) is the technique of directly embossing an image on paper without other patterns or colors, as to show detachment and elegance. Color *gonghua* (as shown in Picture 12.65) is a technique combining watercolor block printing and *gonghua* printing. It delicately integrates the various shapes on colorful paintings with reliefs in multiple layers to give the whole painting more dimension and extraordinary elegance. In terms of procedures, white *gonghua* only needs to be embossed once with a woodblock, while color *gonghua* requires an additional step of embossing the image on the already color-printed paper.

Below, we will look at the *gonghua* printing procedures more closely.

12.3.3.1 Drafting

Gonghua is a colorless printing technique to produce relief, so it is necessary to use paper that is highly transparent while drafting. At present, the common practice is to

Picture 12.64 White *gonghua* in *A Complete Collection of Ten Bamboo Studio Decorative Letter Papers*

first sketch on a transparent polyester sheet that is put over the original work. Once that is done, a second sketch is made on Gampi paper that is put over the previously made draft sheet.

Picture 12.65 Color *gonghua* in *A Complete Collection of Ten Bamboo Studio Decorative Letter Papers*

12.3.3.2 Woodblock Engraving

The wood should be smooth enough to produce exquisite and three-dimensional *gonghua* with distinct layers, yet hard enough to withstand great pressure when printing. Common materials include birchleaf pear wood, jujube wood, and pear wood. Blocks are processed the same way as in standard engraved block printing. The size of the woodblock should be slightly larger than that of the *gonghua* image that needs to be embossed. Even better is to make the woodblock the same size as the whole page, so that the margin space of the paper wouldn't be pressed by the edge of the block and tear. Woodblocks are engraved using the same procedures as watercolor block printing. Incised carving is needed to engrave grooves with smooth surfaces and three-dimensional shapes. In order to engrave the *gonghua* woodblocks this way, it is necessary to use a set of refined carving knives that can engrave different bottom shapes. There is no need to remove unnecessary parts on *gonghua* woodblocks. However, the surface must be smooth without any bump or dent except those of the desired embossed pattern. Otherwise, all these defects will appear without disguise when printing.

12.3.3.3 Printing

A *gonghua* hammer is used for printing along with wool felt. The wool felt is placed between the Xuan paper and the *gonghua* hammer to press the paper into the slender grooves on the woodblock relying on the paper's unique elasticity.

The workers should press hard to print, so the woodblocks must be fixed in place to prevent it from moving. *Gonghua* printing is often carried out on a watercolor block printing table because it is usually combined with this technique.

Unsized *xuan* paper is normally used for *gonghua* printing. To create a color *gonghua*, a pattern is embossed on unsized *xuan* paper that has already been printed with watercolor blocks.

The methods for mounting paper in color *gonghua* and white *gonghua* differ. Color *gonghua* is the combination of watercolor block printing and *gonghua* printing. Therefore, the paper that needs to be embossed is still mounted on the printing table after having been color printed. There is no need to attach additional paper. For white *gonghua*, paper needs to be mounted before printing to make sure the position is relatively stable. The method for mounting it is the same as that of watercolor block printing.

The method for putting the paper onto the *gonghua* block is basically the same as that of block printing, which is to pull the paper above the *gonghua* woodblock with the left hand and then gently put it down. There is no need to worry about color contamination because the woodblocks and paper are all dry and the printing generally occurs after the watercolor block printing has been completed. However, whether in white or color *gonghua*, the woodblocks must be placed in the precise position, as any deviation will impair the overprinting accuracy and quality. Therefore, after covering the paper on the woodblock, the paper is usually held with the left hand without relaxing the grip. Only after the wool felt is pressed onto the back of the paper with the right hand, can the right hand be replaced with the left hand and printing commence.

When printing, the wool felt is pressed with the left hand to prevent it from moving while holding the middle of the *gonghua* hammer in the right hand. One end of the hammer is pressed against the wool felt and the other end against the right shoulder of the worker. When the worker strongly pushes down, the wool felt will sink and press the paper into the grooves of the *gonghua* woodblock. The wool felt can be gently lifted to look at the dents on the paper and check the embossing result. If the result is satisfactory, the paper can be slowly taken off, put aside, and another page added for printing. If not, the wool felt can be gently put back and the same paper printed again. If the craftsmen can alternate flexibly between pressing softly and strongly when printing as well as make reasonable use of the sides of the hammer when pressing the paper into the grooves of the *gonghua* woodblocks, they can deliver vivid paintings with relief-like patterns and distinct layers.

12.4 Conclusion

It is a historical fact that printing originated in China. The resources, technical advancements, and social needs in ancient China provided the necessary conditions for the emergence of printing. Since the invention of printing, people have used various methods to combine materials, develop tools, and integrate techniques to reproduce texts and pictures with high efficiency and quality. Techniques such as block printing, movable type printing, chromatic printing, woodblock New Year picture printing, watercolor block printing, and *gonghua* printing are all exotic flowers that bloomed over the course of history.

In terms of organization and management, three major book publishing systems, namely, governmental printing, private printing, and publishing houses, as well as other important printing institutions in the form of Buddhist monasteries and academies, had already taken shape by the late Five Dynasties period. Government printing started in the Five Dynasties period, gained momentum in the Song and Yuan dynasties, and flourished in the Ming and Qing dynasties. It can be divided into central and local printing institutions. Because of abundant financial resources and strict management, fine paper and ink and efficient binding methods were used to produce high-quality works. They played an exemplary role in the national book printing and publishing industry. Government-printed books are valued by bibliophiles and were titled with different names depending on each printing institution, most of which were set up in different dynasties, such as the Academy Edition, Gongshi Warehouse Edition, Jing Palace Edition, Juzhen Edition, and Bookshop Edition. Private printing began in the Tang dynasty and kept its momentum ever since the Song dynasty. To promote academics and gain fame, officials and powerful families invested in printing certain famous works without profit. Therefore, such books generally are of fine quality and have been carefully revised. Famous private editions include the Shicai Hall Edition of Liao Yingzhong and the Wanjuan Hall Edition of Yu Renzhong in the Song dynasty, the Tianyi Pavilion Edition of Fan Qin and the Jigu Hall Edition of Mao Jin in the Ming dynasty, the Tongzhi Hall Edition of Nalan Xingde, the Zhibuzu Study Edition of Bao Tingbo, and the Shili House Edition of Huang Pilie in the Qing dynasty. Publishing houses, however, included every kind of bookshop. They not only sold but also compiled, engraved, and printed books. Some were even owned by writers themselves. These publishing houses played a substantial role in the production of ancient books and the circulation of commercial books. Many families were engaged in engraving and printing books from the Tang to the Qing dynasty, such as the Yu family in Jian'an, Fujian, and the Chen family in Hangzhou. The books printed by these families were of about the same quality as those printed by governmental and private institutions. Publishing houses aimed to print books that were practical and suited both refined and popular tastes. A large number of books of various types were printed and the technique was in continuous innovation, which boosted the sales and circulation of books. Despite the fact they pursued profit and their prints were of relatively poor printing and collation quality, publishing houses have made an enormous contribution to the spread of culture, meeting people's needs and promoting the development of printing.

The ancient Chinese printing industry was able to flourish and become popular because of the endeavors of craftsmen. One of these craftsmen's names, Lei Yanmei, appeared in an inscription on printed Buddhist images in the Five Dynasties period. By the Song dynasty, the printing industry was highly developed and there was a clear division of labor among transcribers, engravers, printers, and binders. The number of words in each woodblock was engraved on the top of the type page for payroll calculation and the name of the engraver was engraved on the bottom as a symbol of respect for their work. Therefore, the names of many engravers were printed in the books of the Song dynasty. Zhang Xiumin's research found a total of 3,000 engravers' names printed this way during the Song dynasty. In the Liao dynasty, 92 people were once gathered to engrave a scripture at the same time, which shows how many engravers there were. However, in the Jin and Western Xia dynasty, there were less engravers, and only a few of their names were printed. The names of many engravers were also printed on the books in the Yuan, Ming, and Qing dynasties as the printing industry continued to flourish. Many transcribers, printers, and binders have also left their names, but there were much less of them than engravers. Generally, these people were treated as ordinary craftsmen and remained unknown to history, even though they made lasting contributions to the spread of knowledge and culture.

The invention and application of printing not only contributed to the spread, development, and protection of traditional Chinese culture but it also promoted the development of civilization all over the world. The technique first spread to Korea, Japan, and other Southeast Asian countries, then to the Middle East and Arabia through the Central Asia regions, and finally to Europe through Arab and Mongol invasions. In addition to block printing, movable types made of clay, wood, bronze, lead, and iron were also used in North Korea, among which the bronze movable type was the most accomplished technique. In Japan, block printing came up after Chinese engravers voyaged east. The appearance of movable type printing in the country might have been influenced by Korea as well as China. Scholars have also recorded the history of the spread and application of Chinese printing in Europe. From 1440 to 1448, Johannes Gutenberg, a German inventor, printer and publisher, made movable types with metals that were easy to mold. Type cases and matrices were also used, which helped to control the size of the movable types and allowed the mass production of printed books. His use of oil-based ink for printing improved the printing quality and is still used today. In particular, the wooden printing press used by Gutenberg, though simple in structure, greatly facilitated the printing procedures and served as a blueprint for later printing machines. Gutenberg is thus known as the founder of modern printing. Later, the paper matrix, rubber plate, and other plate duplication techniques were improved in Western Europe, and the quality, quantity, and speed of letterpress printing were greatly advanced. In the late Qing dynasty, modern printing technology with high dynamic, efficacy, and precision was introduced from the West and gradually replaced the traditional printing technology in China as the dominant technique.

Contemporary printing technology has bid farewell to the era of movable types and woodblocks. The laser phototypesetting technology initiated by Wang Xuan (1937–2006) has made great contributions to the modernization of Chinese character printing. At the same time, traditional printing techniques, such as woodblock printing and movable type printing at the Guangling Ancient Books Engraving Society, scripture printing at the Jinling Scriptural Press and the Dege Parkhang, and watercolor block printing and *gonghua* printing at the Studio of Glorious Treasures and Duo Yun Xuan, still survive in modern time. However, they are considered more a charming way to remember the technical ingenuity and cultural significance of these techniques in the past. Traditional Chinese printing techniques embody our nation's wisdom and cultural diversity and, thus, should be properly protected and inherited.

Chapter 13
Carving and Painting

Wang Lianhai

13.1 Paper Cutting

13.1.1 History

The traditional Chinese philosophy of "the symbiosis between the actual and the spiritual world" has long existed and serves as the theoretical foundation of paper cutting. The earliest paper cuttings discovered so far were unearthed from the Astana Cemetery of the ancient city of Gaochang in Xinjiang. Among them, the most valuable paper cuttings were the Symmetrical Horses and Flowers, Symmetrical Monkeys and Flowers and two symmetrical chrysanthemum-shaped paper cuttings. The documents unearthed together with these paper cuttings are dated to 541–567. The Symmetrical Horses and Flowers is a pattern of 12 horses cut with a piece of paper folded four times. The Symmetrical Monkeys and Flowers is a pattern of 16 monkeys. The chrysanthemum-shaped paper cuttings are octagonal. These paper cuttings were all used for funerals as condolences for the dead. More than 10 other paper cuttings of silver ingots and *fangsheng* (a piece of jewelry with a pattern of two diagonally overlapping squares) were unearthed together, symbolizing success and good fortune. These objects demonstrate that paper cutting was already quite advanced in the Northern and Southern dynasties (as shown in Picture 13.1).

In the Tang dynasty, paper cutting was more widely used and developed further. Cutting human figures out of paper was very popular then. These paper cuttings were named *rensheng*, "*ren*" meaning "human". Wen Tingyun, a Tang poet, once wrote the lines: "One autumn night, wearing a pale pinkish-grey robe, she held a pair of scissors and cut *rensheng* to wear on her head." The Shosoin, which is the treasure house of Todaiji Temple in Nara, Japan, currently houses a piece of gold foil *rensheng* and a piece of gold foil *chunsheng*, "*chun*" meaning "spring". According

---

W. Lianhai (✉)
Academy of Arts and Design, Tsinghua University, Beijing, China
e-mail: wanglianhai327@163.com

© Elephant Press Co., Ltd 2022
H. Jueming et al. (eds.), *Chinese Handicrafts*,
https://doi.org/10.1007/978-981-19-5379-8_13

Picture 13.1 The paper cutting Symmetrical Horses and Flowers from the Northern Wei dynasty unearthed in Xinjiang

to Shosoin's records, these two paper cuttings were offered to deities or ancestors in the first year of the Tenpyō-hōji era (757), which corresponds to the second year of the Zhide period in the Tang dynasty. The *rensheng* was carved from gold foil with reticulated margins, which is extremely similar to the paper cuttings seen in Foshan, the Guangdong province, today. The *chunsheng* was made by pasting 16 characters cut from gold foil on a piece of light silk. Judging from these characters spelling out festive slogans and well-wishes, it was mainly used for festivals to pray for happiness and safety. Some paper cutting patterns have been mentioned in poetry and articles in the Tang dynasty, such as spring flags, *chunsheng*, spring money (a kind of money used in brothels), spring insects, turkeys, chickens, and phoenixes (as shown in Picture 13.2), all of which were cut and used during festivals. Paper cuttings were also used for funerals, as Feng Yan, a Tang historian, wrote in *Master Feng's Records of Hearsay and Personal Experience*: "Nowadays, large amounts of joss paper are delicately and beautifully cut to be used at funerals." In 1964, joss money, paper shoes, paper belts and paper hats made between the 21st year of the Zhenguan period (641) and the second year of the Kaiyuan period (714) were unearthed in the Khara-khoja Cemeteries of Turpan in Xinjiang. Among them was a paper cutting in the shape of seven people standing side by side, which was identified by archaeologists as "evocation *rensheng*". The six pieces of Decorative Paper Flowers from Cave No. 17 of the Mogao Caves in Dunhuang are now kept in the British Museum. The method for making these paper cuttings is to cut colored paper into the shape of a flower, stack up to nine layers of them on top of each other, and finally paste them together to form a flower. Traces of colored paint have been discovered on it as well. Similar paper flowers are also found on the walls and ceilings of other caves as decoration (as shown in Picture 13.3).

In the Song dynasty, paper cutting was widely used at festivals. For example, on New Year's Day, people cut cyan paper or silk into small flags and pasted over a dozen of them together to fashion headwear called "New Year Streamers". In the

Picture 13.2 *Chunsheng* and *rensheng* from the Tang dynasty, kept in the Shosoin at Todaiji Temple in Nara, Japan

Picture 13.3 Decorative flower paper cuttings from the Tang dynasty unearthed in Dunhuang

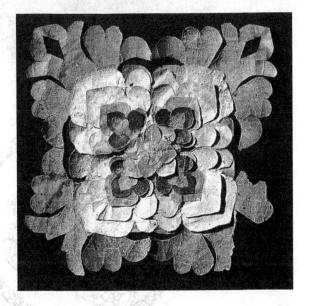

Northern Song dynasty, craftsmen who specialized in paper cutting began to crop up. In *Miscellaneous Notes of Zhiya Hall*, Zhou Mi (1232–1298) wrote: "On the old capital's Tianjie Street, there was a craftsman who could cut a variety of exquisite patterns in any color you wished. Later there was Yu Jingzhi, who was dedicated to cutting a variety of characters and patterns. Then suddenly there was a teenager who could cut characters and flower patterns inside his sleeves without even looking at it, producing paper cuttings even more exquisite and detailed than the former two craftsmen, which caused him to be held in high regard at the time." This shows that in the Northern Song dynasty paper cutting was moving towards specialization and commercialization.

In the Southern Song dynasty, they used paper cuttings to create resist patterns in the black glaze of Jizhou Wares. These patterns included auspicious figures (*siheruyi*), a circular phoenix, a magpie on a plum tree branch, and mandarin ducks. These patterns, which are exactly the same as contemporary paper cuttings, show the great achievements in paper cutting in the Song dynasty (as shown in Picture 13.4).

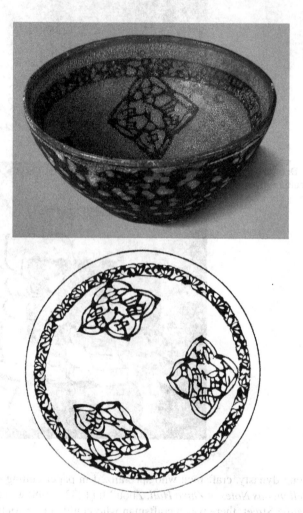

Picture 13.4 A bowl with resist patterns in the glaze (above) made using a pattern (below) cut out of paper

There was a kind of paper cutting called Sunny Doll in the shape of an old woman holding one or two brooms. When it has been raining for a long time, it is said that people should hang a Sunny Doll under the eaves, because it will chase away the clouds and make the rain stop. The poem *Sunny Doll* written by Li Junmin (1176–1260) goes: "With sleeves rolled up, the Sunny Doll holds a broom and keeps on sweeping as soon as it is hung towards the rainy sky." In the Yuan dynasty, paper cutting was also used in revolving lanterns. Even now, such paper cuttings are still called "lantern flowers".

In the Ming dynasty, more varieties of paper cuttings emerged. Apart from the ones used for window and room decoration, festive headwear was also cut from paper. For example, on New Year's Day, people cut patterns of insects and butterflies in Wujin paper called "playful moth" and put them on their heads. On New Year's Eve, people cut red paper in the shape of gourds and pasted them on the doors and windows to exorcise evil spirits. These were called "fighting against plague ghosts". Zhang Dai's (1597–1689?) *Dream Reminiscence of Tao'an* was the first to introduce the use of lantern paper cuttings: "Xia Erjin is a craftsman who can cut colorful paper into extremely exquisite flower patterns. If covered with diamond tulle, these paper cuttings will create an effect of smoke enveloping a peony."

Many paper cuttings from the Qing dynasty have survived to this day, most of which are window paper cuttings with an abundance of patterns. Dyed window paper cuttings and window paper cuttings with contrasting colors also became more and more advanced in the Qing dynasty. The paper cuttings commonly known as *guaqian* (hanging slips), but also known as happy paper, *mencai* (door colors), and *zhaidie* (blessings), can be hung on the lintels of a door or window, or stuck on the niches of Buddhist statues in monasteries and at home. During the Qianlong period, Gu Guang stated in his *New Year's Hundred Chants About Hangzhou Customs*: "Five pieces of colorful paper are used for each room. They are cut into the patterns and characters used on coins and pasted between the beams to expel ghosts and suffering." He also stated that: "*Zhaidie* usually refers to colorful paper that is cut into the shape of a banner with words on it such as 'Wishing for good weather, the country to prosper and the people to enjoy peace'. These banners are pasted on the lintels in Taoist monasteries or Buddhist temples."

In the Qing dynasty, paper cutting prospered the most. At that time, the most famous areas for producing paper cuttings on a large scale and with different artistic styles were Laizhou, Gaomi, Pingdu and Tancheng in Shandong province, Yuxian and Fengning in Hebei province, Fufeng, Qishan and Luochuan in Shaanxi province, Qingcheng in Gansu province, Jinhua in Zhejiang province, Ezhou in Hubei province, and Fenghuang in Hunan province.

From the 1930s to the 1940s, the Second Sino-Japanese War became a popular theme in paper cutting. The paper cuttings in Ezhou, Hubei province, were mainly made to publicize the *New Marriage Law*. After 1949, there were more paper cuttings that focused on current affairs, such as the paper cutting called Equality between Men and Women. This shows that paper cuttings reflect the signs of the times.

Nowadays, paper cutting is undergoing profound changes in terms of its function. With the return to traditional culture and the re-examination of the roots of our culture, traditional paper cutting has been restored and advanced in various aspects. In June 2006, the paper cuttings of 12 cities or counties from 8 provinces were listed in *List of the First Batch of National Intangible Cultural Heritage*, namely, Yuxian paper cutting and Fengning paper cutting in Hebei province, Zhongyang paper cutting in Shanxi province, Manchu paper cutting in Yiwulu Mountain, Jinzhou, Liaoning province, Yangzhou paper cutting in Jiangsu province, Yueqing fine-grained paper cutting in Zhejiang province, Foshan paper cutting, Shantou paper cutting and Chaozhou paper cutting in Guangdong province, Dai-style paper cutting in Luxi, Yunnan province, and Ansai paper cutting in Shaanxi province.

13.1.2 Variety and Distribution

Chinese paper cutting has spread far and wide, and is produced in large amounts and varieties. Most of those engaged in paper cutting are farmers or township residents. Most of them have paper cutting as their hobby and only a few actually specialize in it. The following table shows the 74 regions that produce paper cuttings on a large scale:

Distribution of Paper Cutting	
Heilongjiang province	Hailun
Jilin province	Tonghua, Jilin province
Liaoning province	Jinzhou
Hebei province	Yuxian, Fengning, Xincheng, Chengde
Ningxia Hui autonomous region	Yinchuan, Guyuan, Haiyuan
Shanxi province	Jinnan, Guangling, Yuncheng, Lingqiu, Jishan, Xiaoyi, Xinjiang, Wenxi, Zhongyang
Shaanxi province	Dingbian, Yulin, Suide, Yan'an, Ansai, Fengxiang, Qishan, Weinan, Luochuan, Xunyi
Shandong province	Pingdu, Yexian, Penglai, Yantai, Gaomi, Weifang, Linyi, Tancheng
Hunan province	Wangcheng, Changsha, Fenghuang, Huaihua
Jiangsu province	Yangzhou, Pixian, Nanjing
Zhejiang province	Jinhua, Yueqing, Wuyi, Dongyang, Pujiang, Lishui
Guizhou province	Taijiang, Shidong
Guangdong province	Chaozhou, Foshan, Shantou
Anhui province	Fuyang, Bozhou
Henan province	Lushi, Lingbao, Anyang, Huaiyang, Xinzheng
Hubei province	Xiaogan, Ezhou, Wuhan
Yunnan province	Tengchong, Dali, Luxi
Gansu province	Qingyang

(continued)

(continued)

Distribution of Paper Cutting	
Inner Mongolia autonomous region	Hohhot
Beijing	
Tianjin	
Shanghai	

Varieties and functions of paper cutting are as follow:

13.1.2.1 Window Paper Cutting

Window paper cutting is the most widely distributed and the most popular form of paper cutting, and is produced on the largest scale. It is mainly used decoratively to create a festive atmosphere and express people's desires to bid farewell to the old year and welcome the new, to spread blessings, and to wish for good fortunes. People put up window paper cuttings on holidays, weddings and birthdays, or when they are moving to a new house or opening a new business. This practice is also called "pasting happiness" (as shown in Pictures 13.5 and 13.6).

Window paper cuttings can be divided into four categories: (1) monochrome paper cuttings, (2) colored paper cuttings, (3) ventilation paper cuttings, and (4) paper-plastic paper cuttings. Monochrome window paper cuttings are cut with colored paper, mostly red. These paper cutting are either cut incised, where the paper that remains forms lines and patterns, or in relief, where the blank space expresses the images. This type of paper cutting has an extremely wide variety of subjects covering figures, animals, flowers, plants, landscapes, geometric patterns, and auspicious characters. Paper cutting can realize a kind of artistic vocabulary and expression that cannot be replicated by other artistic means.

Picture 13.5 Window paper cuttings in a farmer's house in Beijing

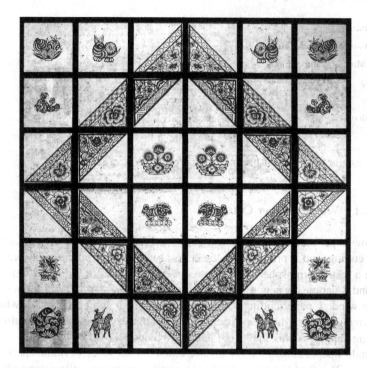

Picture 13.6 Monochrome window paper cuttings in Shanxi by Liu Jinglan

Colored window paper cuttings can be divided into dyed paper cuttings and paper cuttings that use contrasting colors. Dyed window paper cuttings first need to be cut and then dyed. They are usually brightly colored with the colors perfectly integrated into the patterns. The colored window paper cuttings in Yuxian, Hebei province, are famous for their vivid opera figures and are representative of northern paper cutting (as shown in Picture 13.7). At the end of the Qing dynasty, Wang Laoshang, a famous artist, made remarkable contributions to Yuxian's paper cutting collection. In the 1950s, under the influence of a Japanese person, theatrical mask images with colorful and dense patterns appeared in Yuxian's paper cutting scene. Since the 1990s, Yuxian's paper cutting has become famous both at home and abroad, which even caused its export to increase. A number of paper cutting workshops and companies have been established in Yuxian to continuously innovate the custom of paper cutting while still remaining faithful to its traditional characteristics (as shown in Picture 13.8).

13.1.2.2 Decorative Paper

Decorative paper refers to the paper cuttings made to decorate the living environment such as the living room and courtyard. It includes *guaqian*, ceiling paper cuttings,

Picture 13.7 Colored window paper cuttings in Yuxian county

stove paper cuttings, *kangwei* paper cuttings and ventilation paper cuttings. The most popular form is the *guaqian* due to its decorative effect. It usually has one narrow strip hung horizontally on the lintel of a door with five other paper cuttings hanging down from the bottom of the strip, making them flutter in the wind. Auspicious expressions are usually written on it using five Chinese characters, such as *Tianguan chang cifu* (天官常赐福), which means "May heaven bless us often", *Zhonghou chuanjia jiu* (忠厚传家久), which means "May the moral quality of loyalty and kindness be passed on from generation to generation", and *Shishu jishichang* (诗书继世长), which means "Studying hard and learning from predecessors can make the family prosper". When only four characters are used, it is necessary to add one. For example, the expression *Jinyumantang* (金玉满堂), which literally translates into "Gold and jade fill the hall" and means "abundant wealth or many children in the family", should be changed to *Jinyu fu mantang* (金玉福满堂), and *Fushou kangning* (福寿康宁), which means "Good fortune, long life, health and peace", should be changed to *Fushou yong kangning* (福寿永康宁). *Guaqian* with only four paper cuttings fluttering in the wind are solely used for funerals.

Tianjin has the largest number and the highest quality of red *guaqian*. Since the mid-1990s, the custom of putting up *guaqian* in urban areas and surrounding districts during the Spring Festival began to flourish. There were dozens of stalls selling paper cuttings on the Tianhou palace's Menqian square in Tianjin's Ancient Culture Street. Paper cutting markets and specialty stores have also been established at Temple of Great Compassion, Guyi Street, and Gulou East Street. In Tianjian, the locals call

Picture 13.8 1. Yuxian paper cutting. 2. Paper cutting of Qi Baishi, a famous painter in modern China

guaqian "hanging money". There are even colorful ones made with gold paper, which give it a very resplendent look. Nowadays, *guaqian* can still be found in Tianjin, in the rooms and shop windows of modern buildings, with the largest *guaqian* being more than two meters high (as shown in Picture 13.9).

In Tancheng, Cangshan and Linyi in southern Shandong province and Pixian in northern Jiangsu province, the *guaqian* named *huantangzi*, commonly known as *guomenqian*, is very popular. It is made by stacking five pieces of different colored paper on top of each other, then cutting the desired patterns, and finally moving the central parts of the paper to create the effect that each piece of paper has five colors.

Kangwei paper cuttings are paper cuttings that are pasted around the *kang*, which is a kind of heatable brick bed, with a large round flower-shaped cutting in the center

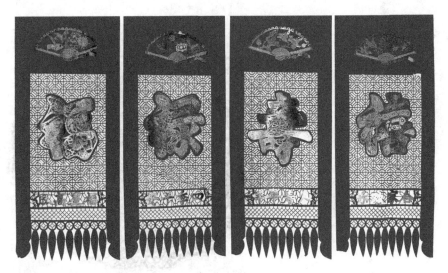

Picture 13.9 Tianjin *guaqian*

and triangular flower-shaped cuttings on the four corners. The blank spaces are used for cutting and pasting comic book-style stories. *Kangwei* flowers are not used purely as decoration, their symmetrical layout revolving around the round flower-shaped cutting in the center also symbolizes the peace and stability of the family.

Ceiling paper cuttings are pasted on the ceiling. Their pattern and function are basically the same as *kangwei* paper cuttings. The pattern is commonly known as "four dishes and one bowl of soup" with a symmetrical flower at the heart and triangular flowers on the corners. Ceiling paper cuttings are put up in newly-built houses or during Chinese New Year, weddings or birthdays. This custom was adopted in the Imperial Palace during the Qing dynasty. The ceiling paper cutting with "four dishes and one bowl of soup" can still be seen on the ceiling of the east corridor of Jiaotai Hall in the Forbidden City in Beijing. The flower at the heart is in the pattern of a phoenix and a dragon, called *Longfeng Chengxiang* (龙凤呈祥), which means "Prosperity brought by the dragon and the phoenix". It is said to be a relic from Emperor Guangxu's wedding (as shown in Picture 13.10).

Stove paper cuttings are popular in provinces in southern China. Stove flowers are paper cuttings of different shapes and sizes that are pasted above or around the stove. It includes patterns related to the Kitchen God.

Ventilation paper cuttings are popular in Shaanxi and Gansu and are pasted on window vents. The square window frames of farmer's houses in northwest China were not entirely closed off with paper so as not to block air flow, so they used ventilation flowers instead. Ventilation paper cuttings have the same shape as the window frame with a mesh pattern carved in the middle. They not only help with ventilation but also prevent mosquitoes and flies from entering the room. The abundant patterns in these paper cuttings also reflect the rich imagination of the region's paper cutting artists (as shown in Picture 13.11).

Picture 13.10 Ceiling paper cutting dating from the Qing dynasty, the Forbidden City, Beijing

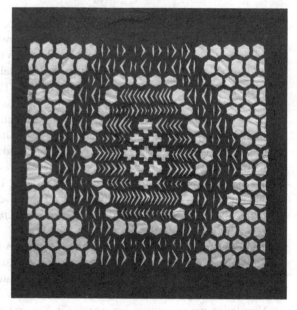

Picture 13.11 Ventilation paper cutting in Fengxiang, Shaanxi province

Wedding paper cuttings, present paper cuttings, and cake flags were also kinds of decorative paper cuttings. Wedding paper cuttings were paper cuttings that were pasted on the dowry at weddings, present paper cuttings were decorative paper cuttings pasted on gifts, and cake flags were small flag-shaped paper cuttings inserted on steamed bread or cakes as decoration. Besides these, there were also offering paper cuttings, which were paper cuttings placed on offerings, and Buddha *diaoqian'er* which were paper cuttings pasted on niches of Buddhist statues. Unfortunately, due to changes in customs, the use of these small decorative paper cuttings gradually died out.

13.1.2.3 Embroidery Patterns

The paper cuttings used to make embroidery patterns lie at the foundation of embroidery. Paper is cut into the patterns to be embroidered and stuck onto the cloth as a draft. After the embroidery is completed, the paper cutting will be completely covered by the threads and becomes impossible to take out, so each paper cutting can only be used once. Zhou Mi's *Miscellaneous Notes of Zhiya Hall* from the Southern Song dynasty states: "There are craftsmen who cut various patterns in different colors", which is the earliest record of embroidery patterns. Embroideries that used gold foil and dyed paper cuttings were unearthed from the tomb of Huang Sheng of the Southern Song dynasty. Guo Mengzhen, a Suzhou woman in the late Qing dynasty, documented her hand-cut embroidery patterns in the *Embroidery Patterns of Jijin Study*, which includes 20 pieces such as "Scrubbing Cattle", "Drinking Horses", "Bathing Children" and "Reading". It is the earliest collection of embroidery paper cuttings to date.

The use and development of embroidery patterns spread far and wide. Below, we will take a look at some of the most famous areas.

Embroidery was first developed in Ezhou, Hubei, where the local artists founded the Embroidery Paper Cutting Union in 1932, which consisted of over 100 people. The union formulated rules and regulations, requiring members (1) not to wave the "embroidery hoop" in front of customers but to lower it to show courtesy when selling embroidery patterns, (2) not to enter the workshop alone to sell paper cuttings, and (3) not to cut or raise prices without permission to ensure fair competition. There are still a large number of embroidery patterns from the early twentieth century in Ezhou today. Some embroidery patterns closely reflect the signs of the times. For example, during the implementation of the *New Marriage Law*, there was a pattern called "New Marriage" (as shown in Picture 13.12).

Zhao Jing'an is an expert in embroidery patterns in Sanhe county, Hebei province. His family has been engaged in paper cutting for three generations. He studied cutting and carving since he was a child and has created an abundance of cuttings, covering colorful window paper cuttings, opera figures, and embroidery paper cuttings. When he was young, he made embroidery patterns for various kinds of embroidery, such as toe caps, insteps, sock bottoms, purses, and undergarments, in Chengde and Zhangjiakou, Hebei, and also in Beijing. His work is characterized by its preciseness and delicacy. Customers always know exactly how to do the embroidery once they see his paper cutting (as shown in Picture 13.13).

Song Yulan from Huaiyang county, Henan province, is another outstanding expert. She also cuts a wide variety of embroidery patterns, but she is best known for cutting tiger-head shoes. She also cuts shoe patterns for embroidering the insteps of the shoes worn as part of a deceased person's graveclothes, with the patterns of toads,

Picture 13.12 The embroidery pattern "New Marriage" on a pillow in Ezhou, Hubei province

Picture 13.13 The embroidery pattern of paper carving "Squirrel Grape" by Zhao Jing'an

geese, and the Golden Boy and Jade Girl, who traditionally serve as the guides to the underworld. There is a local folk song about preparing elders for their end that goes: "One toad and one goose occupy the lotus root, and the Golden Boy and Jade Girl lead the way" and "The rooster sings, the phoenix stops, and the Golden Boy

Picture 13.14 The embroidery pattern for tiger-head shoes in Huaiyang, Henan province by Song Yulan

and Jade Girl carry a red lantern." The embroidery patterns for the graveclothes are designed according to this folk song (as shown in Picture 13.14).

Women of the Miao ethnic group have long been famed for their embroidery. Miao embroidery also uses paper cuttings. In Luxi and Fenghuang in Xiangxi, Hunan province, there are many artisans who create and refine embroidery patterns. Tahu village in Luxi is known for creating the very best of all of them. The embroidery patterns there are known as *yahua* (rolled flowers) or *cuozhi* (filed smooth paper). The method they use is to fix the paper on a wooden board, poke the edges of the pattern with a knife, and then prick the finer inner lines into a dotted line with an awl to show the internal structure of the pattern. In Xiangxi, Hunan province, and Taijiang, Huangping, and Shidong, Guizhou province, the patterns of embroidery patterns include dragons, phoenixes, flowers, and plants. Some animal and plant patterns are mystical and haven't been fully understood until now. The shapes and compositions of these paper cuttings are all large and plentiful (as shown in Picture 13.15).

13.1.2.4 Specialty Paper Cutting

Specialty paper cutting is purely ornamental, with a higher aesthetic value, exquisite workmanship, and elegant styles. They are mounted indoors purely for decorative purposes. Famed artists include Zhang Yongshou, Ku Shulan, Wei Wen'e and Chen Jinhua.

Zhang Yongshou, a native of Yangzhou, Jiangsu province, studied art with his father since he was a child, and was already famous in Yangzhou at the age of

Picture 13.15 The Miao embroidery pattern for cuffs in Guizhou

fourteen. Since the 1950s, he created his first specialty paper cuttings. He was best at cutting chrysanthemum patterns. He wielded the artistic vocabulary of paper cutting fully and effectively. In his works, he was able to create an effect of three-dimensional space despite not using any overlapping techniques (as shown in Picture 13.16).

Picture 13.16 The paper cutting "Chrysanthemum" in Yangzhou, Jiangsu province, by Zhang Yongshou

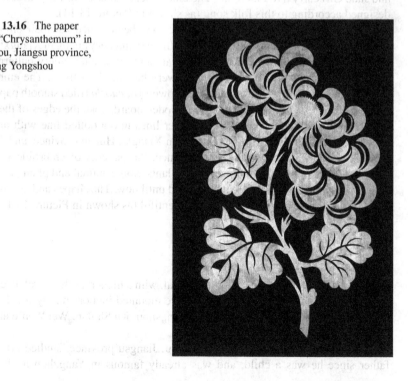

Picture 13.17 The paper cutting "Paper Cutting Lady" by Ku Shulan

Ku Shulan, a native of Xunyi, Shaanxi province, is known as the "Paper Cutting Lady" due to her own distinctive style of cutting and pasting colorful paper cutting patterns. Ku's paper cuttings are characterized by overlapping all kinds of colored paper to form rich and gorgeous lines and shapes (as shown in Picture 13.17).

Wei Wen'e was born in Fengning, Hebei province, in 1963. She studied art with her uncle since she was a child as well and became famous at the age of twenty. Her works, which were mainly relief patterns, absorbed the artistic inspiration of woodblock printed New Year pictures and embroidery, imbued them with local flavor, and combined them with elegant simplicity. The costume characters and opera figures are the most representative of all her paper cuttings, which are vivid, precisely modeled, and show a scholarly style.

Chen Jinhua, a native of Yueqing, Zhejiang province, enjoyed nationwide popularity for creating fine-grained paper carvings. To create a fine-grained paper cutting, she first cut the skeleton of the pattern on a thin piece of paper, then densely arranged geometric patterns on it. The finest parts could contain 51 lines per square inch. To realize the desired artistic effect, she had to make at least 100 cuts per square inch.

Picture 13.18 A fine-grained paper carving by Chen Jinhua

This type of paper cutting is so specialized that it deserves to be categorized separately from window paper cutting and considered as an independent form (as shown in Picture 13.18).

13.1.3 Paper Cutting Process

The production process of paper cutting can be divided into cutting and carving, both of which go through two phases: (1) drafting and (2) cutting or carving. However, colored paper cuttings require an additional dyeing phase.

13.1.3.1 Drafting

Drafting is also called "drawing a model". All folk window paper cuttings, embroidery patterns, and *kangwei* paper cuttings are created by the artists themselves. Most of these artists are rural women, who have not received art training or even attended school, yet they are able to design rich and beautiful images. Although the drafts can be a bit crude, the final paper cuttings can be extremely beautiful.

Another method is called "smoking", which is to brush water on a piece of white paper and stick it on a wooden board or iron plate, then use water to stick the desired paper cutting on the white paper, and finally use the smoke from candles or oil lamps

13 Carving and Painting

Picture 13.19 Smoked paper cutting drafts from Shanxi province

on it. Once smoked, the paper cutting is removed and leaves behind a white pattern on the blackened paper. This stark contrast makes the patterns clear, elegant, and distinctive (as shown in Picture 13.19).

13.1.3.2 Paper Cutting

Women in the countryside are usually taught to cut clothes, vamps, embroidery patterns, and paper from an early age, so they are generally skilled with scissors. When it comes to paper cutting, 3 to 6 sheets of paper can usually be cut at the same time. The first step is to poke holes through the paper's margins and then fix the paper together with paper spills to prevent dislocation. The scissors are held in the right hand and the paper in the left, the paper is cut towards the upper left, and the paper is rotated accordingly. The internal patterns are cut before the outline.

The places where two cuttings intersect need to be dealt with carefully. The edges of the scissors must be cut deeply to achieve the desired effect. Beginners often leave some remnants here, but skilled paper cutting artists can intuitively ensure that the patterns are neat and clear (as shown in Picture 13.20).

13.1.3.3 Paper Carving

Paper carving is mostly used for commercial ends. Products from Yuxian and Fengning, Hebei province, Yueqing, Zhejiang province, Tancheng, Shandong

Picture 13.20 An old woman cutting paper in Lishui, Zhejiang

province, and Pixian, Jiangsu province, labelled as paper cuttings are all actually carved, not cut.

Paper preparation. Make several stacks of 5 pieces of Fenlian paper or thin Xuan paper, sprinkle talcum powder on the surface of each stack and rub it evenly with your hand. Then stack the different stacks on top of each other. Usually 4 stacks are stacked together, but you can stack up to 16 stacks on top of each other. Adhere the cutting pattern on the paper with paper spills and turn over the paper so the front is facing down on the table. Sprinkle water on the paper, cover it with a towel, and sprinkle it again until it is soaked. Then wrap it in waste paper, put it on the slate or the ground, and use your feet to strongly push down on it. Remove the wrapping paper and dry it in the shade. The prepared paper will be as firm and flat as a plate.

Paper carving. The "three treasures" of paper carving are the wax tray, carving knife and grindstone. Artists make their carving knives themselves through forging, quenching and sharpening steel bars or even scrap steel wires. Bamboo pieces are tied on both sides as the handle (as shown in Picture 13.21). A wax board is made by carving a square groove on a wooden board of about 36.7 cm long, 30 cm wide and 1 cm deep, then evenly mixing the powder of hemp stalks and cotton with melted wax and pouring it into the groove. Once cooled, the wax board is ready to use. The proportion of wax and hemp stalk powder will influence the hardness of the wax board, so it is necessary to adjust it to make the board both elastic and hard enough to carve things on. It is also used to polish carving knives.

Carving knives. Carving knives are divided into *lidao* (vertical knives) and *zoudao* (moving knives). The sharp and narrow *lidao* is used for carving details. And the *zoudao* is slightly wider and longer for carving long lines. When carving paper, carving knives must be kept perpendicular to the paper surface and every stack

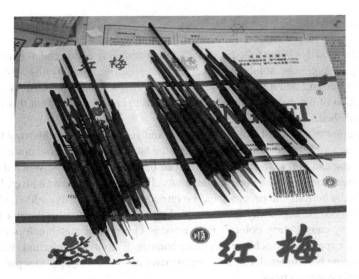

Picture 13.21 Carving knives in Yuxian county

of paper should be carved all the way through (as shown in Picture 13.22). First, carve the internal patterns and remove the paper scraps. Then carve along the outer contour to finish the carving. Finally, recover each stack of five pieces of paper recognizable by the talcum powder on top of each stack. Now it's ready to be dyed.

Picture 13.22 An artisan in Yuxian carving paper

13.1.3.4 Dyeing

The most used dye is the folk dye usually used for dyeing cotton called *pinse*, which is cheap, bright, pure, soluble in water, but fades easily. When dyeing, dissolve the dye in water and add a small amount of high-grade *baijiu* to increase the level of penetration of the dye. Once, arsenic was added to enhance luster, but this tradition was abolished.

The color of dye can be divided into *yin* and *yang*, the former referring to warm colors such as red, yellow and orange, and the latter referring to cold colors such as green, blue and purple. *Yang* colors are dyed before the *yin* colors. When dyeing, it is necessary to ensure that the dye penetrates all the way to the lowest layer. If it fails to reach the bottom, the paper should be turned over and more dye added. Different colors should be kept away from each other to avoid unintentional color mixing, which may cause "dirty colors". Folk artisans choose and match colors based on their rich experience to achieve a delicate balance of strong contrast and rich colors. Once dyeing is completed, roll a rolling pin over the pieces of paper to separate them into single paper cuttings.

13.2 Woodblock New Year Pictures

13.2.1 History

The creation of woodblock New Year pictures sprouted from the Han dynasty's custom of painting directly on a door. It is recorded in *Bibliography of Thirteen Princes* in *History of Han* that: "There is a portrait painting of Cheng Qing, an ancient warrior wearing a short jacket, big trousers and a long sword, on the door of the palace of King Liu Yue." This is the earliest record of a painting on a door. People believed that painting warriors on their doors helped ward off evil and ensure their safety. Tigers were also painted on doors. The legend this stems from was recorded in *Comprehensive Meaning of Customs and Mores* and *Classic of Mountains and Seas*. It speaks of two immortals, Shentu and Yulu, who lived in the peach trees on Dushuo Mountain. They, once they found evil spirits, bound them with reed ropes and fed them to tigers. This is why people often "decorate their house with figures made of peach wood, hang reed ropes and paint tigers on their door" on New Year's Eve as a protection against evil spirits.

In the Han tomb in Yingchengzi, Jinxian county, Liaoning province, there is a 105 by 50 cm painting of a mighty doorman holding a halberd, which served as a rudimentary predecessor of the door gods in China (as shown in Picture 13.23).

Various kinds of door gods were painted on the doors of dwellings, monasteries and palaces in the Tang dynasty, some holding staffs and some flowers. What promoted the popularization of door gods was Emperor Ming of the Tang dynasty

13 Carving and Painting

Picture 13.23 Door paintings in the Han dynasties unearthed in Liaoning, cited from *Tomb Mural* in *The Great Treasury of Chinese Fine Arts*

commissioning Wu Daozi to paint the portrait of Zhong Kui, a Chinese deity traditionally regarded as a vanquisher of ghosts and evil beings. On New Year's Eve in the fifth year of the Song dynasty's Xining period (1072), Emperor Shenzong ordered the painter Liang Kai to also paint the image of Zhong Kui in the Palace Secretariat and the Bureau of Military Affairs. This marks the beginning of the use of New Year pictures in the Imperial Palace.

Printing and selling New Year pictures became popular in the Song dynasty. "When festivals are approaching, the paintings of door gods, Zhong Kui, peach wood charms, the God of Wealth, and other traditional auspicious patterns will be sold on the market." Liu Zongdao, a native of Kaifeng, the capital city, was good at painting *Zhaopen Hai'er*, a picture of children mirroring themselves in a basin. He printed hundreds of copies for every new painting he made himself in order to protect his copyright and prevent others from imitating him. There was another artist known as "Du Hai'er", whose paintings were widely popular during the Zhenghe period (1111–1118). Even the imperial painters would seek help from him to meet the palace's painting needs. After the capital city of the Song dynasty moved south, the shops selling funerary paper horses started to print paintings of Zhong Kui and the horse of fortune. Stores at the Chaotianmenta Square in Lin'an city were also selling the paintings of door gods, peach wood charms, Zhong Kui and tigers. On New Year's Eve, every family would gift and welcome the Six Gods with colorful paper money and put up *Tianxing Tie'er* (streamers put on the door lintel) and the *Door of Fortune*. These images can still be found in today's New Year pictures.

The New Year pictures *Four Beauties of China*, *Throne of Prince of Loyalty and Valor*, and *The Immortal Dongfang Shuo Stealing a Peach* dating back to the Jin kingdom, still exist. The first two are now kept in Russia, and the latter was unearthed in the Stele Forest in Xi'an in 1973. The *Four Beauties of China* (as shown in Picture 13.24) serves as the foundational piece that later New Year pictures depicting beauty are based on.

In the Yuan dynasty, a new kind of New Year picture called *"Nine Times Nine Strokes to Relieve the Winter Cold"* appeared. Before the winter solstice, women would cut 81 plum blossoms and paste them on the windows, and then paint one plum with rouge every morning until, after 81 days, all the blossoms were colored, which heralded the coming of spring. In woodblock New Year pictures, the image

Picture 13.24 *Four Beauties of China* dating back to the Jin kingdom, cited from *Woodblock Printing* in *The Great Treasury of Chinese Fine Arts*

Picture 13.25 The paintings of the auspicious boys called "Wannian" and "Ruyi", currently kept in the National Art Museum of China

of plum blossoms was used in calendars or pasted around the statue of the Kitchen God. Later, it evolved into a rich variety of paintings (Picture 13.25).

The Ming dynasty witnessed woodblock New Year pictures fully flourishing. The engraving technique became more refined under the influence of literary illustrations and the colors of chromatic printing also became brighter. *Red Printed God of Longevity* printed during the Longqing period (1567–1572) and *God of Longevity* during the Jiajing period (1522–1566) are both masterpieces inherited from that time. There were also many hand-painted New Year pictures with festive themes, such as *Shoutian Fulu* and *Niannian Ruyi*, which are all color paints on silk with gold-colored powder expressing good wishes for the coming years. Liu Ruoyu recorded in his *History of the Ming Palace*: "On the 30th day of the twelfth lunar month in the Ming palace, people put peach wood charms beside the door, put up the door god, hung the god of fortune, ghosts resisting judgment and Zhong Kui indoors, and wrote poems on folding screen depicting the changing of the seasons." The door gods of the Ming dynasty were typically in the image of Qin Shubao and Yuchi Gong of the Tang dynasty, the former white-faced and holding a whip, and the latter black-faced and holding a mace. These two far-reaching images are still used in common New Year pictures today.

In the Qing dynasty, each area that produced New Year pictures successively formed its own unique style with a wide range of themes. Printing and painting were undertaken simultaneously, and the output of the pictures also increased dramatically. Therefore, the Qing dynasty was called "The Golden Age of New Year Pictures" by Wang Shucun (1923–2009), a famous contemporary painter of traditional Chinese paintings. There were dozens of kinds of door gods, including Shenshu, Yulu, Qin Qiong and Yuchi Jingde, as well as Hehe Immortals, Liu Hai and Zhang Tianshi. New Year pictures that reflected realistic themes include Yangliuqing's *Burning of Wanghai Building* and *Picture of Liu Tidu's Victory in the Kefu Water-fight*, as well as Taohuawu's *Faren Qiuhe* and *Red River War Victory*. Pictures revealing the corruption of the government include *A Small Investment Brings a Ten- Thousand-fold Profit* and *Tailor as a Straight Line*, which in Chinese literally refers to the magistrate of a county. There are also pictures advocating reform and calling for abolishing old customs, such as *Women Studying* and *Arresting Imported Cigarettes and Poppies*. These New Year pictures reflected the decline of the late Qing society and expressed people's urgent desire for social change (as shown in Picture 13.26).

The theme of welcoming the New Year and praying for blessings remained popular. People used homophones to express wishes with imagery, such as *The Fragrance of the Orchid and Cassia Ascends*, *Auspicious Happiness in Overmeasure*, *Kylin Sending Children* and *Three Rams Bring Bliss* which still exist today (as shown in Picture 13.27). The gods of folk beliefs, such as the Kitchen God, Medicine God, God of Land, God of Wealth, God of Fire, Taiyin Zhenjun and Emperor Conquering the Demon, started to appear more and more frequently in New Year pictures, which

Picture 13.26 A New Year picture which includes the image of a train

Picture 13.27 Yangliuqing New Year picture named *Busy Farmers*, cited from *Chinese Local Handicrafts*

indicated that people were hoping to change their tough reality with the help of the immortals and deities.

After the Republic of China was established, the demand for common New Year pictures began to reduce and their content changed due to social upheavals and the introduction of Western cultures. The development of New Year pictures is entwined with people's daily lives, embodying the aesthetic tastes and pursuits of the common people, which lies at the very core of its long-lasting vitality. In 2006, Yangliuqing New Year pictures from Tianjin, Wuqiang New Year pictures from Hebei province, Taohuawu New Year pictures from Suzhou, Jiangsu province, Zhangzhou New Year pictures from Fujian province, Yangjiabu New Year pictures from Weifang, Shandong province, Puhui New Year pictures from Gaomi, Shandong province (as shown in Picture 13.28), Tantou New Year pictures from Longhui, Hunan province, Foshan New Year pictures from Guangdong province, Liangping New Year pictures from Chongqing, Mianzhu New Year pictures from Deyang, Sichuan province, and Fengxiang New Year pictures from Shaanxi province were all listed in *List of the First Batch of National Intangible Cultural Heritage*.

13.2.2 Production Techniques

There are various carving techniques for woodblock New Year pictures depending on the region. Each expresses its own content and artistic style, but the production process can be divided into roughly the same six major steps: (1) drafting, (2) draft

Picture 13.28 Gaomi New Year picture named *The Central Room at Home* (partial), Shandong province

printing, (3) engraving, (4) ink line printing, (5) chromatic printing, and (6) color painting.

13.2.2.1 Drafting

Drafting is also called "decay" (*xiu*) in Tianjin's Yangliuqing New Year pictures, because the drafts need to be repeatedly revised almost to the degree of decaying. The name for this step is also different in Hebei and Shandong.

Most painters are experienced in drafting and can skillfully create expressive models to use. The first step is to draw an "ink line draft" which is used to specify the size, overall composition and basic contents. The process of drafting is similar to line-drawing. The only difference is that all the points and lines need to be carved. To accommodate the carving and printing process, the drafting style needs to be simple, precise, and clear.

Painters need to first draw the basic images with pencil or charcoal. Once finished, the picture is sprayed with a mixture of alcohol and rosin that forms a protective film. In ancient times, there were no spray nozzles, so their solution was to gradually deepen the initial draft with light ink and once it was finalized, they covered the draft with a thin and soft piece of *xuan* paper and sketched out the final draft with thick ink.

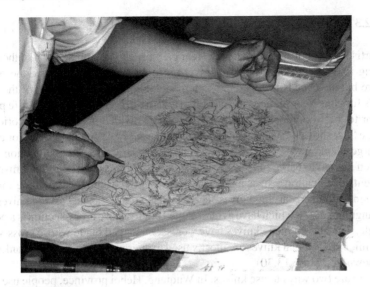

Picture 13.29 Drafting

After the ink line woodblock was carved, the draft was colored in preparation for the color woodblock engraving. There are usually four to six kinds of colors in a picture, and each color had its own woodblock. Therefore, painters needed to prepare four to six color drafts, each with a different color besides the ink line drafts (as shown in Picture 13.29).

13.2.2.2 Draft Printing

For draft printing, wood with a solid texture and delicate grain that do not easily deform are usually used, such as pear wood, jujube wood, pomegranate wood, Chinese rubber wood, lithocarpus fenzelianus wood or walnut wood. In Wuqiang county, Hebei province, people use birch-leaf pear wood as ungrafted pear wood. Break the wood into a three-centimeter-thick plate and cut it into the size of the desired picture. Soak the plate in clear water for 30 days and then dry it in the shade to prevent it from deforming. After it is dried, it needs to be planed and polished. Then, brush glue on the wooden plate and stick the draft face down on it by steadily brushing the paper from right to left to prevent air pockets. After the glue is dry, the ink lines will become unclear, so it is necessary to wipe off some paper scraps with a hard brush dipped in water to make the ink lines more visible again. In Zhangzhou, Fujian province, people use a rough horsetail to rub the plate, while in Wuqiang, Hebei province, sesame oil is applied onto the woodblock after the draft has dried, which can make the paper translucent and the ink line clear.

13.2.2.3 Woodblock Engraving

The artistic effect of woodblock New Year pictures depends on the skill of the person carving it. This is because this engraver not only needs to reproduce the original picture but also needs to make the pattern suitable for printing. Most of the woodblocks for New Year pictures are carved in a vertical direction with the knife perpendicular to the woodblock. Carving knives can be divided into three categories. The first are "flat shovels" with long vertical handles and wide and flat blades for carving out large blank areas. The second are chisels, usually used in conjunction with a wooden mallet, to carve out small blank areas. The third one is carving knives with different specifications, sizes and shapes. There are flat-mouthed, oblique-mouthed and round-mouthed carving knives. There are also some special carving knives, such as triangular knives, which have a right-angled blade whose cross-section is a positive triangle. The curved blade knives have a crescent-shaped blade whose cross-section is a semicircle. Special knives are mostly used to engrave patterns inside blank spaces (as shown in Picture 13.30).

There are two ways to use knives. In Wuqiang, Hebei province, people use hawkbills with a slightly curved blade to carve from the left side of the ink line and then move the knife inward, which means pulling the knife backward. In Yangliuqing, Tianjin province, people carve from the right side of the ink line. No matter from which direction, the engravers always hold the knife with the right hand and assist with the left hand, so as to ensure that the knife moves smoothly. An engraver's skill is reflected in the way he holds and moves his knives. When encountering crossing lines, engravers need to be especially careful not to cut through the intersection, and when multiple lines are juxtaposed, engravers need to guarantee even spacing.

They can rotate the woodblock to find the most suitable position to carve the lines and patterns. Experienced engravers usually divide ink lines into several groups, carving some of them from the left side first, and then rotating the woodblock to

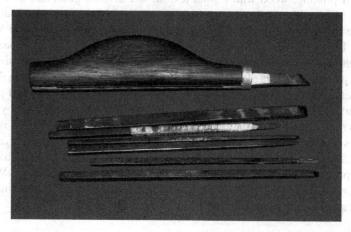

Picture 13.30 Carving tools for engraving

carve them from the other side. When the lines are very thin, engravers should ensure that the cross-section of the engraved wood lines are wedge-shaped with a smaller upper part and a larger lower part. When carving from the left side, the knife should lean slightly to the right, and vice versa. The thinner the line, the greater the inclination angle should be. This ensures that the carved woodblock is strong and durable.

After the ink lines are engraved, the blank spaces are removed with a "picking knife". After all the lines and blocks are engraved, the grooves are deepened and the blank spaces leveled with a flat shovel knife or a flat-mouthed knife. The larger the blank space, the deeper it needs to be engraved, otherwise it might stain when printing. The transition areas between the lines and blank spaces need to be curved, not vertical, to ensure that the lines are solid (as shown in Picture 13.31).

Color woodblocks are engraved using the same method as ink line woodblocks, except that there are simply more blocks to engrave. No matter the kind of woodblock, the borders of the images and lines need to be carved into arcs without any sharp edges, and any blank spaces or unnecessary parts need to be leveled off to prevent staining during printing (as shown in Picture 13.32).

Picture 13.31 Tai Liping, a New Year picture artisan in Fengxiang, Shaanxi province, engraving a woodblock

Picture 13.32 Ink line woodblock and color woodblock

13.2.2.4 Ink Line Printing

New Year pictures are commonly printed with the ink lines first, followed by the colors. Because the ink lines can serve as an alignment tool to the color printing blocks.

The "printing bed" for printing New Year pictures consists of two parts. The right part has a stack of aligned white paper clamped onto it with splints and is aligned with the left side of the platform. The left part is for placing the woodblock on. There is a small gap between the two parts to hang the printed picture.

Before installing the woodblock, ball-shaped glue is stuck on the four corners on the back of the woodblock. The glue, mainly made of rosin, will soften when heated and remain elastic after cooling down. After aligning the woodblock, strongly press it against the printer bed to squash the glue at the four corners and really make it stick. Now, most places use plasticine instead of this special glue.

After the woodblock has been installed, brush ink on the plate with a bristle brush. In the past, people used boiled glue for this step, but now ink is more commonly used. Printers should not apply too much ink on the woodblock to prevent the ink lines from bleeding. Turn the paper on the right side of the printer bed to the left, spread it onto the woodblock, and then brush the back of the paper with a special tool called a *tangzi*. After the ink lines are printed on the paper, carefully lift the paper and hang it in the gap between the two parts of the printer bed. Repeat the above steps for every piece of paper. A *tangzi* is a pressing plate made of palm, which is thick in the middle and thin at both ends. It is placed on a wax plate, so it can also be used to wax the back of the paper.

13.2.2.5 Chromatic Printing

For every colored New Year picture, four to six color woodblocks are commonly required, one for each color. The printing method is the same as that of the ink lines. Light colors are printed before dark colors. When printing, the woodblock must be perfectly aligned to prevent any dislocation or misprints. After printing several pieces of paper, there will be probably a little offset in the position of the woodblock. Experienced printers will promptly hit the edge of the woodblock with a wooden mallet to adjust it back to the original position. The pigments used are generally bright in color but weak in pay-off so that it will not blur the ink lines (as shown in Picture 13.33).

In ancient times, there was a printing method called "woodblock overprinting" which prints pictures with small colored woodblocks without any ink lines printed. Up to now, there are still such unique practices in Zhangzhou, Fujian province and Foshan, Guangdong province.

In recent years, Qi Jianmin, director of the New Year Picture Museum in Wuqiang county, Hebei province, developed a method to print colors before the ink lines. The pigments used are poster colors or acrylic paint with multiple bright, opaque colors. As mentioned before, in traditional printing, the colors could cause the ink lines to blur and destroy the picture. Instead, Qi adopted the method to print color before the ink lines, so that the dark ink lines cover the colors. However, the printing process becomes more difficult since it is necessary to align each colored woodblock precisely without the help of ink lines. Despite the increased cost, the printed pictures have a much better effect when they have brighter colors.

Picture 13.33 Chromatic printing

13.2.2.6 Color Painting

The technique of combining printing and painting for woodblock New Year pictures was invented and adopted in Yangliuqing, Tianjin province, Mianzhu, Sichuan province, Gaomi and Weifang, Shandong province, and in Longhui, Hunan province. Here, printers usually print the ink lines first, then print one or two colors or no color at all, and at last apply color by hand. This technique is mostly used in Yangliuqing, Tianjin province, and these New Year pictures are often very visually appealing. Painters will install plywood or a thin wooden board on a wooden frame and paste Korean paper on it, then install a wooden shaft on one side of the frame so that the board can rotate. This is called the "drawing door", which is similar to a drawing board. Several "drawing doors" will be installed together and can be freely rotated, and together make up the so-called "painting frames". Painters paste the ink line draft on the "drawing doors" and paint colors on it. It involves numerous steps, such as "powder priming", "face dyeing", "face sketching" and "eyes dotting". The color painting process of Yangliuqing New Year pictures is extremely delicate, similar to that of regular artistic color painting, and can create results in exquisite pieces of art (as shown in Picture 13.34).

Mianzhu, Sichuan province is another famous place for painting woodblock printed New Year pictures. Here, the artists paint and dye large areas of the picture with smooth and unrestrained strokes. Artisans in Gaomi, Shandong province pay particular attention to the effect of color transitioning, but color painting is not commonly used. The last possible step to color painting these pictures is to sketch out local parts with gold powder, and then draw patterns with white powder. Such brush techniques create an extremely smooth application of colors and have helped create many stunning paintings.

Picture 13.34 An artisan painting Yangliuqing New Year pictures

13.2.2.7 *Puhui* New Year Picture Techniques

Puhui New Year pictures only exist in Gaomi, Shandong province. This category of New Year pictures that does not use woodblock printing is known as one of the "Three wonders of Gaomi". The process consists of drafting, ash drafting, color painting and powdering.

First, draw the picture on a piece of *xuan* paper with charcoal created from burning willow branches. Then, press the front side of the picture onto draft paper and brush the back of the paper so that the charcoal pattern is printed on it. The draft left on the paper after uncovering the *xuan* paper is called an "ash draft". A charcoal painting can be used to create up to six drafts. Artists can also reapply the charcoal on the same paper to print even more drafts. Then artists usually use big brushes to produce different shades of color and finally trace the picture with ink (as shown in Picture 13.35). The smooth brushwork and color vignetting create a simple and quaint effect (as shown in Picture 13.36).

13.3 Shadow Play

Shadow play, also known as shadow puppetry, is an ancient folk performance art that is popular all over China. Donkey skin, cattle hide, sheepskin or cardboard is cut into figures or animals with movable arms, waists, legs, and other joints. These puppets are manipulated by puppeteers and projected on a translucent screen with lights so that the audience can watch them from the other side. The puppeteers also sing operas during their performance. The singing style has gradually evolved from simple local operas to an independent "shadow opera style".

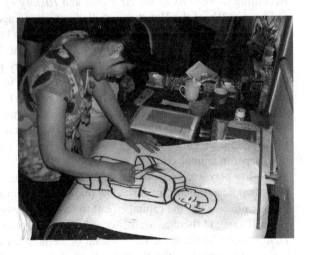

Picture 13.35 Drawing drafts with charcoal

Picture 13.36 *Kicking a Shuttlecock*, a Gaomi *Puhui* New Year picture from Shandong province

13.3.1 History

According to *Records of the Historian* and *History of Han*, Lady Li, a concubine of Emperor Wu of the Han dynasty, died at a young age. One night, Li Shaoweng, one of the ministers, had someone pitch a tent, light a lamp, and control the shadow of a doll in the shape of Lady Li outside of the tent. Emperor Wu, sitting inside the tent, saw the appearance of Lady Li and was very pleased. Researchers regard this story as the origin of shadow play. It is recorded in *A Dream of Lin'an*: "When people first engaged in shadow play in Bianjing city in the Yuan dynasty, they created puppets from normal paper. Later, the production techniques and playing skills were greatly improved. They started to use sheepskin to create puppets, which made it more durable, and decorated them with color." From this, we can see the transition from paper to animal skin.

Shadow play was very popular in the Northern Song dynasty. On the fifteenth day of the first lunar month, small shadow plays were set up at alley entrances to perform shadow plays. After the capital city of the Song dynasty moved south, shadow play gradually spread there as well. According to *Artists of Various Categories* in *Old Things about Wulin*, there were a total of 22 families engaged in shadow play in Lin'an, the capital of the Southern Song dynasty, including the "Three Jia" and

Three Fu", which refer to the families of Jia Wei, Jia Yi, Jia You, Fu Da, Fu Er, and Fu San, and 16 other families such as the families of Shang Baoyi, Shen Xian and Chen Song. During the Lantern Festival, artisans themselves would perform behind the screen as a new form of shadow play. Modern-day Laoting, Hebei province, even has shadow puppets that are as tall as real people.

After the Northern Song dynasty, shadow play spread from the capital city of Bianliang to the whole country. According to research done by professor Wei Liqun, a contemporary shadow play expert, the northern-style shadow play was formed after the invasion of nomads in the late Northern Song dynasty, who captured shadow play puppeteers and learned from them. Some puppeteers moved southward when the capital moved south, gradually forming the south-central-style shadow play popular in the Central Plains and Jiangnan area. Other puppeteers moved across Tongguan in Shanxi province to evade the chaos of the Jingkang Incident in 1127, thus forming the western-style shadow play. These categories have continued to this day. Northern shadow plays are mainly found in (1) Hebei province, which includes the Tangshan style, Leting style, Zhuozhou style, and Hejian style, (2) Shanxi province, which includes the Northern Shanxi style, (3) Shandong province, which includes the Jinan style, Taishan style, Dingtao style, Zaozhuangstyle, and Linyi style, (4) Henan province, which includes the Tongbai style and Luoshan style, (5) Beijing, which has its own Beijing style (6) Liaoning province, which includes the Anshan style in Xiuyan, a Manchu autonomous county, and the Gaizhou style, and (7) Heilongjiang province, which includes the Wangkui style. Western shadow plays are mainly found in (1) Shaanxi province, which includes the Wanwanqiang style of Huaxian (now Huazhou), the Laoqiang style of Huayin, the Daoqing style of the Northern Shaanxi region, the Xianbanqiang style and the E'gong style, (2) Gansu province, which includes the Longdong style, and (3) Qinghai province, which includes the Hehuang style. Shadow plays in central and southern China are mainly found in (1) Hubei province, which includes the Qingzhichuang style of Xiaoyi, the Weipu style of Zhuxi, the Menshenpu style of Jingzhou and the Huanggang style, (2) Sichuan province, which includes the Dengyingxi style of Chengdu and the Northern Sichuan style, (3) Guangdong province, which includes the Lufeng style and Chaozhou style, (4) Taiwan, which has its own Taiwanese style, (5) Yunnan province, which includes the Kunming style and the Tengchong style, (6) Fujian province, which includes the Longxi style, (7) Hunan province, which includes the Hengyang style, and (8) Guangxi's Zhuang autonomous region, which includes the Minnan style. Among them, the shadow play techniques from Beijing, Hejian, Xiuyan, Gaizhou, Wangkui, Taishan, Jinan, Dingtao, Luoshan, Hunan, Sichuan, and Hehuang were listed in *List of National Intangible Cultural Heritage* (as shown in Pictures 13.37, 13.38, 13.39 and 13.40).

Picture 13.37 A female puppet in Tangshan shadow play, Hebei province

Picture 13.38 The Monkey King in Shaanxi shadow play

Picture 13.39 Puppet called *Fanshuai* in Jinnan, Shanxi province

Picture 13.40 The head of a female puppet from Shaanxi province

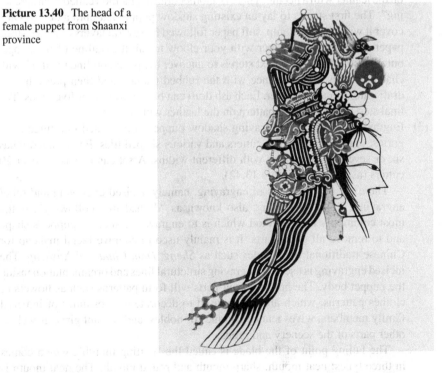

13.3.2 *Production and Performance of Shadow Play*

13.3.2.1 Creating Shadow Puppets

Shadow puppet production includes five processes: (1) leather processing, (2) drafting, (3) engraving, (4) coloring, and (5) assembly.

(1) Leather processing. The skin of the cattle under three years old is the most suitable for making shadow puppets. Cattle are slaughtered in the sixth and seventh months of the lunar calendar, skinned and their hide soaked in clear water. The water should be changed every day. After several days, the hide leather will be taken out of the water, stretched tightly over a wooden frame, and repeatedly scraped with a special scraper to get rid of fur, grease and other imperfections. The darker the cattle, the whiter the leather after processing. Chemicals are not used to make shadow puppets that are used in shadow plays; otherwise, they will become brittle and break easily. Some contemporary handicraft shadow puppets are processed with calcium oxide, sodium sulfide and other chemicals to make the surface smoother and clearer, but they are not firm enough to actually be used for performances (as shown in Picture 13.41).

(2) Drafting. Put the processed leather over a paper draft, and then create the pattern on the leather with a needle. There is another drafting method called "ash spreading". The first step is to lay an existing shadow puppet flat onto a desktop and cover it with slightly damp, soft paper followed by several layers of rough straw paper. Then press the paper with your elbow to rub the outline of the puppet onto the soft paper. The next step is to uncover the paper, outline the draft with charcoal strips in accordance with the rubbed pattern, and then press the ash draft onto the cattle leather. Each ash draft can be reused four or five times. The final step is to create the pattern on the leather with needles.

(3) Engraving. Knives for engraving shadow puppets are divided into three categories: oblique cutters, flat cutters and various-shaped files. Each category has six or seven specifications with different widths. A set can consist of over 20 cutters (as shown in Picture 13.42).

There are two methods of engraving, namely, incised engraving and relief engraving. Relief engraving, also known as "dashed line hollowing", is the most commonly used method which is to engrave the shadow puppet's shape and to remove all other parts. It is mainly used to engrave facial makeup for Chinese traditional opera roles such as *Sheng, Dan, Chou* and *Xusheng*. The incised engraving is used for engraving structural lines and ornamentation inside the puppet body. The hollowed-out parts will form patterns such as flowers or clothes patterns, which are mostly used to decorate the costumes of imperial family members, wives and concubines of nobles, and servant girls, as well as other parts of the scenery and props.

The falling point of the blade is called the "cutting mouth", which comes in three types: neat mouth, sharp mouth and round mouth. The neat mouth is square and straight and is mostly used for square and regular objects, such as

Picture 13.41 Cattle leather processing

Picture 13.42 Shadow puppet cutters

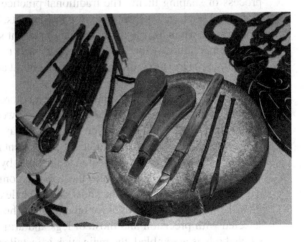

tables, chairs, houses, buildings, and cross-sections of flowers, grass and trees. The sharp mouth is a hollow line pointed at both ends and is mostly used to show the internal structure of hair, beards and flowing clouds displaying the feeling of extension. The round-mouthed hollowed-out lines are semicircular at both ends, which are mostly used at the ends of decorations such as flower clusters and streamers.

Shadow puppets are currently divided into two types: those used for traditional performances and those used solely for decoration. Those used for performances are often made of thick cattle leather, because they must be strong and durable requiring powerful engraving and rough shapes. Those used only for decoration are made of thin materials and treated with chemicals, normally with high transparency and exquisite engraving. However, such puppets are not very durable. Shadow puppet engraving techniques cannot be mastered overnight. Artisans must practice for a long time to achieve the exquisite shapes we know.

(4) Coloring. The engraved shadow puppets can be colored after being polished. In the past, people use a mix of shade and hide glue, but now transparent colors are used. Red, yellow, blue, green, black, and other positive colors are selected more often than mixed colors. Colors are usually painted without fading effects. The painting methods can be divided into line sketching and dyeing. The outline and internal structure of shadow puppets are formed after the engraving. When applying colors, the cutting mouths function as the boundaries to separate different colors. To avoid different colors from mixing with each other, painters must keep some distance between two adjacent colors. Most areas of the shadow puppets that are used for performances are usually painted with dark colors, so that the image is clear when projected. Decorative shadow puppets, on the other hand, require more attention to color matching and vignetting. After coloring, the shadow puppets are ironed flat. Ironing, also known as "firing", is the final process of shaping them. The traditional practice is to heat two special sun-dried mud bricks and use them to iron the puppets. It is vital to keep a moderate temperature when heating, which is an important skill in experienced artisans. If the temperature is too low, the puppets will not be flat enough, and if the temperature is too high, the puppets will be burnt and become brittle (as shown in Pictures 13.43 and 13.44).

(5) Assembly. The various parts that have been colored will then be assembled to become a complete puppet. The connection between the head and the body is called the "insertion knot", which is a double-layer connection with the head of the puppet inserted in the middle for convenient replacement. The arms, legs and feet of the puppet are connected to the body by so-called "fastening nails", which is done by punching holes through the connection part and tying them with hemp ropes or thread. The knots should be flexible and durable at the same time. This method is also used for other joints. The connection points should be selected with precision and those of legs and arms must be aligned. After the whole body is assembled, the main stick is installed on the neck and a movable stick is installed at each wrist for manipulating. In some shadow plays, sticks are also installed at the legs of the puppet. There is a saying that explains the important point of shadow puppet assembly: "Three points and one line produce lifelike shadow puppets. Any small deviation robs the puppets of their vigor."

Picture 13.43 Lu Hai, a shadow play artisan in Beijing, engraving puppets

Picture 13.44 Part of an engraved puppet's head

13.3.2.2 Performance

The stages for performing shadow plays are mostly temporary, as said by puppeteers in Shaanxi province: "Seventeen rafters, eight boards, four ropes and two bowls." The rafters refer to the Chinese fir sticks or bamboo sticks that are used to set up the stage and the boards refer to the tables behind the screen and the fence posts under the screen. The ropes are used to tie the rafters together and the boards to build the stage. The bowls refer to oil lamps. Electric lamps are more often used nowadays, but gas lamps or oil lamps are still used in some underdeveloped areas.

Normally, a shadow play troupe consists of three to ten people. In northwestern shadow play, a performance normally consists of the following performers: *Qiansheng, qianshou, houcao, shangdang,* and *xiadang*. The *qiansheng* and *qianshou* are the two pivotal roles. They are, respectively, responsible for singing and manipulating the puppets.

The *qiansheng* undertakes the responsibilities of singing the major arias for all kinds of puppet roles such as *sheng, dan, jing,* and *chou,* and playing traditional Chinese instruments such as *yueqin* (moon guitar), *shouluo* (hand gong) and *tanggu* (drum). If necessary, they also need to assist the *qianshou* to manipulate the puppets.

The *qianshou,* also known as "under the lamp", manipulates the puppets' movements to perform the plot and also produces various sound effects. Puppeteers use sticks to manipulate shadow puppets with different techniques, including holding, turning, rolling and clamping. It is of vital importance to keep the puppets close to the screen with as little distance as possible, so that the shadow remains clear. There are some fixed patterns to manipulate the puppets to perform certain kinds of actions, such as the *diaomaxian* (hanging horse line), which are horse riding movements that include mounting and dismounting and require cooperation between the *qianshou* and *xiadang*. When performing "carrying water", special skills are needed to achieve the effect of having the shoulder pole and water buckets shake. The techniques used to perform martial arts are divided into two types: hard fighting and soft fighting. The former is a fight with weapons, while the latter is a bare-handed confrontation with punching and kicking. There is also single and multiple manipulations. Single manipulation is to manipulate a single puppet with one hand, while multiple manipulation is to manipulate multiple puppets with one hand, sometimes up to four puppets. The technique of "puppet substitution" can be divided into full substitution and half substitution. The former is to swiftly change a puppet for another one, while the latter means to only change the head, costume or props of a puppet. A good *qianshou* can quickly change their puppets with the help of light effects without being discovered by the audience. The *houcao* is also called the "backstage player", who is responsible for playing instruments including gongs, *wanwan, bangzi,* drums, and *yinluo*. The *shangxian* is responsible for playing instruments including *erxian, shanzi, suona,* and trombone and answering the *"chazihuo"*. The *xiadang* is responsible for playing *banhu* and dismantling puppets.

There is usually more than one artisan for each of the five roles. Sometimes a troupe needs two or more *qianshou* and as many as three *houcao*. It is of vital importance for the five performers to closely cooperate and coordinate with each other to give a successful performance (as shown in Picture 13.45).

13 Carving and Painting

Picture 13.45 Performance of a shadow play troupe in Huanxian, Gansu province

Chapter 14
Special Handicrafts and Others

Yang Yuan and Li Jinsong

14.1 Drilling Wood to Make Fire

The method of drilling wood to make fire is the first great invention of humankind.

Fire was originally a solely natural phenomenon, seen when volcanoes erupt, lightning strikes forests, and coal, oil, and natural gas spontaneously combust. In the beginning, humans, just like other animals, had a deep fear of wildfire. Only later did they realize that fire, despite of its danger, can be used for heating, cooking, lighting, and fending off wild animals. In the early days, human beings used wildfire. Later they gradually invented many different methods to make and preserve fire. In *Five Bookworms* in *Han Fei Zi*, it is written that Suiren "made fire by striking flint as he tried to expel the fishy and foul taste of raw meats". This is the earliest record of manual fire-making in China.

14.1.1 Drilling Wood to Make Fire

At present, the technique of drilling wood to make fire still exists among the ethnic minorities of Li, Gaoshan, Han, Yi, and Lahu in China.

According to ethnologists' research, the Li people's tools for drilling wood to make fire are made of brown kurrajong and consist of two parts. The first part is a rectangular wooden board with a length of about 70 cm, a width of about 5 cm, and a thickness of about 2 cm. One side of the board contains five small holes, each with a groove directly connected to the bottom. Fireweed is placed at the lower part of the

---

Y. Yuan
China National Museum of Women and Childen, Beijing, China

L. Jinsong (✉)
The Institute for the History of Natural Sciences, Chinese Academy of Sciences, Beijing, China
e-mail: lijs@ihns.ac.cn

© Elephant Press Co., Ltd 2022
H. Jueming et al. (eds.), *Chinese Handicrafts*,
https://doi.org/10.1007/978-981-19-5379-8_14

groove to catch fire when heat is introduced. The second part is an over 50-cm-long rod, which is sharp at the lower end so that it can be inserted into the small hole of the board.

When one person makes fire, they can step on the wooden board with both feet, insert the rod into the hole, and twist it between the palms of their hands while pushing down. The head of the rod will rub against the hole wall. Sawdust is produced during this process, which will ignite due to the generated heat. The ignited sawdust then falls onto the fireweed and ignites it. If two people make fire together, there are three methods: (1) one person holds the board and the other twists the rod; (2) one person steps on the board, and both of them twist the rod with one person holding the top end and the other holding the bottom end to keep the rod twisting. The second method is more efficient. (3) A third way is to wrap two ropes around the rod at different positions and have two people hold each end of the rope, pulling it left and right alternatively while pulling them down. This method is very efficient and serves as a transitional form to the bow drill. In 2006, the technique of drilling wood to make firewood of the Li people was listed in *List of the First Batch of National Intangible Cultural Heritage* (as shown in Picture 14.1).

The method of drilling wood to make fire is popular in many ancient civilizations. For example, the Kikuyu people in Kenya used long hardwood to drill holes in wood to generate heat. They then put it in the middle of a pile of dry bark and blew

Picture 14.1 The Li people's technique of drilling wood to make fire

on the embers with fans or mouths. It would take less than a minute for a skillful Kikuyu person to make fire. This technique was also popular in ancient India, Mexico, Tasmania, Southeast Asia, and with the Inuit. There are two methods of using a bow drill. The first is to bite onto wood with the mouth and then push the bow back and forth. The other is to utilize the elasticity of the ropes to rotate the wood at a faster speed.

14.1.2 Making Fire with Bamboo Tubes

The technique of making fire with bamboo tubes is an old method among the Dai, De'ang, Lisu, and Achang people in Luxi (now Mangshi, Dehong Dai and Jingpo autonomous prefecture, Yunnan province). It was still used in the 1950s and 1960s by the Jingpo people living in mountainous areas. The materials include bamboo, brown fuzz (fine fuzz that grows on palm tree), and bamboo sawdust (extremely fine bamboo sawdust scraped out on the surface of bamboo stem with a knife). The first step is to vertically cut a piece of bamboo with a length of about 80 cm and a diameter of 10 cm into two chips, put the mixed brown fuzz and bamboo sawdust into one bamboo chip and cover the other chip onto it to seal the mixture. Then, open a small gap at where the mixture is placed and align the middle of an 80-cm-long and 7-cm-wide blade-shaped bamboo chip with the small gap, so that the chip and the tube are perpendicular to each other. Holding the bamboo tube, the operator makes one end of the bamboo chip against their abdomen and the other end against the ground, while the cutting edge of the bamboo chip faces upward. Then the operator quickly pushes and pulls the bamboo tube to generate friction, which in turn generates heat. After about 3 min, smoke will begin to appear at the gap. Keep pulling until the bamboo chip is worn through and the mixture is ignited. The operator then quickly takes the bamboo chip cover off and blows on the fire to light the hay on fire (as shown in Pictures 14.2, 14.3, 14.4, 14.5, and 14.6).

14.1.3 Making Fire with Paibo

Paibo, in the Jingpo dialect, refers to fire-making tubes that can make the sound of "bo".

The fire-making tube with a length of 15–25 cm and a diameter of about 3 cm is generally made of bamboo, sometimes of ox horn.

The tube is chiseled with a hole of about 1 cm in diameter on one end. A bamboo pole with a thicker end and a thinner end is also needed. The diameter of the thicker end is roughly equivalent to that of the bamboo tube, and the thin part is about 12 cm long, just enough to be inserted into the tube hole. The top of the bamboo pole is concave to hold the mixture of brown fuzz and bamboo sawdust. When making fire, the pole is coated in beef tallow to act as lubricant. After the bamboo pole is loaded

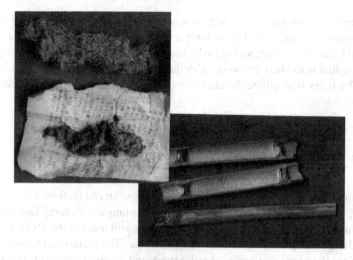

Picture 14.2 Brown fuzz, bamboo sawdust, and bamboo chips

Picture 14.3 Putting the mixed brown fuzz and bamboo sawdust into the bamboo chips

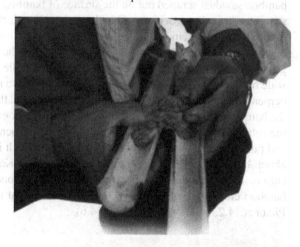

Picture 14.4 Pulling bamboo tubes on bamboo chips swiftly

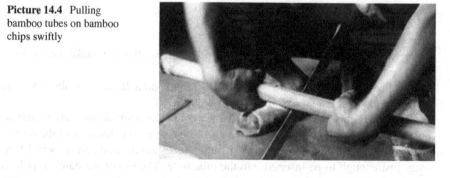

Picture 14.5 Blowing fire

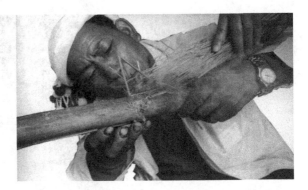

Picture 14.6 Making fire successfully

with the mixture, it is inserted into the bamboo tube until it reaches the bottom. The next step is to pull out the bamboo pole at a very high speed so that with a sound of a "bo", the mixture in the groove produces sparks due to the generated heat. The operator then quickly uses it to ignite the hay to make fire (as shown in Pictures 14.7, 14.8, 14.9, 14.10, and 14.11).

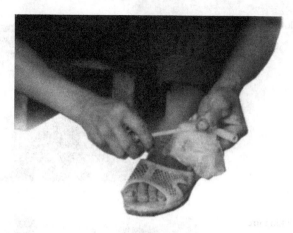

Picture 14.7 The fire-making tube

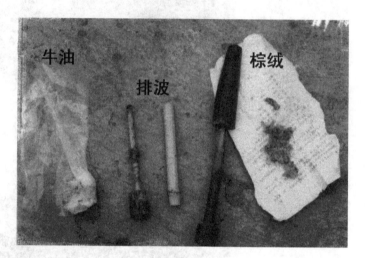

Picture 14.8 Fire-making tools

14.2 The Oroqen Ethnic Minority's Roe Deer Skin Clothing Manufacturing Techniques

This part is based on the materials provided by Yang Yuan.

Picture 14.9 Pulling the bamboo pole

Picture 14.10 Generating sparks

14.2.1 Roe Deer Skin Clothing

Animal skin is one of the earliest natural resources used by humans. As pottery is a product of the agricultural civilization, animal skin products are the symbols of the nomadic civilization.

The Oroqen people have lived a nomadic life in the middle and upper reaches of the Heilongjiang River since ancient times. The favorable ecological environment

Picture 14.11 The fire is successfully lit

gave birth to abundant species of animals and plants. Various wild animals such as roe deer, moose, deer, wild boars, and bears live in the vast forests, creating a natural hunting ground and serving as a source of food and clothing for the Oroqen people. To adapt to the cold climate and nomadic lifestyle, the Oroqen invented various kinds of unique and artistic processing techniques for roe deer skin.

The roe deer, also known as "dwarf deer" and "wild sheep" in China, is the most common wild animal populating the forests of the Lesser Khingan and Greater Khingan Mountain ranges. The color of its fur changes with every season, and it has white hair under its tail. Only male roe deer have horns. This species breeds quickly, likes to live in groups, and is timid and curious by nature. This makes roe deer easy to hunt. The roe deer skin is rough and solid with thick fur, yet very light, which makes it very durable and cold-resistant. The Oroqen are good at making various types of clothing and daily necessities with roe deer skin that suit different seasons. The roe deer's skin in summer, commonly known as "*honggangzi*", is maroon in color with short and thin fur, ideal for making summer clothing. Roe deer skin in autumn is bluish black with short fur, ideal for making spring and autumn clothing and gloves. Roe deer skin in winter is bluish white and light in weight with thick fur, fine hair, and hollow hair stems. Most importantly, it is cold-resistant and thus suitable for making cold-proof clothing. The skin of a young roe deer has brown-red fur and white longitudinal spots on its back, which is soft and beautiful, perfect for making satchels and other ornaments.

The nomadic life of the Oroqen mainly relies on roe deer. With roe deer skins, they make various kinds of products such as *piweizi*, which is used to cover their birch bark buildings, commonly known as *cuoluozi*. Other products include bedclothes, skateboards, bags, divine idols, and handicrafts. Roe deer skin is mainly used to make clothing, such as robes, jackets, trousers, boots, socks, gloves, waistcoats, and hats.

Take roe deer skin robes as an example. Men's robes are called *nilaisu'en* in the Oroqen language, which includes winter robes and summer robes. A winter robe for

14 Special Handicrafts and Others

men is sewn with seven or eight winter roe deer skins with thick fine hair. Such robes are knee-length with the front side of the Y-shaped collar covered from left to right. People can either wear it with the fur inward to keep warm or wear it with the fur outward as camouflage for hunting. Summer robes for men are known as *gulami*, which are hip-length. When wearing such a robe, people need to tie a belt around the waist with the fur outward. The robes are rain-proof and can sometimes serve as camouflage (as shown in Picture 14.12).

Women's robes are called *axisu'en*. Different from men's robes, women's robes are more decorative (as shown in Picture 14.13).

The waistcoats for children are made of skin of young roe deer, which comes with white spots, making the clothing cuter (as shown in Picture 14.14).

The production of *mitaha*, a kind of hat made of roe deer skin, is the epitome of the wisdom of the Oroqen people. It is not only pretty but also cold-resistant and can

Picture 14.12 Men's robe

Picture 14.13 Women' robe

be used for camouflage. It is easier for hunters to approach and hunt animals when they wear these hats.

Wula, also known as *qihami*, are a kind of short boots made of roe deer skin. These boots are so light that people who wear it can walk more silently, making it easier to approach wild animals without being discovered. *Wula* boots, both soft and durable, are also the most commonly worn shoes among the Oroqen (as shown in Picture 14.15).

The Oroqen believe in Shamanism. Shamans are the messengers between people and gods. They generally participate in production activities without receiving remuneration yet are highly respected by hunters and enjoy much prestige. Shaman clothing is usually made of roe deer skin or deer skin with various kinds of decorations, such as iron ornaments in the pattern of the sun, moon, and birds, bronze mirrors, and tassels (as shown in Picture 14.16).

Picture 14.14 Children's robe

Picture 14.15 *Wula* made of roe deer skin

Picture 14.16 Shaman clothing made of roe deer skin

14.2.2 Processing

Oroqen women are experts in processing skin and fur products. The processing consists of several steps, such as skinning, drying, fermentation, scraping, tanning, and dyeing, with each step requiring corresponding tools.

14.2.2.1 Tools

The Oroqen use various tools for tanning. A hunting knife (as shown in Picture 14.17) is used for peeling and scraping tendons and fat. A mallet (as shown in Picture 14.18) is used to smash air-dried roe deer skin. A wooden fodder chopper with serrated upper blade is used to soften the roe deer skin. A leather comb, which is a skin scraping tool with a serrated blade, is used to scrape off fat and tendons to further soften the roe deer skin. Another tool is what is known as a *hedele,* which consists of an arc-shaped wooden handle and an iron blade. This is used to repeatedly tan the roe deer skin after it is scraped, to make it soft and white like cotton cloth.

Picture 14.17 Hunting knife

Picture 14.18 Mallet

14.2.2.2 Techniques

The steps of processing roe deer skin include drying, beating, fermentation, scraping, tanning, and smoking. Generally, the skin is dried to about 80%, beaten with a mallet, and then scraped with a serrated scraper to remove the fat and tendons. These steps are used for primary tanning (as shown in Picture 14.19).

Fermentation. The roe deer's liver is cooked and crushed and then evenly applied on the skin. The skin is then rolled up to let it ferment. Rotten oak bark can also be used instead of roe deer liver, because rotten wood contains many fungi that can help ferment the skin.

Tanning. After the roe deer skin is fermented, it is necessary to scrape off tendons and fat with a scraper to make the skin soft (as shown in Picture 14.20).

Scraping. The residue is scraped off with a scraping knife and then the skin is tanned repeatedly to make it as white as cotton cloth (as shown in Picture 14.21). Finally, the skin is stretched and tanned while being rotated. This way, the skin becomes softer and can be restored to its original size.

Smoking. After the skin is softened, it should be smoked with fire so that it can be shaped and protected from insects (as shown in Picture 14.22).

Picture 14.19 Primary tanning

Picture 14.20 Tanning

Picture 14.21 Scraping

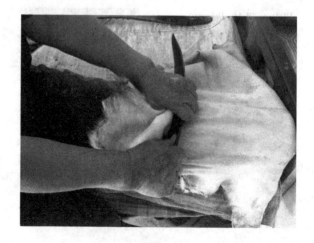

Picture 14.22 Smoking

14.2.3 Sewing

The Oroqen don't use rulers or chalk when making roe deer skin products. Experienced crafters would gently rub and press roe deer skins so that natural creases are formed, which serve as cutting lines. When sewing, special attention should be paid to the direction of the fur. The direction of the fur should be pointing downward.

They also mainly use tendon thread, which is made from the tendons on the back tenderloin of the roe deer, to sew skin products. The roe deer tendons are removed, air-dried, and smashed repeatedly with a mallet. After they've become fine fibers, they are rubbed into threads. Skin products sewn with tendon threads are durable and will not break even when the skin is worn out (as shown in Picture 14.23).

14.2.4 Decoration

When it comes to roe deer skin, the Oroqen's unique decorative skills are largely due to their familiarity with the fur's properties and their proficiency in using natural materials. The roe deer skin products display exquisite craftsmanship and beautiful patterns, which have evolved into an embodiment of their culture, showcasing the wisdom and unique aesthetics of the nomads.

Picture 14.23 Making roe deer tendon threads

14.2.4.1 Dyeing

The dyeing and decoration techniques of Oroqen skin clothing are unique. The robes of the elderly are mostly the original roe deer skin's color, while those of the younger generations are often decorated with yellow and black patterns.

The yellow dye is made from decayed oak bark. The bark is boiled in an iron pan, then *wuluo* grass is dipped in the boiled juice and brushed onto the roe deer skin, which dyes the skin yellow. Clothing that has been dyed will not fade and is durable and insect-proof.

The black dye is made by soaking the black materials formed at the bottom of the pot in hot water, adding a little salt, stirring and waiting for its precipitation. The water is then removed and the remaining black paste can serve as the dye. In addition, black dyes can also be made by boiling walnut bark in water.

To dye the skin black-grey, they burn the residue of rotten oak wood and smoke the tanned skin above the yellow smoke that is generated. This dyes the skin black-grey.

14.2.4.2 Embroidery

The Oroqen have two kinds of embroidery. The first is to embroider patterns on skins with colored silk threads or wool using various embroidery methods, such as the burden stitch, chain stitch, and blanket stitch. The other is to cut the dyed skin into various patterns, embroider them onto the skin products, and then embroider the edges of the patterns with silk thread as decoration (as shown in Picture 14.24).

Picture 14.24 Roe deer skin bag made with the *buhua* technique and blanket stitches

14.2.4.3 Skin Cutting

Skin cutting is a female Oroqen art. Originally, Oroqen women only cut skin, but later they began to cut paper. Most paper cuttings are used to decorate roe deer skin clothing.

14.2.4.4 Inlay

There are two forms of inlay techniques. The first is designed for pure skin, which combines two or three patches of skin with largely different colors (such as black and white, yellow and white, yellow and black, and black, white and yellow). Skin of two different colors is cut into the same pattern and inlaid in an interlaced manner. This is the most distinctive inlay technique for skin. The other technique is called *buhua*, where black skin is cut into various patterns and stuck on the leather. The skin is decorated by the patterns in the center and around the edges (as shown in Picture 14.25).

14.3 The Oroqen Ethnic Minority's Birchbark Products

This part is based on the materials provided by Yang Yuan.

Birch trees are widely distributed in areas between 40 and 70° northern latitudes. Some scholars call this narrow strip the "birchbark culture belt of the sub-Arctic Circle" (as shown in Picture 14.26). China's birchbark culture dates back to the Neolithic Age. Primitive ethnic minorities and nationalities such as the Sushen, Xianbei, Jurchen, Khitay, and Shiwei have all been part of this culture. In this narrow strip in northern China, bark culture has been well-preserved to this day.

Picture 14.25 Roe deer skin bag made with the inlay technique

Picture 14.26 Birch forest in autumn

The Oroqen people have been living a nomadic life in the dense forests of the Khingan Mountains for generations, wearing roe deer skin clothing and using birchbark products. Their birchbark products are beautiful, light, waterproof, and durable, which makes them suitable for the Oroqen hunter-gatherer lifestyle. As recorded in

Summary of Longsha in the Qing dynasty: "The Oroqen like birchbark. They use birchbark to make hats, shoes, utensils, tents, and boats". Due to the wide use and importance of birchbark products, birch trees are also revered as a kind of deity. There are a large number of legends and folk songs about birchbark, such as the *Legend of the Birch Forest*, *Origin of the Five Surnames* and *The Birch Forest*.

14.3.1 Birchbark Products

The Oroqen's birchbark products include barrels, boxes, bowls, chests, baskets, boats, houses, and toys.

Birchbark barrel. The wall and the bottom of the birchbark barrel are both sewn with birchbark. It has a particularly thick bottom. There are holes drilled through the mouth of the barrel, which is inlaid with a thin wood plate so that a stick can be put in and people can hold it easily. Lidded birch barrels are usually delicately made with the lids and walls engraved with patterns (as shown in Picture 14.27).

Birchbark basket. Oroqen women tie birchbark baskets around their waists when collecting wild fruit by hand in the mountain forests. Dried meat, wild fruits, and grain that are kept in such baskets can stay fresh for a long time (as shown in Picture 14.28).

Birchbark boat. Birchbark boats are shuttle-shaped and flat at the bottom with a length of 3–6 m. They can be extremely fast when sailing and are therefore also known as "fast horses". Smaller boats can hold one person, while large ones can hold three to four people. When sailing, the boats are swift and almost silent. The skeleton of the boat is made of pine strips, and the body is made of birchbark sewn

Picture 14.27 Birchbark barrel

Picture 14.28 Birchbark basket

with horsehair thread. Then pieces of birch wood are nailed onto the boat with wooden nails. Finally, all the gaps and nail holes are filled with rosin and evenly scalded with a soldering iron. A birchbark boat can hold 150 kg of cargo, but the boat itself is so light that it can be carried by one single person. When not in use, they should be placed underwater to prevent cracking from exposure to the sun (as shown in Picture 14.29).

Birchbark house. A birchbark house is called "*xianrenzhu*" in the Oroqen language, which means "a house blocking the sun", which is also known as *cuoluozi*. Such houses, made of 30 to 40 tree trunks, are conical with curtains made of reeds. The Oroqen cover the houses with birchbark in warm seasons and roe deer skin in cold seasons to protect the house from the weather and wild animals. These houses are also easy to build and move. The houses are generally built on flat ground near forests, rivers, and pastures with abundant sunshine (as shown in Picture 14.30).

14.3.2 Tools for Birchbark Processing

The Oroqen invent a long sickle-shaped knife specifically made for stripping birchbark. This knife is about 30 cm long with a curved blade. A hunting knife is used for peeling off the bark.

There are two ways to soften birchbark. One is to boil or steam the birchbark in water and the other is to let it soak in water.

Tools such as scissors, needles, thimbles, and tendon thread or twine are used for sewing birchbark. Twine is very strong, especially when it is made by combining hemp and horsetail. When sewing, the Oroqen will dip the twine in oil extracted from an oriental fungus (ganoderma lucidum) to make it smooth and easy to use.

The traditional tool for carving birchbark is the bone awl, which is made of the leg bones of deer or roe deer, the lower leg bones of wild boars, or made by grinding down roe deer horns. A bone awl may have two, three, or four cutter teeth. The

Picture 14.29 Birchbark boat

two-toothed bone awls are used to carve the main patterns, and the three-toothed and four-toothed ones are used to carve the patterns on the side (as shown in Pictures 14.31 and 14.32).

14.3.3 Production Techniques

Every year in the sixth month of the lunar calendar, the Khingan Mountain range is filled with life. During this period, the birchbark is separated from the trunk and thus easily peeled off. Apart from that, the birchbark at that time is also tough, waterproof, and easy to shape. After this "golden season", birchbark will become hard and stick to the trunk, making it difficult to peel.

The work of peeling birchbark is generally undertaken by women. After being peeled, the birchbark is rolled into a tube and placed outdoors to dry naturally.

Picture 14.30 Birchbark house called *cuoluozi*

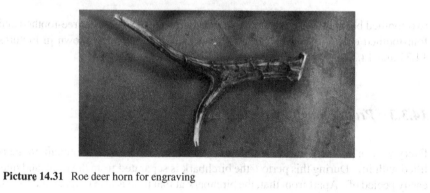

Picture 14.31 Roe deer horn for engraving

Picture 14.32 Engraving with roe deer horn

Picture 14.33 Peeling

Dried birchbark, with the rough skin on the surface removed, will be boiled until it is softened, then stacked until it is flattened, and finally cut as required.

14.3.3.1 Production Process

The production process of birchbark products includes (1) soaking, (2) trimming, (3) cutting, (4) sewing, and (5) engraving. Taking the birchbark box as an example, its general production processes are shown in Pictures 14.33, 14.34, 14.35, 14.36, 14.37, 14.38, 14.39, 14.40, 14.41 and 14.42.

14.3.3.2 Decoration

The production technique of birchbark is passed down from generation to generation among the Oroqen women. The decoration style can be divided into the Heihe style and Tuohe style. Heihe-style decoration is more sophisticated, with delicately

Picture 14.34 Trimming

Picture 14.35 Sewing

Picture 14.36 Combining

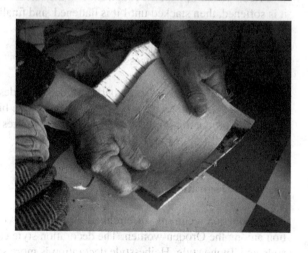

Picture 14.37 Shaping

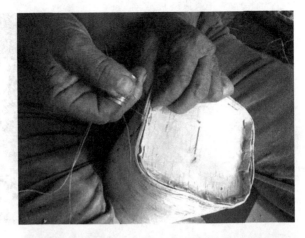

Picture 14.38 Cutting decorative borders

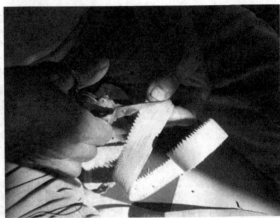

Picture 14.39 Sewing the edges

Picture 14.40 Pattern cutting (to decorate the lid of the box with patterns unique to the Oroqen)

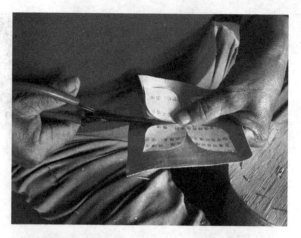

Picture 14.41 Trimming

Picture 14.42 Finished box

engraved patterns and various colors, while Tuohe-style decoration is simpler and less sophisticated, with patterns mostly engraved with bone awls and birchbark color as the main color.

Pattern engraving. A variety of patterns are engraved with toothed tools made of bone. There are various engraving skills such as "line carving", "pattern carving", "stamping", "point piercing", and "point stamping". There are two methods to engrave. The first method is to hold the tool in the left hand with the tool's teeth on the birchbark and then tap the top of the tool repeatedly with a mallet held in the right hand so that an intact pattern will take shape. The second method is to use a bone awl to directly press the patterns into the birchbark by creating dots (as shown in Picture 14.43).

Pattern cutting. The Oroqen's birchbark pattern cutting is simple and vigorous in style. Teenage girls generally learn this art from their mothers (as shown in Picture 14.44).

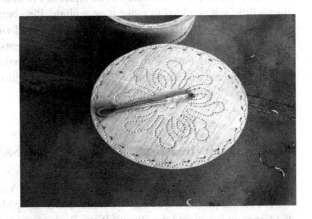

Picture 14.43 Birchbark pattern engraving

Picture 14.44 Birchbark pattern cutting—the cloud pattern

Decorative patterns. There are two ways to make decorative patterns. One way is to cut the patterns first, then stick them on the birchbark products. When using this method, the Oroqen women might cut out the patterns without using a draft, because they actually already have a picture of the desired patterns in their mind. However, sometimes they also stick traditional patterns on the birchbark and then cut those out accordingly. The other way is to use a bone awl to press decorative patterns directly onto the birchbark.

Colors. Decorative colors are mainly applied in Heihe-style decoration. This style is mostly found in Guli township, the Oroqen of Inner Mongolia autonomous region and in the Heihe area of Heilongjiang province. These birchbark products should be dyed after they are settled. The main color is black, but it is also combined with colors such as red, green, and yellow. In the past, the pigments used were mainly extracted from wild plants, including their fruit, leaves, flowers, and bark. *Yageda*, commonly known as red bean, can be used as a kind of raw pigment material. To use it as pigment, the fruit needs to be squeezed to obtain its blue-purple juice and then a little salt is added to make it precipitate. Pine bark can also be used as a raw pigment material. After being boiled and precipitated with salt, it will become dark red. Each color has its specific meaning. Red represents life and is adored by girls. Yellow represents masculinity and symbolizes the male character. White and blue are not frequently used, because they are used by widows in mourning. Black is a supporting color that is combined with other colors.

14.4 The Hezhe Ethnic Minority's Fish Skin Clothing

This part is based on the materials provided by Yang Yuan.

The Hezhe people have lived by fishing and hunting along the Heilongjiang River, the Songhua River, and the Wusuli River Basins for a long time. "Hezhe" is a name invented by local people, which stands for "native", "oriental", and "the lower reaches of the river".

The place where the Hezhe people have lived and thrived since ancient times have always been perfect places for fishing and hunting. This place was described as "a place where you can catch roe deer with sticks, scoop up fish with gourd ladles, and where pheasants fly into your bowl". These unique natural resources are directly linked to the unique production and lifestyle of the Hezhe people. Most of them live along the banks of the rivers and fishing plays a vital role in their lives. Almost every Hezhe family has boats and fishing nets. They have excellent fishing skills. They are even able to judge a fish's species by observing the water ripples. When spearing fish, the Hezhe people always hit their target precisely. More importantly, they primarily spear the fins, so that the fish's skin is kept intact. The Hezhe people have also been historically called the "fish skin tribe" since it is the only ethnic group in China that uses fish skin to create clothing.

They have a long tradition of wearing fish skin clothing. Fish spears and bone needles were unearthed at the Xinkailiu site in Mishan city, Heilongjiang province, which indicates that the Hezhe's ancestors might have started fishing over 6,000 years ago. Bone needles were probably used to sew fish skin clothing. It is recorded in *Classic of Mountains and Seas: East* in *Classic of Mountains and Seas*: "The country of Xuan'gu is located in the north of the country of Heichi. People there wear clothing made of fish skin." It is described in *Annals of Huachuan County* in the Qing dynasty: "Men wear hats made of birchbark, and in winter, they wear mink hats and fox-fur robes. Women wear hats which look like ancient helmets. Their clothing is mostly made of fish skin and decorated with colored cloth and copper bells, which makes it look similar to armor". These historical materials vividly describe and support the Hezhe people's unique custom of wearing fish skin clothing.

By the end of the Qing dynasty, the production technique of fish skin clothing had been an important criterion to measure whether a Hezhe woman was ingenious. When choosing spouses, men not only attached importance to the cooking skill of a woman but they were also concerned about how proficiently the woman could deal with fish skin and sew fish skin clothing. Traditionally, when a girl got married, she had to carefully sew a fish skin coat as her dowry. Fish skin clothing is the epitome of the unique fishing and hunting customs of the Hezhe people.

14.4.1 Fish Skin Clothing

14.4.1.1 Raw Materials

There are three kinds of fish skin used for fish skin clothing.

(1) Scaleless fish skin Scaleless fish skin comes from sturgeon, Soldatov's catfish, and catfish. The scaleless fish skin is thick, which can be used to make leggings, legwraps, uppers (of shoes), and pockets.
(2) Small-scale fish skin small-scale fish skin comes from dog salmon, hucho taimen, sharp-snouted lenok, yellowcheek carp, pike fish, salmon, silver carp, and white fish, which can be used for clothes, trousers, *wula* boots, satchels, and gloves. Silver carp skin and dog fish skin can also be used to make thread for sewing clothes.
(3) Big-scale fish skin Big-scale fish skin comes from carp, grass carp, and black carp. This kind of fish skin is thin and soft, which makes it suitable for coats, trousers, hats, capes, waistbands, and tobacco pouches.

When coated with fish oil, this kind of fish skin can also be used to paste windows. Tanned fish skin will harden when soaked in water, but it will soften again when it is gently rubbed.

14.4.1.2 Fish Skin Clothing

Fish skin is mainly used to make summer clothing such as coats, trousers, leggings, legwarmers, gloves, hats, *wula* boots, bags, and pockets.

There are two kinds of fish skin coats: long-style and short-style. The patterns on the coat are mainly in the shape of clouds, made with exquisite embroidery along the edges of the patterns. The buttons are made of fish bones (as shown in Picture 14.45) and are also used as decoration.

Fish skin trousers are sewn with the skins of Soldatov's catfish, hucho taimen, and pike fish.

Fish skin *wula* boots are also sewn with the skins of Soldatov's catfish, hucho taimen, and pike fish. These boots are light and warm and make no sound on the snow in winter. Some *wula* boots (as shown in Picture 14.46) are made with untanned sturgeon skin. People soak the boots in water to soften them before wearing when catching fish in summer, and then hang them up to dry once the fishing is finished.

There are also fish skin gloves, hats, bags, and sacks.

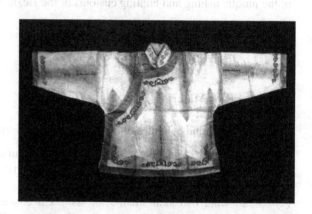

Picture 14.45 Fish skin coat

Picture 14.46 Fish skin *wula* boots

Picture 14.47 Wooden hay cutter

14.4.2 Processing

Fish skin is mainly tanned by women. From the first spring thaw, when men begin to catch fish in the river, women start their work of processing fish skin. They need to finish the processing of all the fish skins needed for the coming year before the river freezes at the end of autumn.

Fish skin processing tools include a skinning knife, wooden saw, wooden hay cutter (as shown in Picture 14.47), mallet, wooden anvil, scraper, and shovel, while fish glue serves as the main auxiliary material.

Sturgeon bladders are steamed into glue. After the glue has cooled down and solidified, it is cut into strips for later use. It is mainly used as an adhesive to stick the decorative patterns onto the fish skin. People stick the pattern onto the fish skin before sewing it into clothes.

Skinning fish requires meticulous work. First the scales are removed before tanning. Then the excess meat tendons are removed with a shovel or scraper and hung on the cob wall to let it air dry. Before being tanned, the fish skin is spread out and sprinkled with maize flour or millet flour to absorb the skin's natural oil.

Then seven or eight pieces of fish skin are wrapped together, tied tightly with ropes, and put in the wooden hay cutter. The tanning process (as shown in Picture 14.48) requires two people working together. One person sits in front of the wooden cutter and constantly turns over the fish skin, while the other one repeatedly lifts and presses down on the wooden cutter. After about 1 day's work, the fibers of the fish skin will have become fluffy and soft. The next step is to scrape the fish skin until it is clean and neat.

14.4.3 Fish Skin Clothing Production Technique

Fish skin clothing is sewn together with fish skin thread, which is mainly made of the skin of silver carp and pike fish. Sewing techniques include filling holes, creating patterns by combining pieces, cutting, sewing, pattern cutting, and decorating. The technique of creating patterns by combining pieces is the most distinctive.

Picture 14.48 Softening fish skin

When the fish is skinned, there will be a hole at the dorsal fin, which needs to be repaired before the fish skin can be put to use. Then any uneven fish skin is cut off and the remaining smooth fish skin is sewn together.

Due to the limited size of fish skin, it is necessary to piece multiple pieces together to make clothing. Combining pieces of fish skin is a critical step with numerous factors to take into consideration. For example, whether the used pattern is beautiful, whether the pieces of skin are smooth, and also what the desired shape and size of the clothes will be. The fish skin should be trimmed into the shape of "△" with smooth edges to facilitate patterned pieces to be combined. First of all, the fish skin is sewn together piece by piece to make the body piece of the coat and then the junctions are sewn once again, but more tightly with fine stitches.

After combining the body pieces of the clothes, the collar is cut out and then the front is cut into two pieces. After comparing the sleeves with the body piece, they are trimmed as needed and sewn to the body, thus finishing the fish skin coat (as shown in Picture 14.49).

Picture 14.49 Creating patterns by combining pieces (The pieced-together fish skin is like a carefully designed and combined work of art, hence the name)

Traditionally, fish skin clothing should be decorated with patterns, which are cut from the skins of pike fish or Soldatov's catfish and attached with fish glue.

Various decorative patterns often appear on the chest, back, collar, cuffs, placket, hem, and corners of the trousers. The cutting patterns have been passed down from generation to generation. When decorating the clothing, the pattern is first attached with fish glue, and then firmly sewn on with fish skin thread.

14.5 Yunnan Ethnic Minorities' Fireweed Cloth Production

This part is based on the materials provided by Yang Yuan.

Some ethnic minorities in Yunnan province still use fireweed to weave cloth and make clothing. This custom is very rare, not only in China but also in the rest of the world. According to research, the people of the Yi, Zhuang, Lisu, and Dai nationalities still produce fireweed cloth.

Xie Zhaozhe, a native of Fujian, served as an advisory officer in Yunnan province from the Ming dynasty's Wanli period to Tianqi period (1573–1627). After collecting statistics for over 10 years, he finished writing the 10 volumes of *A Brief Biography of Yunnan*, which contains the shape of fireweed, the weaving technique, and even the sizes of fireweed cloth in detail. The eighth volume of this book contains *An Unofficial History of Nanzhao* written by Ni Lu from Kunming city, while the second half of this volume describes the Yi people's fireweed clothing. "Luowu men drape themselves in felt and wear fireweed clothing. They have long hair and wear a sword at the waist. Luowu women wear fireweed dresses with shoulder-length braids decorated with seashells and clams. Luowu people do not use bedding. They lie on pine needles and sleep on the ground." The Luowu, an ancient branch of the Yi people, mainly live in Wuding county near Kunming city.

Another branch, known as the "white-clothed people", is living in Heqing county, Dali Bai autonomous prefecture, Yunnan province. They believe that wearing fireweed clothing is an expression of ancestor worship. This makes fireweed clothing an important bond to unite the local people. Legend has it that when the Lotus Mother, the ancestor of the "white-clothed people", was 900 years old, she encountered a huge mountain fire. When it erupted, a swarm of bees carried the Lotus Mother and other survivors to safety. The Lotus Mother was carried to the top of the Morning Glow Mountain, where she blew on a bamboo pipe to gather her people. Day after day, her lip was cut by the pipe and her blood wet the ground. Wherever the drops of blood fell, fireweed sprouted. Later, in order to commemorate the Lotus Mother, the "white-clothed people" designated the 15th day of the 3rd lunar month and the beginning of summer as the Mountain Festival. At that time, they would wear fireweed jackets, climb the mountains, and blow on bamboo pipes to call the Lotus Mother. They would also pick fireweed, spin and weave the white velvet on the back of fireweed leaves into cloth, and wear it to commemorate the Lotus Mother.

Fireweed cloth is actually a kind of blended fabric. People often use hemp or cotton as the warp and use fireweed as the weft. The resulting fabric is stronger, more durable, and more flexible. It is also comfortable, warm in winter and cool in summer.

14.5.1 Preparation of Raw Materials

Fireweed is a wild plant found in the vast area from the central to the western Yunnan. It is called fireweed because, long ago, it was used as kindling. The back of the fireweed's leaves has a thin layer of white velvet, which can be easily separated with your fingers (as shown in Picture 14.50).

The time of collecting fireweed is usually around the Torch Festival on the 24th of the 6th month of the lunar calendar. People only collect the leaves and do not hurt the roots so that the weed will continue to grow (as shown in Picture 14.51).

We will take a look at the Lisu people's fireweed yarn as an example.

Picture 14.50 Samples of fireweed

Picture 14.51 Lisu people collecting fireweed

The fireweed leaves must be washed with clear water to ensure that the velvet is pure white, and put in a basket covered with ferns or wormwood leaves for one night. After this treatment, it will be easier to peel off the velvet layers.

The Lisu will sit around the campfire while they peel off the velvet layer. More specifically, they take the main vein of the leaf as the boundary and first gently tear down the left part of the velvet layer from the leaf's tip with their index finger and thumb. At the same time, they spin it around their fingers to wrap the velvet into a yarn, and continue to spin it over their thigh with their palms. Then they tear down the right part of the velvet layer and splice it with the left part to form a fireweed yarn of about 20 cm long. After repeating this process several times, a large amount of fireweed yarn is obtained. The final step is to wind the fireweed yarn into a ball or wind it on a thread frame.

The preparation of hemp yarn and cotton yarn is generally the same as that in other areas and thus omitted here.

14.5.2 Weaving Fireweed Cloth

The loom used for weaving fireweed cloth has a narrow frame because the fireweed yarn breaks easily.

The warping of fireweed cloth is carried out on the frame. Odd-numbered yarns are lifted up while even-numbered yarns are pressed down to form the so-called "eyes" using the heddle. The yarns should also be threaded through the harness (as shown in Picture 14.52).

Picture 14.52 Lisu people threading the yarns into the heddle and harness

The weaving process of fireweed cloth includes loosening the warp, opening, weft threading, weft pressing, and cloth rolling (as shown in Pictures 14.53 and 14.54).

Due to the different numbers of warp yarns used and different traditional customs, each nationality produces different kinds of fireweed cloth. Generally speaking, the fireweed cloth has a narrow width. Therefore, in order to sew clothes, trousers, skirts, and bags, people must piece together multiple pieces of cloth. For example, a man's jacket's sleeves requires 4 pieces (2 for each sleeve), while a woman's dress requires 12 pieces.

The Yi people in Heqing county mainly use fireweed cloth to make clothes and belts. Some cloth is also used as offerings (as shown in Picture 14.55).

The Lisu people in Binchuan county mainly use fireweed cloth to make women's dresses (as shown in Pictures 14.56 and 14.57).

The Zhuang people in Wenshan city and the Dai people in Chuxiong city mainly use fireweed cloth to make women's dresses or quilts.

Picture 14.53 Lisu weaving fireweed cloth

14.6 Jinuo People's Bark Clothing

This part is based on the materials provided by Yang Yuan.

Bark cloth is composed of natural fibers without any textile processing, and it can only be used to make clothing after being beaten and turned into non-woven fabrics.

The expression of "making bark into cloth" is recorded in *Royal Tribute Painting of the Qing Dynasty*. It is also chronicled in *A Complete Chronicle of Yunnan* written by Ortai during the Yongzheng period (1722–1735) in the Qing dynasty that "the people in Lijiang cover themselves with leaves and live in rock caves instead of houses". These historical records demonstrate that it was common for more primitive ethnic groups in Yunnan to wear leaves and bark.

The custom of making clothes with bark also existed in other areas of the ancient world, such as the Pacific Rim, Southeast Asia, Madagascar, East Africa, even as far away as West Africa. Bark clothing was invented during the germination period of human clothing. The existing bark clothing is of great academic value and can be honored as the "living fossil" of clothing history.

Picture 14.54 Yi weaving fireweed cloth

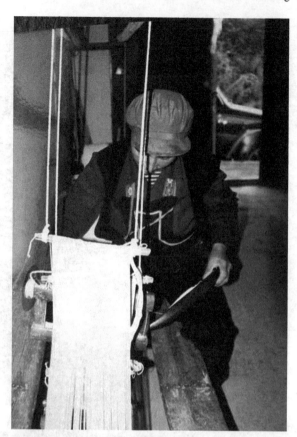

Picture 14.55 Yi fireweed cloth offerings

Picture 14.56 Lisu's fireweed cloth

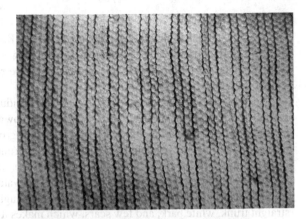

Picture 14.57 Lisu's fireweed dress

Until the beginning of the twentieth century, the ethnic groups who wore bark clothing in China included the Dai, Hani, Jinuo, and Kemu in Yunnan province and the Li in Hainan province. The production methods were basically the same.

In Yunnan, people used upas tree and paper mulberry to make bark clothing (as shown in Pictures 14.58 and 14.59).

The upas tree is a kind of Moraceae and is deciduous with a height between several meters and tens of meters. It is commonly known as the "throat-sealing tree" in China because its sap is highly toxic. Upas trees grow in large quantities in the tropical rain forest in Xishuangbanna's lowlands in Yunnan.

Paper mulberry is also a kind of Moraceae and is deciduous. It can grow to a height of over 10 m. This wild tree can also be cultivated. Paper mulberry is found in large quantities in subtropical areas such as Xishuangbanna. Paper mulberry has a straight trunk, white bark, and few scars, which makes it the most ideal raw material for bark cloth.

Ethnic groups in Yunnan generally cut trees to make bark cloth during the slack season and when there is less rain. The time to cut upas trees is from November to

Picture 14.58 Upas tree

Picture 14.59 Paper mulberry

March, during which the tree would not excrete too much toxic sap. The people who cut the upas trees must be free from wounds, because the tree sap would poison them by entering their blood stream.

The makers of bark clothing are usually old people over 60 years old who possess a high proficiency in the bark clothing production techniques.

The tools they use include an iron axe, iron machete (as shown in Picture 14.60) and mallet.

The whole process of making bark clothing (as shown in Picture 14.61) is as follows: (1) cutting down the trees; (2) peeling; (3) beating; (4) soaking and washing; (5) drying; (6) sewing.

Bark should be beaten before it is dried; otherwise, it will be difficult to soften the bark fibers. The whole bark is soaked in the river, the surface scraped, repeatedly rubbed by hand, and the gum rinsed off until the white fibers become visible. The fiber

Picture 14.60 Iron machetes

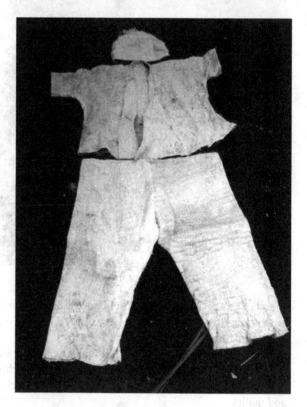

Picture 14.61 Bark clothing of the Jinuo people

without gum is soft and will not rot, which makes it convenient for making clothes. After it has dried, the bark is cut into dedicated shapes and sewn into clothing.

Bark cloth can also be used to make pillows, quilts, and cushions.

14.7 Bronze Restoration Techniques

This part is based on the materials provided by Yang Yuan.

Bronze restoration is closely related to imitation techniques and can be traced back to the Spring and Autumn period. It is recorded in *Essential Criteria of Antiques*, written by Cao Zhao in the Ming dynasty's Hongwu period (1368–1398), that there are two principles for bronze restoration. First, restored bronzes are not forgery; second, new bronzes pieced together with bronze remnants are forgery.

A Collection of Zhonghui's Work, written by Qu Ruwen in the Song dynasty, records that during the Zhenghe period (1111–1118), people would imitate ancient bronze sacrificial vessels such as *ding*, *dou*, *gui*, *xi*, and *zun*, and in the Ming dynasty when the retro style prevailed, some people imitated the bronze mirrors of the Han and Tang dynasties. During the Ming's Wanli period (1573–1620), there were two ancient bronze imitation styles: the northern style and the southern style. The Shi family represented the northern style, and Gan Wentang, a Jinling native who was famous for casting furnaces, and the Cai family from Suzhou represented the southern style. After the Qianlong period (1736–1796) in the Qing dynasty, the folk bronze imitation and reproduction industry could be divided into four styles, which were respectively located in Weifang, Xi'an, Suzhou, and Beijing.

The Weifang style was formed in the Qianlong and Jiaqing periods and took shape in the late Qing dynasty and the early Republic of China. The imitation and reproduction techniques of the Weifang style include: (1) engraving on genuine ancient bronzes or engraving fake inscriptions on genuine antiques; (2) combining the body of a genuine bronze with an imitated cover and adding same inscriptions on both pieces; (3) piecing genuine bronze fragments into a newly shaped bronze; and (4) etching inscriptions using chemicals and the recasting of small bronzes. After the 1950s, the Weifang style's techniques have kept developing till today.

The brothers, Su Yinian and Su Zhaonian, were the representatives of the Xi'an style. Their bronze imitation and reproduction technique was featured in engraving fake inscriptions on genuine ancient bronzes, and they mainly imitated weighing or measuring equipment, such as the bronze square *sheng* made by Shang Yang (Shang Yang Fangsheng), weights of Qin and bronze imperial edicts of Qin. The imitated bronzes were buried underground for years to make them rust.

Since the Ming and Qing dynasties, Suzhou has become a hub of for reproducing ancient bronzes. In the late Qing dynasty and the early Republic of China, representatives of the southern style included Zhou Meigu, Liu Junqing, Jiang Shengbao, Luo Qiyue, and Jin Mansheng.

The founder of the Beijing style was Yu, a eunuch from the Royal Workshop, who was dubbed one of the "Eight Eccentrics" among artisans in the palace. He left the palace at the end of the nineteenth century and since then lived at the Hutong Temple in Qianfu, Qianmen Street, making a living by restoring ancient bronzes for the palace and antique dealers. Yu had seven disciples, and Zhang Tai'en, his youngest disciple, inherited Yu's career and established the Wanlonghe Bronze Bureau. Zhang Tai'en himself also had seven disciples. Yu and Zhang Tai'en, respectively, imparted the techniques to their disciples in the 1930s and 1940s. Both started to work in the museums in the early 1950s, pioneering the restoration of cultural relics. The imitation technique prevailed in the Song, Yuan and Ming dynasties. In the Qing dynasty, Yu and Zhang Tai'en pushed the technique to the restoration stage. Zhang Wenpu and Wang Deshan innovated the technique even further, and Gao Ying, Zhao Zhenmao, and Wang Rongda pushed the technique to its extreme. Finally, the unique Chinese bronze restoration technique was formed.

14.7.1 Restoration Technique

14.7.1.1 Materials and Reagents

Rust removal materials include acid or alkali agents that are either chemically synthesized or found in nature. Materials for filling losses include copper, bronze, and tin-lead alloy. Welding materials include scaling powder and tin solder. Reinforcement materials include epoxy resin, unsaturated polyester resin, cyanoacrylate, and trimethyl resin. Materials for engraving include gypsum, modeling composites, and copper. Molding materials include gelatin, latex, silastic, clay, gypsum, and modeling composites. There are also various coloring materials with different properties.

14.7.1.2 Tools and Equipment

Restoration tools are divided into basic tools and self-made tools. Basic tools are normal tools that are readily available on the market, such as a soldering iron, electric drill, hacksaw, and file. Self-made tools are made by the artisans themselves based on their own specific restoration requirements.

Equipment includes a balance, constant temperature drying oven, sandblasting machine, ultrasonic cleaning machine, and grinding machine.

14.7.1.3 Technique

The unearthed deformed bronzewares are made of copper-tin-lead alloy. They were deformed after being buried underground due to stratigraphic changes, tomb collapses, extrusion and impact. Reshaping means correcting these deformed bronzes. There are different reshaping methods, such as beating, molding, heating, and sawing, which should be selected according to the degree of corrosion and the thickness of the wares.

According to the shapes and defects of the bronze, different methods should be adopted for filling losses. Traditional methods include hammering and casting. Hammering is more difficult and time-consuming, while casting is easier and can be done with a higher accuracy.

Engraving techniques include gypsum engraving, copper engraving, and chemical corrosion.

Gypsum engraving can be divided into transplant engraving and shaping. The former refers to the technique of making a plaster model for the missing parts of the bronze, recasting the model, and then casting the missing accessories by lost-wax method. The accessories would be cast on the body of the bronze after the welding joints are filed. For shaping, artisans must find images of bronze with the same shape and similar patterns as the bronze to be repaired, calculate the ratio between the dimensions of the body and the cover, then make the plaster model and engrave the patterns on different parts. The next step is to cast the missing parts by lost-wax method after the plaster mold is recast and waxed.

Copper engraving poses a challenge for bronze restoration. Take the restoration of the foot ring of the Bronze Vase Decorated with Three Rams as an example.

The artisans have to choose a 2.5-mm copper plate, make it into a foot ring by hammering, and spot weld it onto the bronze.

Kui patterns are painted on the copper plate, which are then connected to the same patterns on the original bronze. The relief pattern has the color of ink, while the intaglio pattern is located between the ink lines.

The copper plate is removed with a soldering iron and fixed on the rosin rubber plate.

The *kui* patterns are engraved along the ink lines with a straight groove chisel. After that, a shovel chisel is used to remove the unwanted parts, except for the ink lines.

Then file the *kui* patterns with a spike iron or a stepping chisel and perfect them with a shovel or a round chisel, so that the patterns become neat and smooth.

Next, design the ground patterns according to the original cloud-and-thunder patterns on the bronze. The ancients didn't specify the number of cloud-and-thunder patterns, so when restoring the patterns, the artisans only need to fill them in the blank space according to the rule.

After engraving the patterns with a small-sized shovel chisel, the artisans should engrave the round patterns with a fisheye chisel at the last point where the second line of the cloud-and-thunder pattern turns inward. The bottom of the patterns should be leveled with a spike iron.

After the above steps, the restoration of the ring foot of the Bronze Vase Decorated with Three Rams is preliminarily completed. But the repaired part should be compared with the original bronze and if the engravings are not deep enough, the artisans have to repeat the above process. Generally, the patterns of the repaired bronzewares of the Shang dynasty need to be engraved seven to eight times, and every single time the bronze needs to be filed again (as shown in Pictures 14.62 and 14.63).

Chemical corrosion. The technique of chemical corrosion is also called "plate rotting", and the word "rotting" refers to corroding the intaglio pattern. This process is suitable for bronzes with a single layer of patterns. After the patterns are corroded, the bronze is wiped clean with solvent, the patterns are trimmed with a steel file and chisel, and groove welded.

Picture 14.62 Bronze Vase Decorated with Three Rams of the late Shang dynasty, before restoration

Picture 14.63 Bronze Vase Decorated with Three Rams of the late Shang dynasty, after restoration

Connection reinforcement. Due to the different times of production, burial environment, and casting technologies, the degree of damage and corrosion vary among bronzewares. Some unearthed bronzewares are well-preserved and have a golden metallic luster, some have yellow-purple profiles, which is indicative of mineralization, and some are red without any metallic luster. Better-preserved bronzewares can be welded simply with scaling powder or rosin. For mineralized bronzes, strong acid is needed for welding. Bronzes with a serious degree of corrosion mostly cannot be soldered and need to be bonded with resin, or even be reinforced first before being bonded together. For special parts, such as tripod legs, mouth edges, and bronze mirrors, a combination of machine processing and welding is required.

Antique finish. The technique of antique finish is a kind of surface processing technique, also called "making rust" or "adding fake rust". It means adding a level of rustiness on the bronzewares during restoration. The antique finish for ancient bronzewares is different from the coloring process of industrial products and metal handicrafts. It is a unique, difficult, and self-contained coloring method that requires a high level of mastery.

There are two general methods to accomplish an antique finish. The first method is permeating or coating the bronzes with a solution made of chemical reagents, which is often used for large pieces or bronze replica. The other method is coloring the welding seams or small pieces with pigments, paints, and binders.

The former method was first used in the Northern Song dynasty. There are related records in *Pure Records of the Cave Heaven* in the Northern Song dynasty, and *Essential Criteria of Antiques* and *A Broad Discussion on the Xuan Furnace* in the Ming dynasty. This method is the first procedure in the whole process of antique finish and can make new bronzewares appear old and dark, colloquially known as "biting old" or "biting black". If treated well, the bronze will have a rusty surface with a suitable color, natural layer and luster, and sufficient feeling of antiqueness, making its appearance differ only slightly from genuine bronzes. The antique finish technique that uses copper carbonate, copper sulfate, ammonium chloride, and vinegar first appeared in the Northern Song dynasty and is still used today.

14.7.2 Rust Removal Technique

The corrosion of bronzes may occur on or under the surface, and the corrosion under the surface is usually called powdery rust, also known as "bronze disease".

If the corrosion of bronzes presents an antique effect without changing the shape and the chemical structure is relatively stable, the corrosion layer can be retained. However, most unearthed bronzes have several layers of rust and if the corrosion keeps expanding, the bronze will rust through and break. Therefore, corrosion inhibition and sealing treatment are needed depending on the degree of corrosion (as shown in Picture 14.64).

Traditionally, plum juice, red fruit puree, ammonium carbonate, acetic acid, nitric acid, and formic acid are used to remove rust.

It is an alternative to using mechanical methods, such as using a hammer, carving knife, and chisel to remove the rust layers on the bronze surface. Other tools to remove rust include sandblasting machines, ultrasonic generators, and grinders.

There are also chemical methods. The reagents used include citric acid solution, ammonium hydroxide solution, basic glycerin, basic potassium sodium tartrate, sodium hexametaphosphate solution, potassium dichromate, and sulfuric acid. When the whole bronze needs to be processed, the artisans may soak the entire object in solution and may even heat the solution properly to accelerate the reaction, conditions permitting. Some bronzewares only need local rust removal. For this, an absorbent piece of cotton is soaked in solution, placed on the rusty part and then, after a certain time, is assisted with mechanical tools to remove the rust. After that, replace the absorbent cotton with new ones and repeat the process until the rust is gone.

Picture 14.64 The corrosion of bronze (The harmful rust is grey)

14.7.3 An Example: The Restoration of the Bronze Chariots

In December 1980, two painted bronze chariots were unearthed from the west side of the Mausoleum of Qin Shi Huang, the first emperor of Qin in Lintong, Shaanxi province, which had been shattered into about 2,000 pieces. The personnel responsible for restoration worked out the restoration plan after participating in the excavation and cleaning, reinforcement of the painting, numbering, and recording, and material analysis of the bronze chariots. They classified the pieces of the bronze chariots according to the size, texture, painting, deformation, location, and stress form, and then determined the restoration process. The restoration principle was to mainly focus on bonding and use welding only as a supplement. They adopted different methods to reinforce the connection between different pieces as necessary.

The steps included piecing, shaping, fracture surface cleaning, connecting, and antique finish (as shown in Pictures 14.65, 14.66, 14.67, and 14.68).

Picture 14.65 The original bronze chariot unearthed from the Mausoleum of Qin Shi Huang, the first emperor of Qin

Picture 14.66 The original damaged bronze chariot number one

Picture 14.67 The restored umbrella cover without antique finish

Picture 14.68 The restored bronze chariot number one unearthed from the Mausoleum of Emperor Shi Huang of the Qin dynasty

14.8 Ceramics Restoration Techniques

This part is based on the materials provided by Zhou Baozhong.

Pottery and porcelain are made of silicate with stable chemical properties, a high level of hardness, and fire resistance. However, due to environmental circumstances, mechanical and chemical damage will occur.

Pottery is porous. In the humid environment underground, soluble salts and other impurities can penetrate its interior, and contaminate or cover its surface with dirt. Due to the changes in the environment, the salts infiltrating the pottery will repeatedly dissolve and recrystallize, thus reducing the strength of the pottery and making it easily degrade into powder. This is why many pieces of pottery are fragmented when unearthed.

Judging from the relics of ancient ceramics, the pottery production technique appeared together with the skills to restore them. For example, a Neolithic-colored earthen pot in Gansu Provincial Museum had once been restored before it was unearthed. It has artificially drilled holes evenly arranged around the edge of the fragments. This shows that the pot continued to be used after the fragments were tied together.

Judging from the porcelain collections of museums and collectors, there is also a kind of restoration technique that repairs the missing parts with foreign materials, such as filling with red sandalwood or rosewood, and inlaying or pasting with gold and silver.

One popular folk technique was joining broken ceramic pieces with a cramp made of copper and iron. Another method was to bond pieces together with shellac and glutinous rice. These were early restoration techniques for ceramics. In the 1930s and 1940s, antique dealers hired artisans to repair ceramics with chemicals in order to make huge profits, and that was when the glaze imitation technique came into being. With this technique, the restored porcelain only shows the smallest differences with the genuine, original wares. At that time, Shanghai had groups of restoration masters who were excellently skilled in this.

In the ceramic restoration industry, artisans passed on the techniques to their sons and male apprentices, but not women unless there were special reasons. Some senior craftsmen refused to pass on the techniques even in the 1970s, and because of this, some skills have been lost. Several ceramic restoration techniques will be introduced through the examples below.

14.8.1 Restoration of Song Dynasty Porcelain with Crack Patterns

Very few porcelains from the prominent Five Great Kilns of Ru, Jun, Ding, Guan, and Ge wares of the Song dynasty were handed down from ancient times, and their prices will sharply once slightly damaged. Therefore, the damaged ones must be carefully repaired so that their artistic value can be maintained. Among them, Guan and Ge porcelain are the most difficult to repair, and the key problem is restoring the crack patterns.

The glaze of Guan wares and Ge wares is thick and crack easily due to the unevenly applied glaze, which caused different shrinkage during the firing process. The cracks differ in size and depth. Small and shallow cracks are golden in color and commonly known as "gold wire", while large and deep cracks are dark brown in color and commonly known as "iron wire". If a porcelain has both kinds of cracks, the cracks are dubbed "gold and iron wires".

When repairing, transparent epoxy resin is used to bond pieces together, so that the joints look just like exquisite crack patterns. Original cracks generally do not need to be repaired.

The following describes how to make crack patterns in missing parts. Porcelain powder and binder were used as filling materials. A flattened model is made with the color and bright luster of glaze and the crack patterns gently engraved with a miniature carving knife. After the patterns were engraved, the same color as the original cracks is mixed, the pigment filled into the engraved cracks, the pigment that spilled on the surface wiped away, and finally it was polished and glazed. Another method was to mix a color similar to the color of the cracks first and paint the crack patterns on the surface with a fine brush. Then, the lines were spray-glazed to recreate the crack effect (as shown in Pictures 14.69 and 14.70).

14.8.2 Restoration of the Blue-and-White Flat Pot Painted with Branches and Sea Waves

The process of restoring the Blue-and-White Flat Pot Painted with Branches and Sea Waves of the Yongle period (1403–1424) in the Ming dynasty, which is kept in the Shanghai Museum, is introduced in the following part. First, the same kind of wares was investigated and the missing parts were molded with clay according to the data. The gypsum model was then recast, reinforced, and polished after it dried, and then the patterns were painted. Then the porcelain pieces were fired using the same glaze formula, tools, kiln, fuel, with temperature as that of the original porcelain. This way, a batch of porcelain pieces with a similar style to the original ones were fired. From those, the most suitable pieces were selected, cut and gently polished with a micro-engraving machine. Then, the technique of antique finish was applied to eliminate the typical luster of new porcelain. Finally, they were pieced together to make the porcelain complete again (as shown in Pictures 14.71 and 14.72).

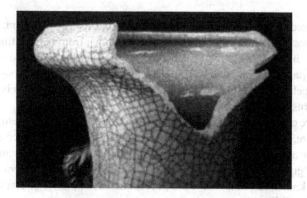

Picture 14.69 Ge porcelain replica with crack patterns of the Qing dynasty, before restoration

Picture 14.70 Ge porcelain replica with crack patterns of the Qing dynasty, after restoration

Picture 14.71
Blue-and-White Flat Pot Painted with Branches and Sea Waves from the Yongle period in the Ming dynasty, before restoration

14.8.3 Restoration of Ming Dynasty Three-Color Glaze

In 1988, a three-color duck-shaped aromatherapy pot with a unique shape was unearthed at the imperial kiln site of Zhusha in Jingdezhen city. It is the only ancient porcelain of the Chenghua period (1465–1487) in existence and is currently kept in Shanghai Museum. The glaze was cleaned before restoration. Artisans also tried to fit the pieces together before bonding them.

Picture 14.72
Blue-and-White Flat Pot Painted with Branches and Sea Waves from the Yongle period in the Ming dynasty, after restoration

After the aromatherapy pot was pieced together, it still had several missing parts in its neck, body, and bottom, which needed to be filled. Ancient porcelains should not be reburned in case they burst, so they needed to be cold-repaired. The gaps between the original and filled pieces were filled with clay. The clay was prepared with adhesive, filler, and mineral pigments; its color was exactly the same as that of the porcelain or glaze.

Nitro varnish, diluent, and porcelain-like pigment were used for mixing the color. When applying the color, the traditional methods such as pen painting, hand painting, and patting with cotton were used together with the machinery method of air brushing. The former was suitable for filling large parts, while the latter was used for slightly damaged parts. After color painting, the glaze and antique finish were still needed (as shown in Pictures 14.73 and 14.74).

Picture 14.73 Fragments of the Three-Color Duck-Shaped Aromatherapy Pot from the Chenghua period in the Ming dynasty, before restoration

Picture 14.74 The Three-Color Duck-Shaped Aromatherapy Pot from the Chenghua period of the Ming dynasty, after restoration

14.8.4 Restoration of Five-color Porcelain from the Ming Dynasty

The restoration of porcelain inscriptions is also difficult. The Five-color Pen Box with Dragon and Phoenix Patterns from the Wanli period (1573–1620) in the Ming dynasty had been repaired before being stored in Shanghai Museum. However, the glaze cracked and it turned yellow later. Therefore, it needed to be repaired again.

First, it was disassembled into different pieces, cleaned, and the small holes of the original cramps were filled with clay. Then the pieces were aligned and re-bonded. The gaps at the joints were filled with clay, then polished and painted with colors.

When restoring the bottom of blue-and-white porcelain, the background color was painted first. Then, it was spray-painted with a small spray pen within a range of 2–3 cm. When restoring the inscriptions, the artisans needed to paint the lines and strokes straight and smooth without any error and make sure that all the details, such as the shadings, were perfect. Finally, the bottom was glazed to achieve the underglaze effect (as shown in Pictures 14.75 and 14.76).

14.9 Painting and Calligraphy Mounting Restoration Techniques

This part is based on the materials provided by Zhou Baozhong.

Picture 14.75 Five-Color Pen Box with Dragon and Phoenix Patterns from the Wanli period of the Ming dynasty, before restoration

Picture 14.76 Five-Color Pen Box with Dragon and Phoenix Patterns from the Wanli period of the Ming dynasty, after restoration

14.9.1 Development of the Mounting Techniques

Painting and calligraphy mounting techniques have a long history. A large number of precious calligraphic works, paintings, and inscriptions have survived through vicissitudes precisely because they have been mounted. Mounted paintings and calligraphic works also have better artistic effect. The popular saying of "A good painting is made up of three-tenths painting and seven-tenths mounting" indicates the close relationship between painting and calligraphic works, and mounting. In short, a piece of Chinese calligraphic works or painting is considered incomplete if it is not properly mounted.

In 1973, the *Silk Painting Depicting a Man Riding a Dragon*, unearthed from the Chu tomb of the Warring States period, Hunan province, provides us with valuable materials for understanding the origin of the mounting art. This silk painting is wrapped with a thin bamboo strip and tied with a brown silk rope. The T-shaped silk painting unearthed from the Han tomb No.1 at the Mawangdui site in Changsha also has a bamboo stick at its top tied with a brown ribbon, and the middle edges and lower corners of the painting are decorated with a tubular silk braid woven with blue fine linen thread. These relics are of great value to the study of the origin of painting and calligraphy mounting.

Xu Ai, during Emperor Wu's reign, and Yu He and Chao Shangzhi, during Emperor Ming's reign, were all skilled at mounting in the Song dynasty of the Southern Dynasties period. Emperor Wu of Liang once ordered Zhu Yi and Xu Sengquan to mount and thoroughly protect the paintings and calligraphic works collected in the palace. After he acceded to the throne, Emperor Yang of Sui divided the paintings and calligraphic works kept in the Imperial Household Department into three classes: top, middle, and lower. It is recorded that "top works should be mounted with red glaze rollers, middle works with cyan glaze rollers, and lower works with lacquer rollers". Zhuge Ying and Jiang Zong were sent to manage and supervise the mounting process. Emperor Taizong of Tang also had Chu Suiliang and Wang Zhijing supervise the mounting progress. Zhang Longshu and Wang Xingzhen were skilled workers at mounting at that time. During the Tang dynasty's Kaiyuan period (713–741), *Complete Library in Four Sections* was decorated with rollers, ribbons, and labels of different textures and colors.

Before the Tang dynasty, most of the paintings and calligraphic works were horizontal, so scrolls were the main mounting form. Examples include *Nymph of the Luo River* painted by Gu Kaizhi in the Jin dynasty, *Spring Excursion* painted by Zhan Ziqian in the Sui dynasty, and *Five Oxen* painted by Han Huang in the Tang dynasty. Bamboo slips were the main media for writing documents in China before the introduction of paper, which influenced the invention of silk books and paper mulberry bark books. Books constitute a high percentage of the mounted objects in the beginning.

The first dedicated recording for the mounting techniques in Chinese history is an article in the *Notes of Past Famous Paintings* written by Zhang Yanyuan of the Tang dynasty. The article introduces the methods of making paste, cleaning, filling losses, and installing rollers with concise language, making the article a valuable material to this day. It also lists several scrolls inscribed with the names of the mounting artisans in the Sui and Tang dynasties. Inscribing the name of the mounting artisan on the paintings or calligraphic works is similar to the painters or calligraphers signing names on their own work. This process enhances their sense of responsibility and therefore should be advocated.

Scrolls were the main mounting form before the Tang dynasty. Scroll paintings and calligraphic works could only be laid flat for appreciation due to its length. Later, the form of screen painting appeared, which generally refers to paintings fixed on one or several connected wooden frames. It later became an important indoor decoration due to its practical and decorative value.

Screen paintings are hard to move, so the form of vertical scroll of paintings and calligraphic works appeared later. *Mount Kuanglu* by Jing Hao in the Five Dynasties period is an early vertical scroll painting. The two silk paintings, unearthed from the Liao tombs at Yemaotai Cemetery in Faku county, Hebei province, in 1974, were the earliest vertical scroll works of art found so far. Despite their simple mounting technique, the paintings have all the necessary mounting components such as a header mount, footer mount, and rollers.

The Song dynasty witnessed the heyday of Chinese painting history, and that was also when the mounting industry flourished. Mi Fu and his son, who had unique insights on mounting, created the horizontal scroll, the well-known Xuanhe mounting style, and the ingenious fish scale mounting style, pushing the mounting art to a new stage. The butterfly mounting style appeared later because people often painted on small-sized silks and round fans in the Song dynasty.

Calligraphy of the Shaoxing Reign in the Southern Song dynasty, written by Zhou Mi, introduces the application of the mounting technique in the imperial palace in detail, which indicates that the mounting of paintings and calligraphic works mainly took shape in the Song dynasty.

The Ming dynasty saw an important stage of the development of mounting art. Tang Han, Ling Yan, and Shu Sou were all famous mounting masters in that period. The elegant and unique Suzhou mounting style had a particularly great reputation because of its delicate workmanship and exquisite materials. In terms of mounting theory, Zhou Jiazhou introduces the whole process of mounting techniques with 42 sections in his *Records of Mounting*, making the book a valuable reference book.

Vertical scrolls were the main mounting form in the Ming dynasty. After mid-Ming, the practice of inscriptions was quite popular, and the frames of the painting or calligraphic works would also be preserved when re-mounted, forming a nested structure that included the original frame with a new frame. The mounting artisans, in order to conform to the trend, would inlay an extra piece of plain white paper above the painting called *Shitang* to leave marginal space for inscription. This tradition became a popular form by the late Ming dynasty and served as the origin of the couplet mounting style, which is popular today.

The development of the mounting techniques made steady progress on the basis of inheriting the existing achievements in the Qing dynasty. During the Kangxi and Qianlong period (1662–1796), the mounting industry developed in leaps and bounds, and gradually formed a simple and solemn style. The new style, the Beijing mounting style, took its place next to the Suzhou mounting style, to become one of the two major mounting styles.

The common styles of vertical scrolls in the Qing dynasty include: one color, two color, three color, *Xuanhe*, *Shitang*, half silk, brocade brow, and paper-inlaid silk. The paintings were always inlaid or dug when mounted. The scroll style includes edge bumps, edge turns, and edge covers. The styles of album decoration include push-awning, butterfly, folding, five-inlay, edge covers, and edge turns. There are various types of mounting components such as horizontal scrolls and couplets, large and small, long and short. *Records for Chinese Calligraphy and Painting Mounting*, written by Zhou Erxue, elaborates on the mounting technique and is as valuable as *Records of Mounting* (as shown in Pictures 14.77 and 14.78).

Picture 14.77 Pasting the fan in the Qing dynasty

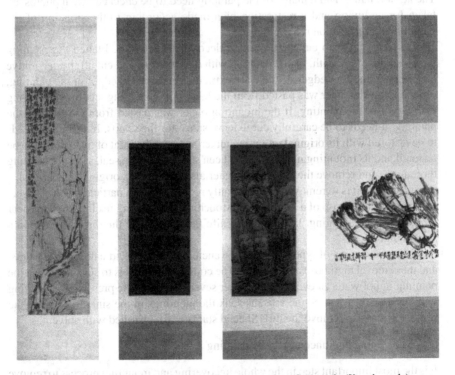

Picture 14.78 Left to right: one color, two colors, three colors, Song style (*Xuanhe* style)

14.9.2 Painting and Calligraphy Restoration Techniques

14.9.2.1 Ancient Painting and Calligraphy Restoration

Ancient paintings and calligraphic works are always dirty or damaged. Generally, the older the work, the more serious the damage. If a mounted painting is damaged, the original decorations, the backing paper, and the supporting paper should be removed and the painting need to be washed, cleaned, completed, and then remounted. This process of "replacing the old appearance with a new one" is called uncovering and mounting, and its purpose is to restore the original appearance of painting and calligraphic works.

Uncovering and mounting is a complicated process that requires meticulousness. Zhou Jiazhou of the Ming dynasty once made an in-depth exposition on the topic stating that "restoring ancient relics is like seeing a doctor", in which he proposed that "a good doctor could easily cure a patient, while a bad doctor would accidentally kill a patient". Mounting artisans are like doctors who diagnose and treat the disease of the paintings and are thus called the "doctors of paintings".

(1) Preparation

The author, name, and damage of the painting need to be checked. Then photos are taken for archiving and a restoration plan made according to the situation of the painting and restoration requirements.

The first step is to cut the unwanted decorative materials. Flatten the painting with a board underneath, fix the painting with ruler boards, and cut off the decorative materials on the four edges. If there are extra materials on the edges of the painting, or the mounting paper was pasted from the back of the painting, cut them off along the margin of the painting. If the mounting paper was pasted from the front of the painting, it needs to be carefully cut to leave space for the seams. If a painting needs to be restored with its original mounting materials, or if the joint of the painting to be restored and its mounting materials have been stamped with a seal, soak the painting thoroughly and remove the decorative materials to restore its original appearance.

The next step is to remove the dirt. Gently remove foreign particles such as insect excrement with the tip of a knife without touching the painting itself. For stains near the holes of the painting, lay a cutting knife flat and scrape the dirt away within a small range.

There are dozens of types of stains for ancient paintings and calligraphic works, and the removal methods vary for each. The common method is to soak and wash the painting in hot water or even boiling water several times while pressing the painting with grouped brushes. Some artisans soak the paintings in the sink and change the water regularly to remove the dirt. Special stains can be washed with solvent.

(2) Moistening and uncovering the painting

It is the most important step in the whole uncovering and mounting process to remove the backing paper and supporting paper, which have been pasted on the back of the painting in the mounting process.

The first step is to moisten the painting. Lay the painting face down on the platform and sprinkle clear water on it. Then use a brown brush to flatten it and make it tightly adhere to the platform. Before restoring damaged or crisp paintings, a layer of silk is laid on the mounting platform. After moistening, flattening, and drying, lay the painting face down, then brush and paste the mounting material on it. The silk or cotton painting, whether it is damaged or not, needs to be fixed with oil paper before restoration to ensure that the painting is straightened. This process is called "turning over oil paper". If it is torn, the painting first needs to be pieced back together, and then fixed with oil paper.

When uncovering the painting, the backing paper needs to be removed before the supporting paper. Uncovering the supporting paper is more difficult and a little risky, because it fits tightly with the painting layer. If the painting has been mounted with a great amount of paste and thus got rotten, it will be more difficult to restore.

(3) Filling the losses

It is recorded in *Records of Mounting*: "To fill the losses of a painting, it is a must to obtain the same paper and silk as the painting and calligraphic work itself. If the new material has a different color, it can still be dyed. But the thickness of silk and paper, if different, cannot be changed. Therefore, even if there is a goddess that can patch up the sky, she will first need to refine the five-color stones. The silk and paper to be used to fill the losses should be completely the same as the original material." This passage indicates that the key to filling the losses of a painting is the selection of materials.

(4) Supporting the painting

The painting layer of ancient paintings is different from that of new ones. It is recorded in *Records of Mounting*: "The restoration of paintings is finished after the painting is supported. The painting is removed of its rottenness and restored to its liveliness, as if cured by the famous physicians Hua Tuo and Bian Que. The process not only restores the appearance of the painting but also endows it with charm and elegance." In other words, restoring a painting involves the critical process of moistening, uncovering, and filling losses. Its function can be compared to the medical skills of Hua Tuo and Bian Que, two famous physicians in ancient China.

(5) Completing the painting

After the above-mentioned process, the incomplete parts of the painting must be drawn and colored to restore its integrity. Completing the painting refers to the process of completing both the lines and the color. Normally, the lines are completed first, then the color. But there are also practices of completing the color first. This process is a comprehensive inspection of all the steps before and also the epitome of the uncovering and mounting technique.

14.9.2.2 Restoration of Unearthed Paintings and Calligraphy

Most of the unearthed paintings and books were buried underground as funerary objects. Some of them have been soaked in water for a long time and thus rotted, some have been folded, pressed, bonded to lacquer wood, or connected with clay and lime into blocks, some have been eroded by dirty liquids, and some have been eaten by insects and rats. Therefore, it is necessary to take special measures to protect these precious cultural heritages.

(1) Cleaning and uncovering

Dry uncovering. This method can be used for books, paintings, and calligraphic works with relatively intact textures, which have not been soaked in water. For a book stacked together, first remove the external dust with a wool brush. Then, put it on mulberry bark paper, remove the binding line and paper spill, pick up one corner of the page with a needle cone, insert a thin bamboo opener, and slowly uncover the page. Every time a page is uncovered, unfold it and place it on a piece of flat and clean *xuan* paper. Stack the pages in order and press them firmly with wooden boards.

Wet uncovering. Paintings and calligraphic works that have been soaked in water cannot be uncovered immediately due to their rotten texture. They can only be processed when they are half dry.

Water uncovering. Immerse the gauze or plastic board with dense holes into the basin bottom, put the painting and calligraphic work into the water in the basin, and utilize the floating force of the water to divide pages. After a period of time, use the bamboo opener to gently divide the pages so that water can penetrate the inner layer quickly. A little alcohol can also be added to accelerate the progress. After the books that are stuck together are divided into several small stacks, take out the books one after another with the gauze or plastic board, put them flat on a foam plastic board padded with mulberry bark paper to dry, and uncover the pages after the paper has regained its toughness.

(2) Post-processing

After disinfection, reinforcement, cleaning, and uncovering, the unearthed paintings and calligraphic works still need to be processed, which includes the restoration of paintings and the binding of books (as shown in Pictures 14.79 and 14.80).

There are various special techniques all over China, and some of them are not included in this book. Queshan county in Henan province is rich in gold, silver, copper, iron, and other mines, and used to be famous for smelting and casting. There is a unique skill of striking "iron firework" in this area, which has been passing down for generations. The first step is to melt iron with a tilting cupola. Then make a stick with willow wood and dig out a pit with a diameter of about 3 cm at the front end of the stick. The craftsman, who wears a gourd ladle on his head, injects the molten

Picture 14.79 Liao dynasty scriptures, before restoration

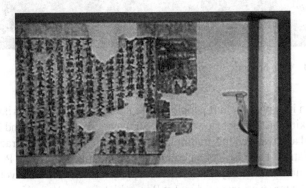

Picture 14.80 Liao dynasty scriptures, after restoration

iron into the pit and strikes the stick to make the molten iron splash into the air and become a dazzling "iron firework". On holidays, people build a two-story shed with a height of about 2 *zhang* (6.7 m) with willow branches and give performances of striking "iron firework". The lively view of the "iron firework" lighting up the sky is quite splendid (as shown in Pictures 14.81 and 14.82). As a grand event, it is as popular as dragon dance and setting off fireworks. There is a proverb that goes "Taoist priests and craftsmen are a family, hitting 'iron fireworks' together". It is said that the formula of molten iron is modulated by Taoist priests and is kept secret.

Picture 14.81 "Iron firework" all over the sky

Such molten iron will not hurt people. There is also a formula of "striking white but not red, striking fast but not slowly". In recent years, Yang Jianjun and Zhao Yongfu have studied and restored this traditional skill to keep it alive. Such practice deserves recommendation. Here is another example. After retiring in 2000, Professor Feng Leyun of Renmin University of China, out of personal interest, has devoted himself to improving the tenon art invented by Lu Ban, which is popular in Yangzhou, Jiangsu province. In the decade from 2000 to 2010, he created over 30 wooden sculptures in the shapes of animals, such as lions and bulls (as shown in Pictures 14.83 and 14.84), which were well received by sculptors such as Ren Shimin. Ren held the view that such projects could be included in the curriculum of primary and secondary schools to cultivate students' intelligence and manual dexterity. This technique has been listed in *List of the Intangible Cultural Heritage* in Beijing's Changping district.

14 Special Handicrafts and Others

Picture 14.82 The craftsman striking "iron firework" (The bottom left corner of the picture is the molten iron ladle)

Picture 14.83 The wooden sculpture called Bullfighting created by Professor Feng

Picture 14.84 Professor Feng disassembling the wooden bull

Chapter 15
Protection, Inheritance, and Revitalization of Traditional Crafts

Hua Jueming

In recent years, with the progress of China's modernization, the growth of national strength and people's awareness of cultural protection, especially after the State Council promulgated *List of the First Batch of National Intangible Cultural Heritage* in 2006 (among the 518 items in the List, there are 89 traditional handicrafts, accounting for more than 1/4 of the total), the necessity and urgency of rescuing and protecting traditional crafts have become a social consensus and received wide attention. This chap will be focused on the protection forms, inheritance mechanisms, and development and revitalization of traditional crafts.

15.1 Forms of Protection

China, well-known as a country rich in traditional crafts, possesses abundant handicraft resources, which should be protected in different forms according to their humanistic, scientific and technological value, mode of existence, and prospect of future development. Five forms should be adopted in the authors' opinion.

15.1.1 Autonomous Protection

According to the *Convention for the Safeguarding of the Intangible Cultural Heritage* made by the UNESCO in 2003, "safeguard", or protection, in a broad sense means "measures aimed at ensuring the viability of the intangible cultural heritage". As far as traditional handicrafts are concerned, no matter whether the mode of their

H. Jueming (✉)
The Institute for the History of Natural Sciences, Chinese Academy of Sciences, Beijing, China
e-mail: huajueming@163.com

© Elephant Press Co., Ltd 2022
H. Jueming et al. (eds.), *Chinese Handicrafts*,
https://doi.org/10.1007/978-981-19-5379-8_15

existence is individual, family, workshop, or enterprise, most of them rely on market development, and the maintenance and development of their vitality are closely related to their business conditions. Craftsmanship is created and inherited by artisans and the society they live in; therefore, artisans and their economic entities should be considered as the main body of traditional handicraft protection and the source of their inner vitality. Among the 89 handicrafts included in *List of the First Batch of National Intangible Cultural Heritage,* 49 can achieve the goal of protection and inheritance by their own strength, which we call autonomous protection. With the continuous growth of the national economy, the improvement of people's living standards, and the growing demand for handicrafts, as well as the gradual improvement of intangible cultural heritage protection, the vitality of various craft economic entities is expected to be maintained or even strengthened, and the proportion of autonomous protection may increase or maintain the current level, which will be a great boon of traditional handicrafts in China. Meanwhile, literature, material, and physical objects related to traditional handicrafts should be systematically collected, collated, and protected in the form of archives or databases by means of interviews, investigations, transcripts, audio, and video recording. Those are the basic works that must be done for all items included on the national, provincial, municipal, and county-level lists of traditional handicrafts (as shown in Picture 15.1). Some handicrafts, such as tinfoil paper making (formerly used as funeral offerings), are often regarded as outdated and related to superstitious activities. However, as a historical existence, it contains some religious culture and has certain scientific and technological value, so it is worth protecting.

Picture 15.1 *Records of customers' shoe sizes,* a booklet recording the making of Beijing nei lian sheng multi-layer sole cloth shoes

15.1.2 Protection for Cultural Memory

It is inevitable that the traditional crafts, especially the production technology, are replaced by modern technology. Some items, such as pig iron smelting and casting technology in Yangcheng county, Shanxi province, have played an extremely important role in history. The annual output of the plowshare, which was Yangcheng's famous brand product, was over 300,000 at its peak. They were sold to many countries including Nepal and North Korea. The manufacturing technology can still be used for reference, so it was listed in *List of the First Batch of National Intangible Cultural Heritage* (No. 385). However, such an ancient production technology has been replaced by modern iron-making technology in practical application. After the mid–1980s, with the popularization of machine tillage and new agricultural machinery, the demand for plowshares decreased greatly; moreover, charcoal is used as fuel and reducing agent in plow furnace to make the plowshare, which is harmful to the environment. The last plowshare-making workshop in Yangcheng county closed in 1994, and the pig iron smelting and casting technology that had been passed down for thousands of years dropped out with it. Apparently, it is unnecessary for such items to be protected in their original forms. The more appropriate way is to protect the relevant material and physical evidence for display and education in the form of cultural memory, as is what the Yangcheng government has been doing and achieved good results (as shown in Picture 15.2).

Picture 15.2 Exhibition of pig iron smelting and casting technology in Yangcheng county, Shanxi province

15.1.3 Government Protection

Traditional crafts are of strong vitality. Despite the impact of modernization and foreign technology, many traditional Chinese handicrafts are still operating well or have no danger of dropping out, and some even have great potential and broad prospects for expansion, such as Yixing purple clay (*zisha*) (No. 376), tie-dyeing of the Bai nationality (No. 376), Nanjing gold foil (No. 386), Suzhou Ming-style furniture (No. 395), Tiantai dry ramee-lacquer (No. 403), wine and vinegar brewing (No. 407–412), and *xuan* paper (No. 415). Those skills can be inherited autonomously at present; however, it is still necessary for the government to take protection measures as policies based on laws and regulations. For example, the government should ensure that the origin of raw materials is not polluted (as shown in Picture 15.3), resources are not destroyed or over-exploited, fake and shoddy products are cracked down upon, brands and trademarks are established, vicious competition and over-exploitation are prevented, and preferential measures such as tax reduction and exemption are given to enable the traditional handicrafts to grow normally and healthily in market economy. At the same time, it is also of vital importance for the inheritors to make necessary efforts to ensure the inheritance and sustainable development of the original skills.

15.1.4 Supportive Protection

For some traditional crafts, businesses must be shut down due to poor management or lack of successors. Take Song brocade (No. 364) as an example. As one

Picture 15.3 Long-stalk straw planting base in Jingxian county, Anhui province

of the three types of famous brocade in China, it is not only of important historical value but also has rich scientific and technological connotation and unique artistic style, with great economic potential. Such items are expected to be transformed into an autonomous inheritance with the financial support from the government, communities, and enterprises.

15.1.5 Maintenance Protection

Some traditional skills, such as drilling wood to make fire of the Li nationality (No. 437) and bark cloth (No. 434), are relics of ancient technology, and a historical witness and living fossil of the wisdom and creativity of ancestors and have important academic value in ethnology and folklore (as shown in Picture 15.4). It will be of great significance if they are revitalized and maintained with the financial support of the government, communities, and tourist enterprises.

We have estimated and analyzed the endangered status and protection forms of 89 traditional handicrafts included in *Lists of the National Intangible Cultural Heritage*, which are detailed in the following table.

Our analysis and estimation may be slightly different from the actual facts due to the incompleteness of information obtained, but they can still provide insight into the current situation of traditional crafts and the proportion of various forms of protection. It can also be known from the table that most of the traditional crafts will still be the living culture for a long time as long as the protection work is done properly. With the passage of time and the replacement of social economy and culture, mode of existence of many skills will change accordingly, and there will be ups and downs, so the protection forms should be adjusted accordingly. It is exactly this kind of dynamic protection pattern that puts high demands on the actual work. Only by being familiar with the grass-roots situation, diligent inspection, and timely change of the protection forms can the predetermined protection objectives be achieved.

Protection forms of 89 traditional Chinese handicrafts

Category	Number of items	Protection form				
		Autonomous protection	Protection for cultural memory	Government protection	Supportive protection	Maintenance protection
Making tools and machinery	8	5	2	6	1	
Agricultural and mineral processing	9	8	1	8		
Construction	6	1		4	3	2

(continued)

(continued)

Category	Number of items	Protection form				
		Autonomous protection	Protection for cultural memory	Government protection	Supportive protection	Maintenance protection
Weaving, dyeing, embroidery, and sewing	14	6	1	12	6	2
Ceramics	12	4	2	10	6	2
Metallurgy and metalworking	9	7	1	7		1
Lacquering	7	5		7	2	
Furniture making	1	1		1		
Calligrapher's tools	12	8		11	3	1
Printing	4	2		4	2	
Special handicrafts and others	7	2	2	3	1	2
Total	89	49	9	73	24	10
Percentage (%)		55	10	82	27	11

15.2 Inheritance Mechanisms of Traditional Crafts

Oral instruction and mentoring of masters to apprentices are the long-standing and extremely effective skill inheritance mechanism in China's traditional handicraft industry, and the usual mode is family inheritance and father-son succession. *Record on the Subject of Education* in *Classic of Rites* says, "The son of a good smelter is sure to learn how to make a *qiu*; that of a good bowyer to learn how to make a *ji*". The Chinese idiom "Ke Shao Ji Qiu" (meaning "follow in one's father's footsteps") derives from this sentence. *Ji* refers to the dorlach made of *ji* wood; *qiu* refers to *fengqiu*, i.e. the oxhide bellow made of animal skin. They are the necessities for the bowyer and the smelter. Only after he learns how to make the *ji* and the *qiu* from his father can the son inherit the family skills passed down from generation to generation and become a skillful craftsman who is familiar with the whole process of craftsmanship. As the old saying goes, "the son of a worker is always a worker", which is a major feature and advantage of the ancient Chinese handicraft industry. The son learns the craft from an early age and gets quite skillful by continual practice. That's one of

Picture 15.4 Cutting raw materials for bark cloth

the most important reasons why Chinese handicraft workmanship is wonderful and superb.

Admittedly, the practice of family and master-apprentice mentoring inheritance of skills has its inherent defects, such as excessive technical secrecy, sticking to the beaten track, and extinction of the skill because of the death of inheritors. However, there must be some merits about such a tradition since it can pass on from ancient times to modern times, showing tenacious vitality for 3,000 years. A case in point is Yang Wentong and Yang Fuxi, the ninth and tenth generations of Beijing Ju Yuan Hao Bow and Arrow Workshop. Yang Yi, the son of Yang Fuxi, dropped out of school to study the craft with his father after the death of his grandfather, Yang Wentong. After several years of study, he has made great progress and is expected to become a capable inheritor of the family's handicraft (as shown in Picture 15.5). Another example is Li Changhong, a famous purple clay master, and his wife, Shen Juhua; both are the apprentices of Gu Jingzhou, a great master of the craft. Li Xia, Li Qun, Chu Tingyuan, Li Ni, Li Ming, and Xu Ping, Li and Shen's children, and daughters–in–law have all studied ceramic art since they were teenagers. They also learn the skill from some famous masters such as Xu Xiurong and Tan Yaokun besides their own fathers. Their two ceramic art companies, Changhua and Mingchen, have produced many excellent works because of their family background and skill foundation, which are cherished by collectors. For another example, Beijing Arts and Crafts Association and Niulanshan Erguotou Winery recently invited a number of famous craftsmen to recruit apprentices and held a grand ceremony to honor the masters in order to ensure the inheritance of precious skills such as engraved lacquer, jade carving, ivory carving, cloisonné enamel, and *baijiu* brewing, which had a great impact in the industry (as shown in Picture 15.6). There are many similar examples, which show that this traditional way of inheritance can continue to play an important role for a long period of time if it can develop its strong points and avoid its weaknesses. It is in accordance with the requirement of the *Convention for the Safeguarding of the Intangible Cultural Heritage*, i.e. intangible cultural heritage can be effectively inherited through non-formal education.

At the same time, we should also see that since the early 1950s, with the transformation of social and economic system and industrialization, many traditional crafts

Picture 15.5 Family inheritance of Juy Yuan Hao Bow and Arrow Workshop, Yang Fuxi and his son Yang Yi

Picture 15.6 Apprenticeship ceremony of Niulanshan Winery

have experienced the transformation of ownership from individual and family workshops to collective and state-owned enterprises. In this case, especially in enterprises adopting a modern management system, the original inheritance mode of mentoring from masters to apprentices has been replaced by new modes, such as leader–member relationship. For example, positions of general technologist, senior technologist, technologist, workshop director and sect chief, etc. are set up in many large-scale wine-making enterprises including Maotai, Luzhou Laojiao, and Xinghuacun Fenjiu. They are masters of distilled liquor brewing skills, and the inheritors of traditional crafts as well. There is not necessarily a clear and strict relationship of mentoring between them, nor is there a traditional mater–apprentice relationship of "being a teacher for one day and being a father for a lifetime". Their learning experience is quite diverse. Some learnt the craft from their masters through practical work; some are taught and trained by teachers in training classes, technical schools, technical secondary schools, junior colleges, and colleges of higher learning. They belong to different generations because they learn the skill at different times, thus distinguishing the generations of skill inheritance. For example, the blending technicians of Wuliangye Group constitute the inheritance pedigree according to their founders and training classes held over the years (as shown in Picture 15.7).

To sum up, there are two ways of inheritance of traditional crafts, namely, master–apprentice and family inheritance, and social inheritance, both of which, with their own merits, will undoubtedly coexist and be a complementary to each other for a long time. This is of great benefit to the protection and sustainable development of traditional crafts in China.

It is worth noting that under the new historical conditions, the inheritance of traditional crafts has been undergoing gratifying changes. Take purple clay pottery

Picture 15.7 Master–apprentice inheritance of wine–making skills in Luzhou Laojiao Winery

as an example. The skill is prosperous and has a promising prospect, which is related to the industry's emphasis on the cultivation of new generations of artisans since the 1950s. Young people with good qualifications, high academic qualifications, and promising prospects are sent to colleges and universities for further study, which ensures the stamina of sustainable development of the industry. Li Changhong, who is mentioned earlier, studied at the Central Academy of Arts and Crafts in his early years. All his six children studied at the same school as their father or at Jiangsu Light Industry School with a formal education background. Another example is Liu Jing, director of Anhui Chaohu Duoyingxuan Stationery Factory, who graduated from Hefei Union University in 1994, and then engaged in the restoration of traditional letter paper with his father. He successfully restored the making technique of hand–painted powdered and waxed paper in 1999; he also succeeded in restoring the making technique of gold and silver on paper of the Ming dynasty by collaborating with Professor Zhang Binglun and Dr. Fan Jialu of the University of Science and Technology of China in 2000. Later, he invented gold-painted paper, a new variety of better and more exquisite quality, by using modern high-tech gold-like material, and developed cinnabar paper, transparent paper, and colorful silk paper. He was appointed as a researcher of folk-art creation, the youngest among 30 distinguished researchers, by the Chinese National Academy of Arts in 2005, and was also invited to teach and give live demonstrations at the Central Academy of Fine Arts (as shown in Picture 15.8).

The inheritance of traditional crafts, which are representative of the living culture, is of great significance. The above examples undoubtedly point out the direction for the cultivation of inheritors, which is worth vigorously promoting.

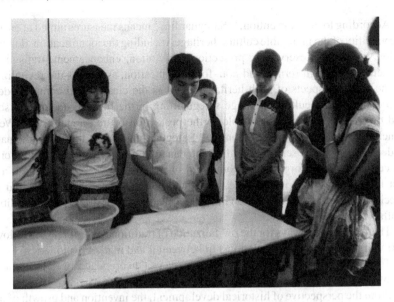

Picture 15.8 Liu Jing teaching at the Central Academy of Fine Arts

15.3 Development and Revitalization of Traditional Crafts and Market Exploitation

At present, there are two diametrically opposed ill practices in China regarding the development and revitalization of traditional crafts and market exploitation.

One is over–exploiting the traditional crafts in harmful ways driven by interests, the results of which are cutthroat competition, a deformed production process, and deteriorated quality, finally leading to the sharp decline in their reputation. Such kind of so–called "protection" is destruction. A case in point is Beijing cloisonné enamel in the mid–1980s when its production was blindly expanded to many township enterprises. Due to poor skills and shoddy production, cloisonné enamel, which had enjoyed a good reputation at home and abroad and ranks among the best in foreign exchange earning handicrafts, was degenerated as low-grade stall goods. As a result, the whole industry was greatly weakened, and it wasn't until in recent years that it gradually recovered. Similar examples are not uncommon. Some projects with good market trends fall into the abyss in a twinkling of an eye, which needs our close attention and strong countermeasures.

The other tendency, on the contrary, is the neglect of government officials and scholars, who lack confidence in the development and revitalization of traditional crafts in view of the above unhealthy phenomena.

The authors hold that these two opposing practices and tendencies are one–sided, which are contrary to *Convention for the Safeguarding of Intangible Cultural Heritage* of the UNESCO. They are harmful in many ways, especially the former.

According to the Convention, "'Safeguarding' means measures aimed at ensuring the viability of the intangible cultural heritage, including the identification, documentation, research, preservation, protection, promotion, enhancement, transmission, particularly through formal and non–formal education, as well as the revitalization of the various aspects of such heritage". This is the clearest and complete definition of intangible cultural heritage safeguarding in the form of documents so far, and of course, it is also applicable to the protection of traditional crafts. We can conclude from the above definition that "safeguarding" is not limited to the narrow understanding of preservation and maintenance but refers to a complete working process and system organically composed of multiple links. Revitalization, as the last link, is obviously inseparable from the whole safeguarding process and is of special significance to ensure the vitality and sustainable development of traditional crafts.

There are various ways for the revitalization of traditional crafts, and the following part will be focused on the relationship between it and market exploitation.

The market exploitation of traditional crafts is a theoretical as well as a practical problem.

From the perspective of historical development, the invention and growth of traditional crafts have always been related to social needs and markets. At the very beginning, the purpose for our ancestors to make pottery, weave cloth, smelt iron, and brew wine was to meet the needs of their own production and daily life. Later, with the division of labor and its expansion, pottery making, weaving, even iron smelting, and winemaking gradually became professionalized and commercialized, which is described as "exchanging millet for machinery does not harm the potters or blacksmiths" in *Meng Zi*. This kind of craft–making was for the needs of others, which appeared during the Warring States period at the latest. Even making bronze gifts and jade articles for the royal family was a kind of social demand. From ancient times, medieval times, near ancient times to modern times, and contemporary times, the form, pattern, connotation, and rise and fall of traditional crafts have always been related to and under the influence of the development of social economy and humanities.

There are still some crafts made purely out of personal hobbies even today. However, another fact is that the survival, inheritance, development, and revitalization of traditional crafts heavily depend on market exploitation. Without the market, craftsmen can't survive, and craftsmanship can't be passed down, let alone developed and revitalized. The main body of craftsmanship is the craftsmen and their communities, which should consciously shoulder the responsibility of protection. Of course, government support is necessary, but it is also limited.

Take purple clay pottery as an example. Its survival and development are based on the custom of tea drinking in China. The craft has been developing steadily over the years because of the prosperity of tea culture and the lasting demand for high–grade tea sets, especially purple clay teapots (as shown in Picture 15.9).

Another example is batik and tie–dyeing, the survival and development of which are based on people's love for their artistic value and practical use. It is precisely because of the improvement of people's living standard, the active development of

Picture 15.9 Purple clay teapots and flowerpots, top–grade tea sets popular both at home and abroad

folk artisans, and the continuous renovation of product varieties and patterns that the demand for batik and tie-dyed cloth has been increasing in recent years, which helps to create a good environment for the sustainable development and revitalization of such handicrafts, such as Nantong blue calico (as shown in Picture 15.10).

Traditional crafts have tenacious vitality of their own. Even in some international metropolises with a high degree of modernization like Beijing, there is still room for old businesses such as scissors and knife grinding. We make an investigation on

Picture 15.10 Drying yard for blue calico

some knife-grinding craftsmen and find that many of them are farmers living in the urban fringe area or suburbs. All their working equipment includes a grindstone tied to the back seat of the bicycle, a small bucket, and an iron fork beside the seat. Since they have farm work to do at home, they only work 20 days a month and many live in basements to save money and travel time. They wander about the streets and lanes to grind knives of common households and only earn 30 yuan every day, board and lodging expenses excluded, but they can still make a living on it. They only charge 2–3 yuan for each knife, which is ground well; and their door–to–door service is quite convenient for the customers. Sometimes when they are too busy, they take the knives back home and deliver them to customers in a few days. The experience of those artisans can help us understand the internal relationship between crafts and market to some extent. The sustainability of traditional crafts and their role in solving the problems of agriculture, rural areas, and farmers and the development of the central and western regions in China should not be underestimated.

The examples of purple clay, batik, tie-dyeing, blue calico, bamboo weaving, and straw weaving (as shown in Picture 15.11) show that the survival of traditional crafts is inseparable from the market. It is a historical necessity that inheritance and revitalization of them depend on market exploitation. We should find our own way to protect traditional crafts by converging the wisdom and initiative of the people and also learn from the successful experience of Japan and South Korea. Excessive exploitation for quick success and instant benefit must be resolutely stopped, and neglect or giving up for fear of risks is also highly undesirable.

In the final analysis, the inheritance and revitalization of traditional crafts are all to ensure their vitality. Inheritance is a prerequisite for revitalization, and revitalization is a reliable guarantee for inheritance. They need the joint efforts of the government, communities, artisans, and experts.

The government should do its duty of leadership. Administrative departments should make the protection of traditional crafts part of their work and take measures such as formulating protection laws and regulations, providing artisans spiritual and material awards, tax reduction and exemption, and special fund support, cracking down on counterfeiting, etc., to effectively promote their inheritance and revitalization.

Picture 15.11 Bamboo weaving in Qionglai bamboo weaving market of Sichuan province, and straw weaving in Xuxing, Shanghai

As for the artisans themselves as well as their communities, they should play the main role in the sustainable development of the crafts by improving their own skills, abiding by professional ethics, actively exploring markets, making innovations, establishing industry organizations, and formulating regulations, training apprentices, and encouraging young people to study at colleges and universities.

What's more, experts should help artisans and communities better understand the value of traditional crafts, improve their awareness of cultural protection, and provide consultation and guidance for the improvement of product design and quality, product promotion, as well transformation of operating mechanisms.

Handicrafts are eternal. Their protection is a cause that benefits the present generation and will continue to do so for future generations as well. Handicrafts encounter new opportunities and new missions in the new era. It is believed that with the cooperation of the government, communities, artisans, and experts, the revitalization of traditional crafts will provide a strong guarantee for its inheritance and sustainable development, contribute to the maintenance of China's culture and national characteristics, and play an important role in the development of the central and western regions and the modernization of urban and rural areas.

Conclusion: Destiny of Traditional Crafts in Contemporary Times

Crafts are basically technology, which is dynamic and changes over time. Each era has its own crafts and technology. For example, modern weaving and dyeing technologies are different from those in the recent past, in the Middle Ages, or in ancient times. Keeping pace with the evolution is one of the characteristics of traditional crafts.

During the long period of pre-modern times, people were seldomly aware of cultural protection. The cultural tradition of valuing humanities over technology in China leads to the fact that many great inventions and creations, such as softening of cast iron and engraved block printing, are often absent from historical records or only recorded in fragments.

Since the second half of the twentieth century, the importance, necessity, and urgency of cultural protection have become the consensus of the international community and different measures have been taken. At the turn of the twenty-first century, the protection and inheritance of intangible cultural heritage, including traditional crafts, has been put on the agenda of the international organizations. China also initiated its protection of intangible cultural heritage in 2003. In 2006 and 2008, the State Council successively promulgated *List of the First and Second Batch of National Intangible Cultural Heritage*. A total of 299 items of traditional handicrafts plus folk arts such as sculpture, weaving, tying, and engraving were included in the lists, accounting for more than a quarter of the total of 1,175 items.

Previously, due to the misunderstanding and lack of relevant mechanisms, the protection and inheritance of traditional crafts have long been neglected and ignored, and some precious skills have been lost without noticing. Nevertheless, many of them are still preserved in their original state among the folk people due to China's vast territory and unbalanced social and economic development. Handicrafts have strong vitality, and craftsmen are dedicated, cheerful, and folk-rooted (as shown in Picture A.1). With the promotion of intangible cultural heritage protection, the traditional crafts on the lists are expected to be protected under the leadership of the government and through the joint efforts of artisans, communities, enterprises, and experts. And those that have not yet been on the nation's intangible cultural inheritance lists will hopefully be included in the provincial and national lists step by step and be protected and passed on to future generations. In this sense, the destiny

Picture A.1 Craftsmen in the teahouse (Photo by Shao Guangwei)

of traditional crafts in contemporary times is much better than in modern and ancient times, which is owing to the progress of the society and is fortunate for them (as shown in Pictures A.2 and A.3).

However, their protection is still in its infancy at present, and the standardized, orderly, and effective protection target can only be realized after a period of running-in. In this case, the conscious actions of artisans and communities are of particular importance. They, as the inventors and inheritors of traditional crafts, should be the main subjects of protection, inheritance, development, and revitalization, with the necessary support from the government. Traditional crafts have their own characteristics, with diverse functions and modes of existence, and therefore their protection forms and inheritance mechanisms are also different. For example, the wood roller oil-extracting skill driven by animal power is unlikely to sustain due to technological improvement, so it is advisable to protect it only for cultural memory, while Shanxi's mature vinegar and Shanghai's Laofengxiang gold and silver fine craftsmanship are in benign development and should be inherited autonomously (as shown in Picture A.4). Another example is Beijing's *zha caizi* (making colorful flowers with silk as decoration), which is expected to expand and develop with the support of the cultural administrative department (as shown in Picture A.5). Some skills that do not

Conclusion: Destiny of Traditional Crafts in Contemporary Times

Picture A.2 The old workshop of Ju Yuan Hao Bow and Arrow Workshop is less than 8 m^2, in which only one person can operate, and it is impossible to recruit apprentices

Picture A.3 Ju Yuan Hao moved to a new site, and its working conditions have been improved significantly since it was included in the national list

Picture A.4 The demonstration of Shanxi mature vinegar brewing skills, which attracts many tourists, and the benefits of enterprises increase accordingly

meet the requirements of the times are bound to be replaced by new technologies, but they should be properly protected and maintained by the government as living fossils of ancient relics if they are of important historical and academic value, such as the large windmill in Shuyang county, Jiangsu province and the fish skin processing technology in Heilongjiang province.

Since the 1950s, a new pattern of coexistence of family inheritance and social inheritance for traditional Chinese crafts means that they are inherited both through formal and non-formal education. This kind of practice should be vigorously advocated and promoted since it is beneficial to the protection and revitalization of traditional crafts (as shown in Pictures A.6 and A.7). As in the Yangtze River, the waves behind ride on the ones before: The new generation excels the old. In recent years, it is not uncommon for young people with high educational qualification to inheritor traditional crafts, and on the other hand, many on-the-job young artisans are also receiving academic education, striving to improve their cultural level and skills. Traditional crafts will surely be further developed and revitalized when a new generation with high education level and good skills takes over the industry.

Handicrafts are eternal, and human beings have instinctive interest for crafts. With the improvement of modernization, craft activities and products will be more widely integrated into people's daily life and become an integral part of modern life; the protection, inheritance, development, and revitalization of crafts will also get more

Conclusion: Destiny of Traditional Crafts in Contemporary Times

Picture A.5 The decorated archway at Beijing Longtan Temple Fair

Picture A.6 Jiading bamboo carving techniques taught in primary schools

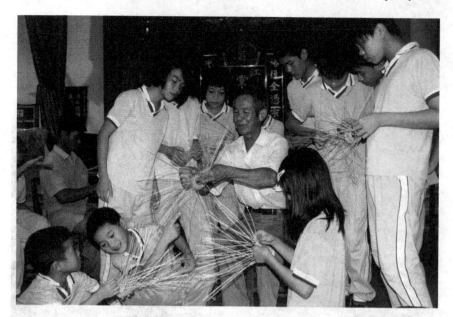

Picture A.7 Huang Yanxing, the inheritor of sea catfish (locally known as mud fish) catching, teaching the skill of fish cage weaving in Doumen district, Zhuhai city, Fujian province

attention and become an indispensable part in modernization. We have reasons to be optimistic about the destiny of traditional Chinese crafts in contemporary times, although there is still a long way to go.

Handicrafts are a treasure. Please regard them with respect and develop alongside them.

Postscript

With the implementation of intangible cultural heritage protection in China in recent years, the rich cultural connotation and important value of Chinese traditional crafts have attracted more and more attention. Since 2009, Elephant Press in Zhengzhou city, Henan province has published in succession a series of *Chinese Handicraft* in Chinese, covering all fourteen categories of traditional crafts, which are widely welcomed.

Based on the materials provided, we write this book to concentrate on the essence of Chinese traditional handicrafts in 14 chapters, to help our readers get a holistic understanding of each field in a short time. The Introduction part is entitled "Re-introducing Handicrafts", which summarizes the core characteristics and value of traditional Chinese crafts, emphasizing that "handicrafts are a treasure; please regard them with respect and develop alongside them", which is the theme and expectation of this book. The 14 chapters encompass 14 different categories, namely, (1) making tools and devices, (2) agricultural and mineral processing, (3) construction, (4) weaving, dyeing, and embroidering, (5) ceramics, (6) metallurgy and metalworking, (7) sculpture, (8) weaving and tying, (9) lacquering, (10) furniture making, (11) calligrapher's tools (12) printing, (13) carving and painting, and (14) special handicrafts and others. The 15th chapter is about the protection and sustainable development of traditional crafts.

Most of the content in this book is taken from the *Chinese Handicrafts* series, the authors of which are Feng Lisheng and Guan Xiaowu (making tools and devices), Zhou Jiahua, Li Jinsong, Zhu Xia and Guan Xiaowu (agricultural and mineral processing), An Peijun and Yang Rui (construction), Qiu Gengyu (ceramics), Hua Jueming, Li Xiaocen and Tang Xuxiang (metallurgy and metalworking), Ren Shimin (sculpture), Wang Lianhai (weaving and tying), Qiao Shiguang (lacquering), Lu Jun (furniture making), Fan Jialu and Fang Xiaoyang (making calligrapher's tools), Fang Xiaoyang and Wu Dantong (printing), Yang Yuan and Li Jinsong (special handicrafts and others), who are also the authors of this book.

They take on the task of writing respectively as follows.

Hua Jueming: Introduction, Chaps. 5, 6, 9, 10, 11, And 15, conclusion and postscript;
Li Jinsong: Chaps. 2 and 14;
Wang Lianhai: Part of Chaps. 4, and 8 and 13;
Guan Xiaowu: Chaps. 1 and 12;
Zhao Hansheng: Part of Chap. 4; Luo Xingbo: Chaps. 3 and 7.

The book is edited by Hua Jueming and reviewed by Liao Yuqun. Li Xiaojuan, director of the editorial board office, Li Jinsong, academic secretary, and Bai Yunyan, special editor, have done a lot of work for the completion and revision of the book.

This book is a sub-project of "Investigation and Comprehensive Research of Traditional Crafts" of Chinese Academy of Sciences (CAS), sponsored by the Institute for History of Natural Sciences. Many thanks to the leaders of the Comprehensive Planning Bureau of CAS and the Institute for their support, the leaders, and responsible editors of Elephant Press for their great work during the process of publishing this book.

Suggestions are welcome from readers.

Hua Jueming

March 12, 2012

Appendix I

Chinese dynasties		
Dynasty		Time of Beginning and Ending
Xia		2070–1500 B.C
Shang		1500–1046 B.C
Western Zhou		1046–771 B.C
Eastern Zhou	Spring and Autumn period	770–476 B.C
	Warring States period	475–221 B.C
Qin		221–207 B.C
Western Han		202 B.C.–8 A.D
Xin		8–23
Eastern Han		25–220
Three Kingdoms period	Wei Kingdom	220–265
	Shu Kingdom	221–261
	Wu Kingdom	222–280
Western Jin		265–316
Eastern Jin and 16 Kingdoms	Eastern Jin	317–420
	Sixteen Kingdoms	304–439
Northern and Southern dynasties	Song	420–479
	Qi	479–502
	Liang	502–557
	Chen	557–589
	Northern Wei	386–534
	Eastern Wei	534–550
	Northern Qi	550–577
	Western Wei	535–557

(continued)

(continued)

Chinese dynasties

	Northern Zhou	557–581
Sui		581–618
Tang		618–907
Five Dynasties period		907–960
Ten Kingdoms period		902–979
Liao		916–1125
Song	Northern Song	960–1127
	Southern Song	1127–1279
Western Xia		1138–1227
Jin		1115–1234
Yuan		1271–1368
Ming		1368–1644
Qing		1644–1912
Republic of China		1912–1949
People's Republic of China		1949-

Appendix II

English	Chinese PRC
A Brief Biography of Yunnan	《滇略》
A Brief Introduction to the Scenery of Emperor Capital	《帝京景物略》
A Brief Map of Mines in Southern Yunnan	《滇南矿厂图略》
A Broad Discussion on the Xuan Furnace	《宣炉博论》
A Classificatory Compilation of Literary Writings	《艺文类聚》
A Collection of Stories of Model Women	《古今女范》
A Collection of Transmitted Plays Published by the Warm-Crimson Studio	《暖红室汇刻传奇》
A Collection of Zhonghui's Work	《忠惠集》
A Complete Book on Agriculture	《农政全书》
A Complete Chronicle of Yunnan	《云南通志》
A Complete Collection of Ten Bamboo Studio Decorative Letter Papers	《十竹斋笺谱》
A Dream of Lin'an	《梦粱录》
A Dream of Red Mansions	《红楼梦》
A New Account of the Tales of the World	《世说新语》
A Rare Ming Dynasty Big Table	《一件珍奇的明代大案桌》
A Selection of Metric Poem and Fu from Finest Blossoms in the Garden of Literature	《文苑英华律赋选》
A Song in Slow Time	《长歌行》
Album of Erotic Paintings	《竞春图卷》
Along the River During the Qingming Festival	《清明上河图》
An annotation of the Diamond Sutra	《金刚般若波罗密经注解》
An Unofficial History of Nanzhao	《南昭野史》
Analects on Women	《女论语》

(continued)

(continued)

English	Chinese PRC
Analysis on Proofreading of Sphragistics and Four Examples	《校补金石例四种》
Ancient Jade Catalogue	《古玉图》
Anecdote Novel of Tang Dynasties	《唐语林》
Annals of Lu'an Prefecture	《潞安府志》
Annals of Dongchuan Prefecture	《东川府志》
Annals of Fuyang County	《富阳县志》
Annals of Gusu	《姑苏志》
Annals of Huachuan County	《桦川县志》
Annals of Huzhou Prefecture during the Tongzhi Period	《同治湖州府志》
Annals of Jiajiang County	《夹江县志》
Annals of Jiaxing Prefecture	《嘉兴府志》
Annals of Jingde County	《旌德县志》
Annals of Liangshan County	《梁山县志》
Annals of Luxian County	《泸县志》
Annals of Nanxun Town	《南浔镇志》
Annals of Shexian County	《歙县志》
Annals of Taiding Prefecture during the Daoguang Period	《道光太定府志》
Annals of Weining County	《威宁县志》
Annals of Wuxi County	《无锡县志》
Annals of Wuxian County	《吴县志》
Annals of Xia in Records of the Historian	《史记·夏本纪》
Annual Customs and Festivals in Peking	《燕京岁时纪》
Appreciation and Collection of Four Treasures of the Study	《文房四宝鉴赏与收藏》
Appreciation of Famous Chinese Inkstones	《中国名砚鉴赏》
Art of War from Shi Shangong	《施山公兵法心略》
Art of War From Zhuge Liang	《诸葛武侯兵法心要》
Atlas of Chinese Farm Tools	《中华农器图谱》
Atlas of Science and Technology through the History of China	《中国古今科技图文集》
Avatamsaka Sutra	《华严经》
Awakening Zhuang Zi	《唤庄生》
Axis Painting	《轴画法》
Bai Annotation	《百论疏》
Beiping Letter Papers	《北平笺谱》

(continued)

Appendix II

(continued)

English	Chinese PRC
Bibliography of Thirteen Princes in History of Han	《汉书·景十三王传》
Biographies of Exemplary Women	《列女传》
Biography in the Southern Ming Dynasty	《南疆绎史勘本》
Biography of Cai Lun in History of Later Han	《后汉书·蔡伦传》
Biography of Du Shi in History of Later Han	《后汉书·杜诗传》
Biography of Huitong in Rong Chuntang Set	《容春堂集·会通君传》
Biography of King Gesar	《格萨尔王传》
Biography of Su Qin	《苏秦列传》
Biography of Su Qin of Records of the Historian	《史记·苏秦列传》
Biography of Yue Fei in History of Song	《宋史·岳飞传》
Biography of Zhang Heng in History of Later Han	《后汉书·张衡传》
Biography of Zhenhai in History of Yuan	《元史·镇海传》
Birthday Celebration of Guo Ziyi	《郭子仪拜寿》
Block Printing in Tang and Song Dynasties	《唐宋时期的雕版印刷》
Book of Agriculture	《农书》
Book of Anthroposcopy according to Animals and Numbers	《演禽斗数三世相书》
Book of Changes	《易》
Book of Diverse Crafts	《考工记》
Book of Documents	《书》
Book of Guqin Music at Yuguzhai	《与古斋琴谱》
Book of Jin	《晋书》
Book of Later Han	《后汉书》
Book of Lei and Si	《耒耜经》
Book of Luochuang	《萝窗小牍》
Book of Rites	《礼记》
Book of Songs	《诗经》
Bow-making Craftsman in Book of Diverse Crafts	《考工记·弓人》
Calligraphy of the Shaoxing Reign	《绍兴御府书画式》
Casual Records of Mo Zi and Zhuang Zi	《墨庄漫录》
Catalog of Paper Money	《楮币谱》
Ceremonies and Rites	《仪礼》
Changqing Selection of Yuan Zhen	《元氏长庆集》
Characters of Five Classics	《五经文字》
Characters of Nine Classics	《九经字样》
Cheng Weishi Annotation	《成唯识论述记》
Chicken Rib Compilation	《鸡肋编》

(continued)

(continued)

English	Chinese PRC
China at Work	《劳作中的中国》
China at Work: An Illustrated Record of the Primitive Industries of China's Masses, Whose Life is Toil, and Thus an Account of Chinese Civilization	《手艺中国: 中国手工业调查图录》
Chinese Ancient Rare Books Bibliography	《中国古籍善本书目》
Chinese Handicrafts	《中国手工技艺》
Chinese Handicrafts: Four Treasures of the Study	《中国手工艺·文房四宝》
Chinese Handicrafts: Metallury and Metalworking	《中国手工艺·金属采冶和加工》
Chinese Local Handicrafts	《中国乡土手工艺》
Chongning Tripitaka	《崇宁藏》
Chunguan in Rites of Zhou	《周礼·春官》
Classic Chinese Furniture: Ming Dynasty	《明代家具珍赏》
Classic of Mountains and Seas	《山海经》
Classic of Regions Beyond the Seas: East in Classic of Mountains	《山海经·海外东经》
Classics and History in Donglai	《东莱经史论说》
Classification of State Offices	《职官分纪》
Code of Great Ming Dynasty	《明会典》
Collected Commentaries on Elegies of Chu State	《楚辞集注》
Collected Essays of Miaoxiang Pavilion	《妙香阁文稿》
Collected Poetry	《诗藁》
Collected Talks of a Drunkard	《醉翁谈录》
Collected Works of Weinan	《渭南文集》
Collected Writings of Cai Yong	《蔡中郎集》
Collection of Chinese Buddhist Temple Gazetteers	《中国佛寺志丛刊》
Collection of Civilization: the Traditional Techniques of Papermaking and Printing	《典藏文明——古代造纸印刷术》
Collection of Important Essays on Sericulture	《蚕桑萃编》
Collection of Laws and Regulations of the Han Dynasty	《汉官仪》
Collection of Suichu Hall	《遂初堂集》
Collections of Instruments	《器准》
Collections of Qing Dynasty	《清秘藏》
Colorimetry	《色彩学》
Commentaries of Dongpo on the Book of Changes	《东坡易传》
Commentaries to the Classics from the Hall of the Free Mind	《通专堂红解》
Commentary of Gongyang	《公羊传》
Commentary of Guliang	《榖梁传》

(continued)

Appendix II

(continued)

English	Chinese PRC
Commentary of Zuo	《左传》
Compilation of Martial Arts	《武编》
Compilation of Regulations in the Song Dynasty	《宋会要辑稿》
Compilation of Unofficial History of the Wanli Period	《万历野获编》
Complement of Dumen Bamboo Branch Poems	《续都门竹枝词》
Complete Collection of Illustrations and Writings from the Earliest to Current Times	《古今图书集成》
Complete Collection of Traditional Chinese Handcrafts: Papermaking and Printing	《中国传统工艺全集·造纸与印刷》
Complete Library in the Four Branches of Literature	《四库全书》
Complete Works of Traditional Chinese Handicrafts: Lacquer Art	《中国传统工艺全集·漆艺卷》
Complete Works of Traditional Chinese Handicrafts: Weaving, Dyeing and Embroidery	《中国传统工艺全集·织染刺绣卷》
Complete Works Translated by Master Xuan Zang	《玄奘法师译撰全集》
Comprehensive Examination of Literature	《文献通考》
Comprehensive Meaning of Customs and Mores	《风俗通义》
Comprehensive Mirror to Aid in Government	《资治通鉴》
Concise Title Catalogue of Siku Quanshu	《四库简明名录标注》
Convention for the Safeguarding of the Intangible Cultural Heritage	《保护非物质文化遗产公约》
Court Ladies Preparing Newly-Woven Silk	《捣练图》
Crimson Silk Threaded Carpets	《红线毯》
Daily Notes on Green Bamboo	《留青日札》
Daya Building Studio	《大雅楼画室》
Decoration Technology of Lacquering	《国漆髹饰工艺学》
Diamond Sutra	《金刚经》
Discourses of the States	《国语》
Discovery and Early Spread of Chinese Movable Printing Technology	《中国活字印刷的发明和早期传播》
Discrimination on Sound and Meaning	《音义辨同》
Discussing Writing and Explaining Characters	《说文解字》
Dongba Scripture	《东巴经》
Dream Pool Essays	《梦溪笔谈》
Dream Reminiscence of Tao'an	《陶庵梦忆》
Dream Talk by the Pine Window	《松窗梦语》
Dreams of Splendor of the Eastern Capital	《东京梦华录》
Duke Wen of Teng I in Meng Zi	《孟子·滕文公上》
Eight Treatises on Following the Principles of Life	《遵生八笺》

(continued)

(continued)

English	Chinese PRC
Elementary Learning	《小学》
Embroidery	《绣谱》
Encyclopedia Britannica	《不列颠百科全书》
Erotic Color Prints of the Ming Period	《秘戏图考》
Essays on the Emergence of Pillars with Buddhist Inscriptions	《涌幢小品》
Essential Criteria of Antique	《格古要论》
Essential Principles of Mathematics	《数理精蕴》
Essential Techniques for the Peasantry	《齐民要术》
Essentials of a New Method for Mechanizing the Rotation of an Armillary Sphere and a Celestial Globe	《新仪象法要》
Essentials of Categorized Matters Like Joint Jade Circles Quoted From Old and New Literature · Volume One	《古今合璧事类备要·前集》
Essentials of Fire Attacks	《火攻挈要》
Exposition of the Great Learning	《大学衍义》
Extensive Records of the Taiping Era	《太平广记》
Facts of the Song Dynasty	《宋朝事实》
Fang's Ink Collection	《方氏墨谱》
Fine Workmanship: A Brief History of Chinese Furniture	《精工细巧——中国家具史略》
Finest Blossoms in the Garden of Literature	《文苑英华》
First Collection of Elegant Poems in Xiuye Hall	《修业堂初集肆雅诗钞》
First Collection of Xianping Bookstore	《仙屏书屋初集》
Five Books on Phonology	《音学五书》
Five Bookworms in Han Fei Zi	《韩非子·五蠹》
Five Oxen	《五牛图》
Five Thousand Years of Chinese Science and Technology	《中华科技五千年》
Flying on a Crane at Yaotai	《瑶台跨鹤图》
Fortification of the City Gate in Mo Zi	《墨子·备城门》
Four Beauties of China	《四美图》
Four Treasures of the Study	《文房四宝谱》
Full Literature of the Tang Dynasty	《全唐文》
Funeral Records	《丧大记》
Furniture Making in Chinese Handicrafts	《中国手工艺·家具制作》
Gems of Ancient Chinese Technological Inventions	《中国古代科技文物展》
Genealogical Annals of the Emperors and Kings	《帝王世纪》

(continued)

Appendix II

(continued)

English	Chinese PRC
Genealogy of the Zhai Family in the East of Jinghe River	《泾川水东翟氏宗谱》
General Principle to the Art of War	《武经总要》
Gengzhu in Mo Zi	《墨子·耕柱篇》
Geographical Works	《地工开物》
Girl Dreams of a Rich Dowry	《十里红妆女儿梦》
Great Master in Zhuang Zi	《庄子·大宗师》
Great Overview on Brush Note Style Novellas	《笔记小说大观》
Great Treatise in Book of Changes	《易经·系辞》
Guidelines by Lu Ban	《鲁班经》
Guodian Chu Slips	《郭店楚墓竹简》
Guqin	《古琴》
Handy Essentials for Medical Prescriptions for Use in the Army	《军中医方备要》
Heaven and Earth in Zhuang Zi	《庄子·天地》
Heavenly Creations	《天工开物》
Historical Records	《史鉴》
History of Flowers	《花史》
History of Han	《汉书》
History of Ink	《墨史》
History of Science and Technology in China: Textile	《中国科学技术史·纺织卷》
History of Science and Technology in China: Papermaking and Printing	《中国科学技术史·造纸与印刷卷》
History of the Ming Palace	《明宫史》
History of Yuan	《元史》
Hongwu Tripitaka	《洪武大藏经》
Hua Ji	《画继》
Huai Nan Zi	《淮南子》
Hundred Battles Strategy From Liu Bowen	《刘伯温先生百战奇略》
Hundred Pictures of Ancient Chinese Prints	《中国古代版画百图》
Hundred Rivers Reach the Sea of Learning	《百川学海》
Hydraulic Engineering in the Annals of Shaoxing	《绍兴府志·水利》
Illustrated Handbook of Lacquering Decoration Record	《髹饰录图说》
Immortals' Birthday Celebration	《群仙祝寿图》
Imperial Anthology of Tang and Song Poetry	《御选唐宋诗醇》
Imperial Facts Encyclopedia	《皇朝事实类苑》
Industry of Hangzhou Annals	《杭州志·工业篇》
Ink Spectrum	《墨谱法式》

(continued)

(continued)

English	Chinese PRC
Inkstone History	《砚史》
Inkstone Record	《砚记》
Inkstone Spectrum	《砚谱》
Instructions for a Grand Nation in Huai Nan Zi	《淮南子·泰族训》
Instructions of the Ancestor of the August Ming	《皇明内训》
Instructions to the Factory and Library of the Ministry of Industry	《工部厂库须知》
Interior Decoration and Furniture of the Ming Dynasty in China	《中国明代室内装饰和家具》
Interpretation of Lacquering Decoration Record	《髹饰录解说》
Interpretation of Ships in Interpretation of Names	《释名·释船》
Investigation and Research on Chinese Traditional Machinery	《传统机械调查研究》
Istanbul Declaration	《伊斯坦布尔宣言》
Jade Book	《玉篇》
Jiajiang, the Hometown of Chinese Painting and Calligraphy	《中国书画之乡——夹江》
Jindao Tripitaka	《金道藏》
Journey of Shadow Play	《皮影之旅》
Kaicheng Stone Classics	《开成石经》
Kangxi Dictionary	《康熙字典》
Karma Purifying Mantra	《无垢净光大陀罗尼经》
King of Sea in Guan Zi	《管子·海王》
Lacquer Art Theory	《漆艺论》
Lacquering Decoration in Liangshan Yi Nationality	《凉山彝族的漆艺工艺》
Lacquering Decoration Record	《髹饰录》
Ladies' Classic of Filial Piety	《女孝经》
Lao Zi	《老子》
Laoxuean Notes	《老学庵笔记》
Law on the Protection of Cultural Property	《文化财保护法》
Le Machine	《机械》
Leisure	《闲情偶寄》
Lessons for Women	《女诫》
Li Lou I in Meng Zi	《孟子·离娄上》
Liao Tripitaka	《辽藏》
Lie Zi	《列子》
Listening to Someone Pounding Cloth at Night	《夜听捣衣》
Litang Daoting Lu	《里堂道听录》

(continued)

(continued)

English	Chinese PRC
Local Officials in Rites of Zhou	《周礼·地官》
Lotus Sutra (Saddharma Pundarika Sutra)	《妙法莲花经》
Luoxuan Biangu's Letter Paper Album	《萝轩变古笺谱》
Luxuriant Dew of the Spring and Autumn Annals	《春秋繁露》
Lyrics to the Melody of Sandy Creek Washers	《浣溪沙》
Ma Weidu Talking about Collection: Furniture	《马未都说收藏·家具篇》
Mahayana Mahaparinirvana Sutra	《大般涅经》
Master Feng's Records of Hearsay and Personal Experience	《封氏闻见记》
Master Yin's Guide to Letter Papers	《殷氏笺谱》
Meaning of Rituals in Classic of Rites	《礼记·祭义》
Memorials of the Ministers Collected in the Song Dynasty	《宋诸臣奏议》
Meng Zi	《孟子》
Mengqiu	《蒙求》
Method of Making Ink	《墨法集要》
Mirror of Craftsmanship and Guidelines by Lu Ban	《鲁班经匠家镜》
Miscellaneous Notes of Zhiya Hall	《志雅堂杂钞》
Miscellaneous Records at Xishan Mountain	《西山杂志》
Miscellaneous Records of the Western Capital	《西京杂记》
Mo Zi	《墨子》
Models of Ink Stick	《墨苑》
Mount Kuanglu	《匡庐图》
Mountain Official of Earth Official in the Rites of Zhou	《周礼·地官·山虞》
National History Supplement	《国史补》
New Theories of Huan Zi	《桓子新论》
New Year's Hundred Chants About Hangzhou Customs	《杭俗新年百咏》
Notes and Examinations of Thirteen Classics	《十三经注疏》
Notes of Ceasing Farm Labors	《辍耕录》
Notes of Kui Tan	《愧郯录》
Notes of Past Famous Paintings	《历代名画记》
Notes of the Jade Hall	《玉堂杂记》
Notes on Comprehensive Mirror to Aid in Government	《注<资治通鉴>》
Notes on Physics	《物理小识》
Notes on Sacred Edict of Emperor Kangxi	《圣谕广训注》
Notes on the Antiquity and Present Days	《古今注》
Nymph of the Luo River	《洛神赋图》

(continued)

(continued)

English	Chinese PRC
Object Map and Proportionality	《物图与比例》
Ode to Brush Pen	《笔赋》
Ode to the Capital of Shu	《蜀都赋》
Ode to the Hydraulic Wheel	《水轮赋》
Ode to the Water Wheel	《水车行》
Odes to Three Capitals	《三都赋》
Office of Earth in Rites of the Zhou	《周礼·地官司徒》
Office of Spring	《春官》
Old Dreams Reappear	《旧梦重现》
Old Things about Wulin	《武林旧事》
On Boardland in Xun Zi	《荀子·疆国篇》
On Imperial Testing Policies	《御试策》
On Lacquering and Painting	《谈漆论画》
On Phonology	《音论》
On Salt in Sichuan	《蜀盐说》
On Sundries	《上杂物疏》
On Ten Types of Paper	《十纸说》
Origin of the Five Surnames	《五姓的由来》
Original Phonics in Book of Songs	《诗本音》
Outlaws of the Marsh	《水浒传》
Outline of the Finest Blossoms in the Garden of Literature	《文苑英华纂要》
Painting History of Wise Men	《智者绘画史》
Painting of Chetian Lantern Market	《车填灯市图》
Painting of Emperor Xianzong of Ming Enjoying Lantern Festival	《明宪宗元宵行乐图》
Painting of Joyous Matters of Approaching Peace	《升平乐事图》
Painting of Planting and Harvesting	《耕获图》
Painting of Sericulture and Weaving	《蚕织图》
Painting of the Panche at the Water Gate	《闸口盘车图》
Paper and Printing	《造纸和印刷》
Paper and Printing	《纸和印刷》
Part I of General Principle to the Art of War	《武经总要·前集》
Pattern of the Family in Classic of Rites	《礼记·内则》
Peacock Flies Southeast	《孔雀东南飞》
Phonetic Explanation of Nine Classics	《九经韵览》
Phonetic Explanation of Spring and Autumn Annals	《音释春秋》
Photos of Old Peking	《京城旧影》

(continued)

Appendix II

(continued)

English	Chinese PRC
Picture of Little Ducks in the Lotus Pond	《莲塘乳鸭图》
Picture of Washing Horse	《洗马图》
Pictures of Ancient Chinese Civilization	《中华古文明大图集》
Pictures of Farming and Weaving	《耕织图》
Plea to Send Officials to Jiangyan County	《奏为乞差京朝官知井研县事》
Preliminary Compilation of Ceramic Movable Type Trial Printing	《泥版试印初编》
Preparation against Tunnelling in Mo Zi	《墨子·备穴篇》
Principles and Stitching of Chinese Embroidery	《雪宦绣谱》
Process of Printing in Wuying Dian Using Movable Types	《钦定武英殿聚珍版程式》
Proportionality Thoery	《比例学》
Proposal on Farming and Sericulture	《敬陈农桑四事疏》
Prosperous Suzhou	《盛世滋生图》
Pure Records of the Cave Heaven	《洞天清录集》
Questions of Tang in Lie Zi	《列子·汤问》
Quick Approaches of Chinese Characters	《急就篇》
Quiet Girl in Book of Songs	《诗·静女》
Readings of the Taiping Era	《太平御览》
Recommendation on the Safeguarding of Traditional Culture and Folklore	《保护民间创作建议案》
Record of Cai Weng's Monument	《蔡翁碑叙录》
Record of Copper Seismograph	《地动铜仪经》
Record of Folk Life Customs	《云间据目抄》
Record of National Geography	《广志绎》
Record of Rebuilding Cai Hou Temple for Longting Marquis in the Han Dynasty	《重修汉封龙亭侯蔡公祠记》
Record of the Pure and Uncommon	《清异录》
Record of Wan'an Bridge	《万安桥记》
Record on the Subject of Education in Classic of Rites	《礼记·学记》
Records for Chinese Calligraphy and Painting Mounting	《赏延素心录》
Records of Carriage and Costumes in History of Song	《宋史·舆服志》
Records of Carriages and Costumes in Book of Jin	《晋书·舆服志》
Records of Customers' Shoe Sizes	《履中备载》
Records of Industries in China	《中国实业志》
Records of Inkstone, Lei and Jade Tablets	《砚耒圭绪录》
Records of Laws and Systems of the Qing Dynasty	《大清会典事例》

(continued)

(continued)

English	Chinese PRC
Records of Mounting	《装潢志》
Records of Relations Between Rulers and Officials in Past Dynasties	《历代君臣事迹》
Records of Silk Embroidery in Cunsutang	《存素堂丝绣录》
Records of the Historian	《史记》
Red Printed God of Longevity	《朱拓寿星》
Reflections on Things at Hand	《近思录》
Research on Furniture in the Ming Dynasty	《明代家具研究》
Research on Joining Technology of Joiner Tenon in Warring States Period	《战国细木工榫接合工艺研究》
Review on Tao Te Ching	《道德经广圣义》
Revision on Phonetic Notation of Book of Changes	《校正音释济经》
Rites of Zhou	《周礼》
Royal Tribute Painting of the Qing Dynasty	《皇清职贡图》
Rustic Talks from the East of Qi	《齐东野语》
Salt Industry and Trade	《盐法通志》
Second of the Six Poems on the Night of Lantern Festival	《上元夜六首之二》
Secret Books on Water and Land Attack and Defense Strategy	《水陆攻守战略秘书》
Selections from Lu's Commentaries of History	《吕氏春秋》
Sequel Manuscript of Jiannan Poems	《剑南续稿》
Shu Lin Qing Hua	《书林清话》
Silent Night Thinking	《静夜思》
Silk Painting Depicting a Man Riding a Dragon	《人物御龙帛画》
Slang Dictionary	《通俗文》
Sleeve Records	《袖中记》
Songs of Chu	《楚辞》
Splendid Flower Valley	《锦绣万花谷》
Spring Excursion	《游春图》
Statue Measurement Sutra	《造像度量经》
Stone Inkstone Song	《杨生青花紫石砚歌》
Strategies of the Warring States	《战国策》
Study on Tools Obtaining Well Salt in Ancient China	《中国古代井盐工具研究》
Summary of Longsha	《龙沙纪略》
Superfluous Things	《长物志》
Supplemental Annals of Panyu County: Industry	《番禺县续志·实业篇》
Supplementary to the History of the Qing Palace	《清宫史续编》

(continued)

Appendix II

(continued)

English	Chinese PRC
Taiping Heavenly Sun	《太平天日》
Talking about Mahogany Furniture in the Ming and Qing Dynasties	《漫话明清红木家具》
Talking about Mahogany Furniture in the Ming and Qing Dynasties	《漫话明清红木家具》
Talks on Salt and Iron	《盐铁论》
Tathagatagarbha Sutra	《宝箧印陀罗尼经》
Ten Bamboo Studio Painting Spectrum	《十竹斋书画谱》
Ten Defects in Han Fei Zi	《韩非子·十过》
Ten Volumes on Lady's Demeanor	《闺范十集》
Textual Criticism on Calendar	《星历考原》
Textual Research on Lacquerware Production in the Qing Dynasty	《清代造办处漆器制作考》
The Book of Song	《宋书》
The Fifth Collection of Rongzhai Essays	《容斋五笔》
The Golden Lotus	《金瓶梅》
The Great Treaties II in the Commentaries on Book of Changes	《易·系传下》
The History of Chinese Sculpture	《中国雕塑史》
The Hong-Sueh Sketches	《鸿雪因缘图记》
The Master Who Embraces Simplicity	《抱朴子》
The Material Culture of Yunnan: Farming	《云南物质文化·农耕卷》
The Night Revels of Han Xizai	《韩熙载夜宴图》
The Nineth Revision of the Genealogy of Xu's Family in Piling	《九修毗陵徐氏宗谱》
The Orchid Pavilion	《兰亭序》
The Overall Survey of the Ocean's Shores	《瀛涯胜览》
The Travels of Marco Polo	《马可·波罗游记》
The Universal Declaration on Cultural Diversity	《文化多样性世界宣言》
Third Madam Teaches Son	《三娘教子》
Tian Zifang in Zhuang Zi	《庄子·田子方》
Tomb Mural in The Great Treasury of Chinese Fine Arts	《中国美术全集·墓室壁画》
Tongzhitang Jingjie (Commentaries to the Classics from the Hall of the Free Mind)	《通专堂红解》
Treasure Theory	《宝藏畅微论》
Treaties of Foods and Commodities in History of the Ming Dynasty	《明史·食货志》

(continued)

(continued)

English	Chinese PRC
Treaties of Foods and Commodities of History of the Han Dynasty	《汉书·食货志》
Treatise of Five Elements in History of Later Han	《后汉书·五行志》
Treatise of Food and Loan in History of Song	《宋史·食货志》
Treatise of Geography in History of Han	《汉书·地理志》
Treatise on the Huayan Samadhi	《华严三昧章》
Tribute of Yu	《禹贡》
Tribute of Yu in Book of Documents	《尚书·禹贡》
Tripitaka in Kaibao	《开宝大藏经》
Trivial Talking about Painting	《绘事琐言》
True Doctrines of Music	《律吕正义》
Tune: Endless as the Sky	《应天长》
Twenty-Four Histories	《二十四史》
Utensils in the Study Room	《考槃余事》
verse July in Lessons of Bin State in Book of Songs	《诗·豳风·七月》
Vimalakirti-Nirdesa-Sutra	《维摩诘所说经》
Water-powered Trip-Hammer under the Willow	《柳阴云碓图》
Weaving in Code of Great Ming Dynasty	《明会典·织造》
Weights and Measures Law of the Republic of China	《中华民国度量衡法》
Well-balanced Records of Guihai	《桂海虞衡志》
What I See and Hear in the Imperial City	《酌中志》
Woodblock Printing in The Great Treasury of Chinese Fine Arts	《中国美术全集·版画》
Words in Lu Garden	《履园丛话》
Writing Brush made of Brownish Rabbit Hair	《紫毫》
Writing of a Scholar	《书林别话》
Wu You's Picturesque Treasure	《吴友如画宝》
Wuliangshou Jing Youpotishe Yuansheng Ji Zhu	《无量寿经优婆提舍愿生偈注》
Xifang Jile Shijie Yizheng Zhuangyan Tu (Image of the Magnificent World of Ultimate Joy in the West)	《西方极乐世界依正庄严图》
Xu Zi	《许子书》
Yangzhou Original Boat Record	《扬州画舫录》
Yinming Ruzhengli Annotation	《因明入正理论疏》
Yongle Encyclopedia	《永乐大典》
Yongxing	《咏兴》
Youyang Notes	《酉阳杂俎》
Yueman Qingyou Book	《月曼清游册》
Yunxian Miscellaneous Notes	《云仙散录》

(continued)

Appendix II

(continued)

English	Chinese PRC
Za A-pi-tan Xin Lun (Samyuktabhidharma-hrdaya-sastra)	《杂阿毗昙心论》
Zhaocheng Jin Tripitaka	《赵城金藏》
Zhong Annotation	《中论疏》
Zhuang Zi	《庄子》
Zifu Tripitaka	《资福藏》